DATABASE SECURITY AND AUDITING: Protecting Data Integrity and Accessibility

By Hassan A. Afyouni

THOMSON
COURSE TECHNOLOGY

Australia • Canada • Mexico • Singapore • Spain • United Kingdom • United States

Database Security and Auditing: Protecting Data Integrity and Accessibility

By Hassan A. Afyouni

Executive Editor:
Mac Mendelsohn

Senior Product Manager:
Eunice Yeates-Fogle

Senior Acquisitions Editor:
Maureen Martin

Editorial Assistant:
Jennifer Smith

Development Editor:
Gretchen Kiser

Production Editors:
Brooke Booth, Cecile Kaufman

Senior Marketing Manager:
Karen Seitz

Associate Product Manager:
Mirella Misiaszek

Quality Assurance Testing:
Chris Scriver, Serge Palladino, Burt LaFountain

Compositor:
GEX Publishing Services

Cover Designer:
Laura Rickenbach

Printed in Canada

1 2 3 4 5 6 7 8 9 WC 09 08 07 06 05

For more information, contact Thomson Course Technology, 25 Thomson Place, Boston, Massachusetts, 02210.

Or find us on the World Wide Web at: www.course.com

ISBN 0-619-21559-3

Brief Contents

Table of Contents

Chapter Six

Part II

Chapter Seven

Chapter Nine

Part III
Chapter Ten

Introduction

THE COST OF DATA LOSS IS RISING progressively every year. Companies are losing data due to malicious attacks and improper implementation of database security and auditing. Data integrity and accessibility must be protected in order to ensure the company operability.

Yesterday's DBAs were equipped with all sorts of technical skills that empowered them to manage the database for optimal efficiency and use. Today's DBAs must have in addition the ability to implement security policies and auditing procedures in order to protect one of the most valuable assets of an organization—data. Data has become so mission critical and indispensable an asset that an organization could become paralyzed and inoperable if data integrity, accessibility, and confidentiality is violated.

This book is designed to provide the reader with an understanding of security concepts and practices in general and those specific to database security in a highly detailed implementation. Not only will the reader gain a good understanding of database security, the reader will be shown how to develop database applications embedding from simple to sophisticated security and auditing models using Oracle10*g* and Microsoft SQL Server 2000.

Intended Audience

This book is intended for any person who is involved with database applications. The reader could be a developer, system analyst, business analyst, data architect, database administrator, or a systems development manager working with database applications. This book can be used as a textbook in colleges and universities, in career training schools, or as training material for companies with IT professionals. The book assumes the reader has a basic knowledge of database concepts. The book's pedagogical features are designed to provide a learning experience to equip the reader with all the tools necessary to implement database security and auditing in order protect data. Each chapter includes a case study that places the reader in the role of problem solver, requiring you to apply concepts presented in the chapter to achieve a successful solution.

Book Organization

This book is divided into three logical parts: the first part (Chapters 1 to 6) introduces you to general concepts related to database security; the second part (Chapters 7 to 9) discusses topics related database auditing; and the third part (Chapter 10) provides practical case projects covering all the information covered in this book.

Chapter Descriptions

Here is a summary of the topics covered in each chapter of this book:

Chapter 1, "Security Architecture" presents fundamental security concepts that serve as the building blocks to data and database security. This chapter covers important concepts such as information systems components, database management systems functionalities, and major components of information security architecture. These concepts and others are presented and explained from a database security perspective.

Chapter 2. The first line of defense is the network that connects users to the database and the second line of defense is the operating system of the server where the database resides. "Operating System Security Fundamentals" provides a quick but essential tour of the operating system functionalities from a security perspective. The focal points of the chapter are an explanation of the components of the operating system security environment, operating system vulnerabilities, and password policies.

Chapter 3, "Administration of Users" is a step-by-step walkthrough on how to create, drop, and modify user accounts in Oracle10*g* and SQL Server 2000. This chapter outlines all the various security risks related to user administrations that administrators must be aware of. Finally, this chapter concludes with best practices adopted by database administration experts.

Chapter 4, "Profiles, Password Policies, Privileges, and Roles" covers the security components of a database account. This chapter provides full description and instruction on how to administer these components and details on how to grant and revoke these components to and from database user accounts. This chapter concludes with best administration practices related to these topics. Administration instructions are provided for both Oracle10*g* and SQL Server 2000.

Chapter 5, "Database Application Security Models" presents concepts that are the core of database application security. It presents different application security models that can be adopted and implemented for most business models.

Chapters 6, "Virtual Private Databases" explains in detail the concept of virtual private databases and presents step-by-step implementation using views in SQL Server 2000 and Oracle10*g*. In addition, this chapter explores and implements virtual private database functionalities such as Application Context and Policy Manager provided by Oracle10*g*.

Chapter 7, "Database Auditing Models" is a fundamental chapter for understanding database auditing. The chapter explains the differences between and the interdependence of security and auditing. This chapter defines the role and responsibilities of the auditor and presents a full discussion of the auditing environment and auditing benefits and side effects. Another important part of this chapter is discussion of various database auditing models to be implemented in Chapters 8 and 9.

Chapter 8, "Application Data Auditing" presents an implementation of application data changes auditing. Step-by-step instruction shows the reader how to develop and implement the database auditing models presented in Chapter 7 in both SQL Server and Oracle10*g*. The fine-grained auditing feature provided by Oracle10*g* is outlined in this chapter along with detailed explanations on how it can be implemented as part of the database administration procedures.

Chapter 9, "Auditing Database Activities" provides the reader with various demonstrations and illustrations on how to implement database auditing using Oracle10*g* and SQL Server 2000. Subtopics of database auditing include database events, data control statements, and data definition operations. A full description and implementation of Oracle10*g* AUDIT statement is provided in this chapter. SQL Server 2000 and Oracle10*g* tracing facilities are also covered.

Chapter 10, "Security and Auditing Project Cases" is implementation only. No new database concepts are presented. However, this chapter presents the most important phase of the learning process—implementation in practical business situations. This chapter presents five different cases that encompass all the major topics and materials covered in this book.

Appendix A, "Security Checklist" presents a checklist of security dos and don'ts that impact database security.

Appendix B, "Auditing Security" presents a checklist of auditing dos and don'ts for database auditing.

Features

To help you in fully understanding database security and auditing, this book includes many features designed to enhance your learning experience.

- **Chapter Objectives.** Each chapter begins with a detailed list of the concepts to be mastered within that chapter. This list provides you with both a quick reference to the chapter's contents and a useful study aid.
- **Illustrations and Tables.** Numerous illustrations of security and auditing concepts and models are presented supplement text discussion. In addition, the many tables provide details of database objects presented in this book as well as some of the practical and theoretical information.
- **Hands-On Projects.** Although it is important to understand the database and auditing concepts, it is more important to improve and build your knowledge with real life implementation of hands-on projects. Each chapter provides includes several Hands-On Projects aimed at providing you with practical implementation of concepts and scenarios covered in each chapter. These projects cover both Oracle10*g* and Microsoft SQL Server 2000.
- **Chapter Summaries.** Each chapter's text is followed by a summary of the concepts introduced in that chapter. These summaries provide a helpful way to review the ideas covered in each chapter.
- **Review Questions.** The end-of-chapter assessment begins with a set of review questions that reinforce the ideas introduced in each chapter. These questions help you evaluate and apply the material you have learned. Answering these questions will ensure that you have mastered the important concepts of database security and auditing.

- **Case Projects.** Located at the end of each chapter are Case Projects. In these extensive exercises, you implement the skills and knowledge gained in the chapter through real design and implementation scenarios.
- **Glossary.** For easy reference, a glossary at the end of the book lists the key terms in alphabetical order along with definitions.
- **Step-by-step demonstrations.** These are essential components of the book illustrating in detail how to implement most of the concepts presented in each chapter.
- **Scenarios.** Embedded within the chapter text, scenarios are very helpful in relating the concepts to real life situations.

Text and Graphic Conventions

Wherever appropriate, additional information and exercises have been added to this book to help you better understand the topic at hand. Icons throughout the text alert you to additional materials. The icons used in this textbook are described below.

The Note icon draws your attention to additional helpful material related to the subject being described.

Each Hands-On activity in this book is preceded by the Hands-On icon and a description of the exercise that follows. The Hands-On icon also appears in the chapter to identify which Hands-On Project provides practice for the current topic.

Case Project icons mark Case Projects, which are scenario-based assignments. In these extensive case examples, you are asked to implement independently what you have learned.

Special icons highlight information in the book that explain how you can secure information using people, products, or procedures.

Instructor's Materials

The following additional materials are available when this book is used in a classroom setting. All of the supplements available with this book are provided to the instructor on a single CD-ROM. You can also retrieve these supplemental materials from the Course Technology Web site, *www.course.com*, by going to the page for this book, under "Download Instructor Files & Teaching Tools."

Electronic Instructor's Manual. The Instructor's Manual that accompanies this textbook includes the following items: additional instructional material to assist in class preparation, including suggestions for lecture topics; recommended lab activities; tips on setting up a lab for the Hands-On Projects; and solutions to all end-of-chapter materials.

ExamView Test Bank. This cutting-edge Windows-based testing software helps instructors design and administer tests and pretests. In addition to generating tests that can be printed and administered, this full-featured program has an online testing component that allows students to take tests at the computer and have their exams automatically graded.

PowerPoint Presentations. This book comes with a set of Microsoft PowerPoint slides for each chapter. These slides are meant to be used as a teaching aid for classroom presentations, to be made available to students on the network for chapter review, or to be printed for classroom distribution. Instructors are also at liberty to add their own slides for other topics introduced.

Figure files. All of the figures and tables in the book are reproduced on the Instructor's Resource CD, in bitmap format. Similar to PowerPoint presentations, these are included as a teaching aid for classroom presentation, to make available to students for review, or to be printed for classroom distribution.

Lab Requirements

To the User

This book is divided into three parts and each part is designed to be read in sequence from beginning to end. Each chapter in the Database Security part builds on preceding chapters to provide a solid understanding of all the necessary concepts and practical implementations of security in database applications. Also, each chapter in the Database Auditing builds on preceding chapters to provide a comprehensive understanding of auditing from a database perspective. The last part of the book is designed to provide the reader with practical cases using all concepts learned in previous parts of the book.

Hardware and Software Requirements

The following are the software requirements needed to perform cases and code presented in the chapter and the end-of-chapter materials:

- Oracle10*g* (10.1.0.2.0)
- Windows SQL Server 2000
- Oracle Enterprise Manager
- Oracle Policy Manager
- Windows 2000

Please note that SQL Server will not install on Windows XP. For more information on the SQL Server 2000 system requirements, please refer to the Microsoft™ Web site at: *http://www.microsoft.com/sql/evaluation/sysreqs/2000/default.asp*

Specialized Requirements

The code presented in this book requires a good understanding of Oracle PL/SQL language and Microsoft SQL Server 2000 Transact-SQL language. In some instances knowledge of Oracle and Microsoft SQL Server administration may be required.

Special Acknowledgments

I would like to thank Jason Penniman for his contribution of writing the Microsoft SQL Server code presented in this book. His talent and skills never failed to amaze me.

Acknowledgments

The completion of this book is attributed to every member of the team that worked diligently on this project. My thanks to Mac Mendelsohn, Vice President, Product Technology Strategy, for giving me the opportunity to write this book, Maureen Martin for her support, and Eunice Yeates-Fogle for patiently managing this project. The Quality Assurance Team, Chris Scriver, Serge Palladino, and Burt LaFountain did a terrific job, as did Brooke Booth and Cecile Kaufman who shepherded the manuscript through the production process. Special thanks to my friends for their support: Robert Payne, Garry Boyce, Sou Chon Young, Bob Hurley, Vinnie Falcone, Barbara Griffin, and to the reviewers and production team. I am indebted to the following individuals for their respective contributions of perceptive feedback on the initial proposal, the project outline, and the chapter-by-chapter reviews of the text:

- Randy Weaver, Everest College
- Barbara Nicolai, Purdue University Calumet
- Anthony Dashnaw, Clarkson University
- Michelle Hansen, Davenport University
- Kenneth Kleiner, Fayetteville Technical Community College
- G. Shankar, Boston University
- Dan Rafail, Lansing Community College
- Ylber Ramadani, George Brown College
- Debbie Rasnick, Virginia Highlands Community College
- John Russo, Wentworth Institute of Technology
- Arjan Sadhwani, San José State University
- Ningning Wu, University of Arkansas at Little Rock

Dedication

I dedicate this book to my beautiful, beloved, and devoted wife whose love and support is never ending, and to the pearls of my life: my daughter, Aya, and my sons, Wissam and Sammy.

About the Author

Hassan A. Afyouni has been working in the information technology field as a consultant for over fifteen years as database developer, database architect, database administrator, and data architect. He has been an instructor at several universities in Canada and the United States, a corporate trainer for some major corporations, and a curriculum developer for various courses and programs.

PART ONE

Security Architecture

1

LEARNING OBJECTIVES:

Upon completion of this material, you should be able to:

- Define security
- Describe an information system and its components
- Define database management system functionalities
- Outline the concept of information security
- Identify the major components of information security architecture
- Define database security
- List types of information assets and their values
- Describe security methods

Introduction

A quick look at security statistics reveals that security violations and attacks are increasing globally at an annual average rate of 20%. Statistics show that virus alerts, e-mail spamming, identity theft, data theft, and other types of security breaches are also on the rise. Rising at a faster rate are the related costs for preventive and protective measures. In response to this situation, organizations are focusing more heavily on the security of their information. This book places you in the role of a database administrator who is responding to this increasing focus on security by strengthening the security of your organization's database. The first part of this book deals with topics that enable you to implement security measures on your database to protect your data from a variety of violations. To prepare for the technical discussions in the chapters to follow, this chapter presents an introduction to concepts such as general security, information systems, database management systems, and information security—all of which act as the basis for database security.

To gain an understanding of the issues you would face as a database administrator trying to implement increased security, consider the following scenarios. They give you a feeling for the types of security topics covered by the first half of this book.

- A prominent institution hires you to manage a team of skillful database developers, architects, and administrators. Your first challenge on the job is to design and implement a new database security policy to secure data and prevent data integrity violations.
- You are a database administrator for a small startup company. Your company just won a contract from a large, reputable organization to implement a new database application. One of the requirements is to enforce a stringent security policy, which was never before a priority for your company.
- You are a database developer assigned to a new project, which involves the latest technology. As you read the functional specification of the module you are to build, you discover that the data to be stored must be encrypted.

These are a few of the many scenarios you're likely to encounter as you progress through the world of work. This chapter covers both security principles and implementation, in general, and database security, more specifically.

Not long ago, most companies protected their data simply by preventing physical access to the servers where the data resided. This practice was sufficient until several incidents occurred in which data was jeopardized, compromised, and hijacked. Corporations quickly moved to enforcing security measures via operating systems, which prevented data violations by requiring the authentication of the identity of computer users. This approach was successful until new vulnerabilities and new threats brought different types of risks to database systems and applications.

Database management systems that depend on operating systems cannot survive without the implementation of security models that enforce strict security measures. Most database management systems did not have a secure mechanism for authentication and encryption until recently, when serious research and development was initiated to add security components that enable database administrators to implement security policies.

Yesterday's DBA was equipped with all sorts of technical skills that empowered him or her to manage a database efficiently. Today's DBA is required to have an additional skill—that of implementing security policies that protect one of the most valuable assets of a company—its data.

Regardless of your job title, as a team member of a corporation that employs database applications, you must be prepared to protect your company from a variety of security threats. This chapter is designed to increase your expertise and knowledge so that you will be prepared for your database security responsibilities. The chapter presents an overview of several fundamental concepts essential to implementing the security of a database environment.

Security

You have just arrived at your office after a restful vacation. The minute you open the office door, you are shocked to see that all the locked drawers are open, your work files are missing, and your computer has disappeared. You immediately start to list in your head the most sensitive confidential files that are missing. You begin to panic as you consider what would happen if the information within these files were leaked to the public. You remind yourself that the new project you have spent months developing is gone and someone could be selling it to other vendors. Your anxiety rises when you consider what could have happened to you personally if the incident had happened while you were working late in your office. A few minutes later, your manager steps into your office to tell you that the company had been forced to conduct an unexpected audit, and that all the sensitive information in your office had been temporarily moved to an area where the auditors were working. This scenario involves the sense of personal security, which is best described as the level and degree of being free of danger and threats.

The subject of this book is database security. As you begin this book, it is important to know just what that is. **Database security** is the degree to which all data is fully protected from tampering or unauthorized acts. However, this definition is not entirely complete. To fully understand the definition, you need to take a quick tour of various information systems and information security concepts. The following sections dip into these topics to build a foundation for defining and understanding database security.

Information Systems

In today's global market, corporations all over the world are competing to gain a portion of market share. In some cases, corporations are striving to dominate a sector of the market, and in other cases they are just trying to stay afloat and survive. Regardless of the goals of these businesses, their success is usually attributed to the wise decisions of the CEOs. Wise decisions are not made without accurate and timely information. At the same time, the integrity of that information depends on the integrity of its source data and the reliable processing of that data. Data is processed or transformed by a collection of components working together to produce and generate accurate information. These components are known as an **information system**.

An information system can be the backbone of the day-to-day operations of a company as well as the beacon of long-term strategies and vision. Information systems can be categorized based on usage. Figure 1-1 illustrates the typical management pyramid showing the category of information system used in each level of management. For example, lower-level management uses information systems that assist management and employees with operational tasks, such as inventory systems or point-of-sale (POS) systems.

Middle-level management uses systems that deal with midterm goals, such as a forecasting systems that project sales for the following quarter. Upper-level management works with systems that assist with long-term goals, such as business model simulation and reasoning.

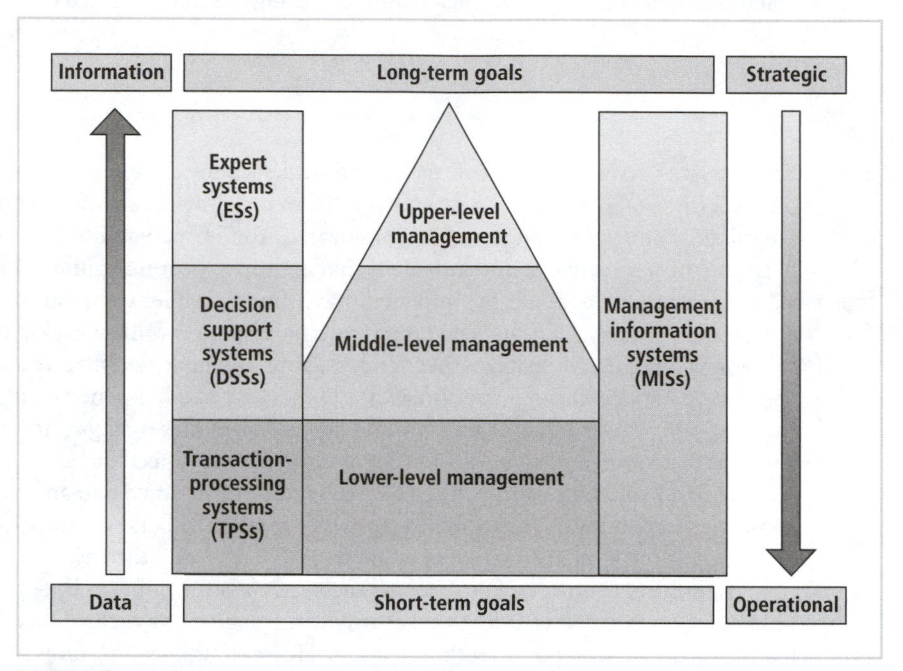

FIGURE 1-1 Typical use of system applications at various management levels

As illustrated in Figure 1-1, information systems are classified mainly into three distinct categories: transaction-processing systems, decision support systems, and expert systems. Table 1-1 describes the characteristics and typical applications for each type of system.

TABLE 1-1 Characteristics of information system categories

Category	Acronym	Characteristics	Typical Application System
Transaction-processing system	TPS	▪ Also known as online transaction processing (OLTP) ▪ Used for operational tasks ▪ Provides solutions for structured problems ▪ Includes business transactions ▪ Logical component of TPS applications (derived from business procedures, business rules, and policies)	▪ Order tracking ▪ Customer service ▪ Payroll ▪ Accounting ▪ Student registration ▪ Car sales

TABLE 1-1 Characteristics of information system categories (continued)

Category	Acronym	Characteristics	Typical Application System
Decision support system	DSS	▪ Deals with nonstructured problems and provide recommendations or answers to solve these problems ▪ Is capable of performing "What-if?" analysis ▪ Contains a collection of business models ▪ Is used for tactical management tasks	▪ Risk management ▪ Fraud detection ▪ Sales forecasting ▪ Case resolution

Expert system	ES	▪ Captures reasoning of human experts
		▪ Executive expert systems (ESSs) are a type of expert system used by top-level management for strategic management goals
		▪ A branch of artificial intelligence within the field of computer science studies
		▪ Software consists of:
		  ▪ Knowledge base
		  ▪ Inference engine
		  ▪ Rules
		▪ People consist of:
		  ▪ Domain experts
		  ▪ Knowledge engineers
		  ▪ Power users
			▪ Virtual university simulation
			▪ Financial enterprise
			▪ Statistical trading
			▪ Loan expert
			▪ Market analysis

Regardless of the type of information system and purpose, an information system consists of the following components (see Figure 1-2 for an illustration of a typical information system):

- *Data*—Collected data and facts used as input for system processing, and data stored in the database for future reference or processing
- *Procedures*—Includes manual procedures, guidelines, business rules, and policies implemented in the system or used as part of the system
- *Hardware*—Computer systems and devices such as disks, chips, faxes, scanners, and printers
- *Software*—Application code, languages used to develop code, database management system, operating system used, and any other utilities or tools
- *Network*—A communication infrastructure to connect client processes to the system
- *People*—Users, managers, business analysts, programmers, system analysts, database administrators, and system administrators

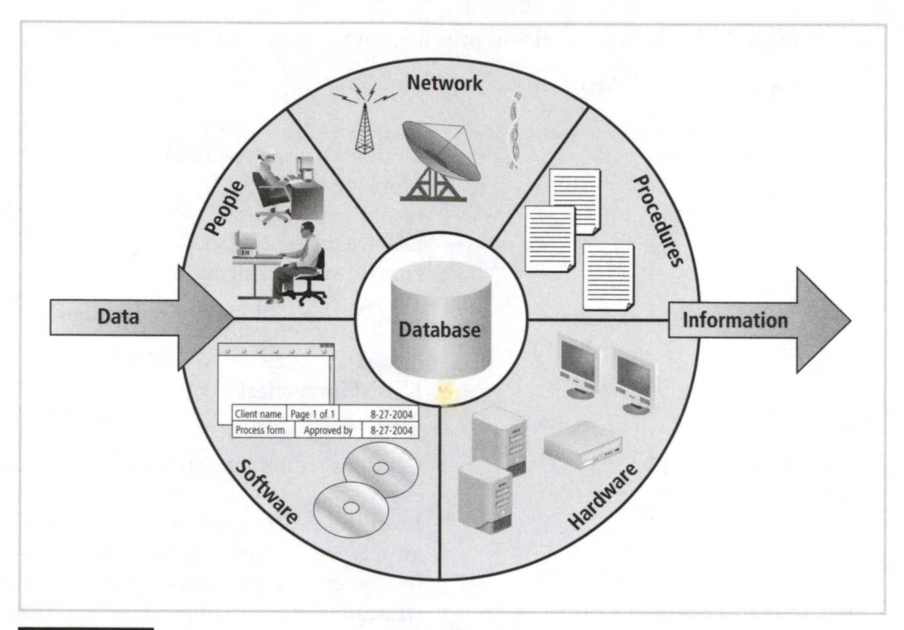

FIGURE 1-2 Information system components

Figure 1-2 shows that data is entered into the system to be processed immediately or to be stored in a database to be processed later when needed. The database is a core component in today's most commonly used system architecture, which is a form of the client/server architecture that was introduced in mid-1990s. The success of client/server architecture is due to the flexibility and scalability that it offers to system architects.

The concept behind a client/server application is based on the business model of a customer ordering a service or product and the representative of a business granting that request. In the client/server environment, you can think of the client as the customer and the server as the representative of a business granting the client's request. The client/server architecture can be implemented as one-tier, two-tier, and *n*-tier designs. A tier is a logical or physical platform. From a physical point of view, single-tier architecture is characterized by the client and server components residing on the same hardware platform. From a logical perspective, single-tier architecture is characterized by the client and the server coexisting as one component. A component is a logical (software) module such as a function, process, or a program. Figure 1-3 illustrates one-tier, two-tier, and three-tier client/server architecture. For example, a two-tier architecture can be composed of a front-end module used to validate data and to submit requests to the database server that processes and responds to the client-submitted requests.

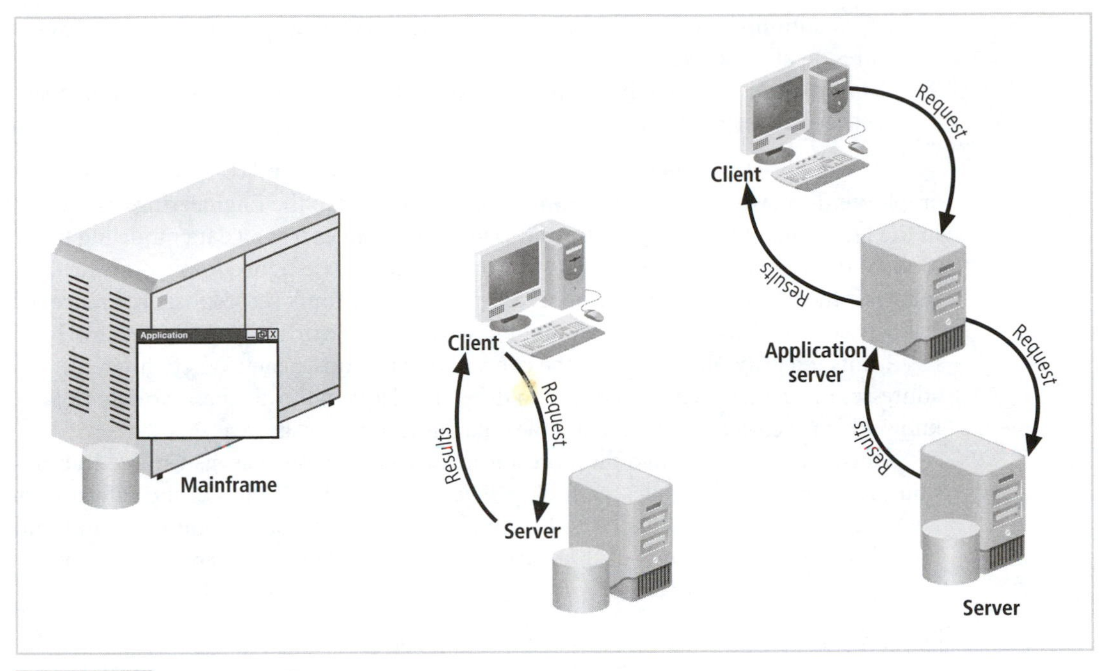

FIGURE 1-3 Examples of different client/server tier design

The client/server architecture is composed of three layers: the first is the user interface, which is typically the client; the second is the network layer, which is the backbone of the application architecture; and the third is the core of the client/server architecture, which responds to all requests submitted by the client (this third is the database server layer). In other words, all applications use some sort of a database server. The database is managed by a collection of programs whose main purpose is to allow users to store, manipulate, and retrieve data efficiently. The collection of programs that manage the database is known as a database management system (DBMS). The next section presents an overview of the architecture and functions of database management systems.

Database Management Systems

As the database is an integral part of an information system, the need for reliable and efficient programs to manage the database becomes essential to the success of the information system. Although many corporations develop DBMSs in which each DBMS has a distinct implementation and architecture, they all have the following basic common functionalities:

- Allow developers and administrators to organize data in an orderly fashion.
- Allow users to store and retrieve data efficiently.
- Allow users to manipulate data (update and delete).
- Enable developers and administrators to enforce data referential integrity and consistency. Data is considered to have referential integrity when a relationship between two tables is always maintained (never broken).

- Allow administrators to enforce and implement data security policies and procedures on all database levels.
- Allow administrators to back up data in case of a failure and provide a mechanism to recover and restore data.

Here is a brief example of how data can lose its integrity and consistency. An employee, Tom, who is in the Employee table, is assigned to the Engineering department in the Department table (Employee and Department tables from a data modeling point of view are related 1 to 1. This means that one employee is assigned to only one department). If the Engineering department record is deleted, Tom's record loses reference to the department. This means you do not know Tom's department, and the data for Tom has lost its integrity. Now examine the concept of data consistency. When different addresses for the same employee exist in different places in the database, you do not know which is correct, and therefore the data has lost its consistency.

Of course, a DBMS can offer more advanced functions such as distributed transactions, replication, and parallel processing. Figure 1-4 provides a view of the database and DBMS environment that illustrates the similarity between those environments and information systems. Both consist of the same components—data, hardware, software, networks, procedures, and database servers.

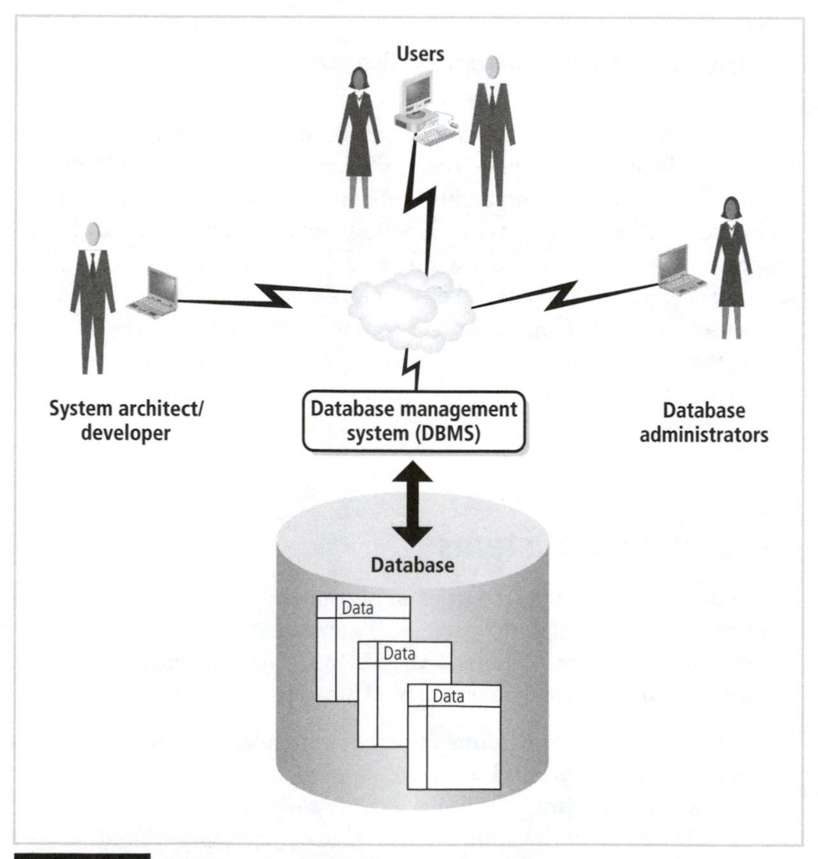

FIGURE 1-4 Database and DBMS environment

By this point, the chapter has presented a quick overview of security, information systems, and database management systems. Now that you have some understanding of those topics, it is time to tie the topics together.

Information Security

Security, as defined earlier, is the sense of feeling safe or protected from deliberate or accidental threats. So what about information security? Information is one of an organization's most valuable assets, and many companies have an Information Security department that protects the information and assures employees and managers that the information is safe. Information is safe if it is protected from access by unauthorized users. At the same time, to be useful, information must be accessible at all times to authorized users. **Information security** consists of the procedures and measures taken to protect each component of the information systems involved in producing information. This means protecting data, hardware, software, networks, procedures, and people—all the components of the information system.

According to the National Security Telecommunications and Information Systems Security Committee (NSTISSC), the concept of information security is based on the **C.I.A. triangle**, in which "C" stands for Confidentiality, "I" for Integrity, and "A" for Availability. The C.I.A. triangle is a framework for protecting information. The C.I.A triangle should guide your efforts to enforce information integrity and shield data from being tampered with by unauthorized persons, being modified accidentally by employees, or losing consistency because of the incorrect coding of business requirements and rules. Ensuring that the information system is available when it is needed and at the same time protected from downtime caused by external or internal attacks or threats can be a difficult balancing act. To achieve this balance, you must establish security policies that are not so stringent as to make data inaccessible. Finally, you should not lose sight of confidentiality. Sensitive data and information should be kept secret and only divulged based on data and classification. Figure 1-5 illustrates the C.I.A. triangle.

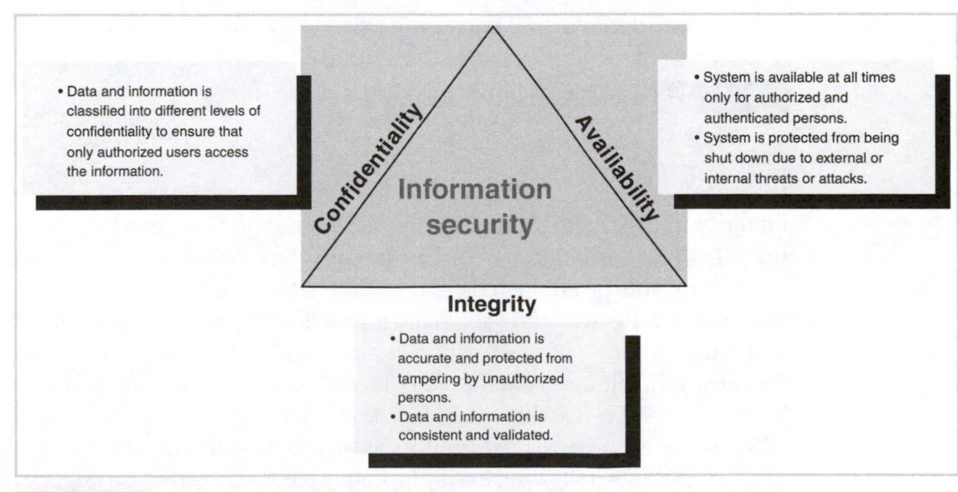

FIGURE 1-5 Information security C.I.A triangle

Confidentiality

As indicated in the previous section, **confidentiality** is one of the three principles of the C.I.A. triangle. Confidentiality addresses two aspects of security that have subtle differences. The first aspect is the prevention of unauthorized individuals from knowing or accessing secret information. The second aspect is the process of safeguarding confidential information and disclosing secret information only to authorized individuals by means of classifying information. If either of these two factors is violated, the confidentiality principle of the C.I.A. triangle is breached and information security is at risk.

From this discussion, you may conclude that this balancing act is hard to achieve, if not impossible. It *is* difficult to implement but not impossible if you properly classify your information and design a process to implement and enforce confidentiality. You should classify your company's information into different levels—each level having its own security measures. To devise an effective classification system, you need to understand that classification schemes vary with different companies, government agencies, and other institutions. What determines classification is the type of business and its policies and procedures. However, companies usually classify information based on the degree of confidentiality necessary to protect that information. Figure 1-6 presents a model that can be adapted to implement controls for each level.

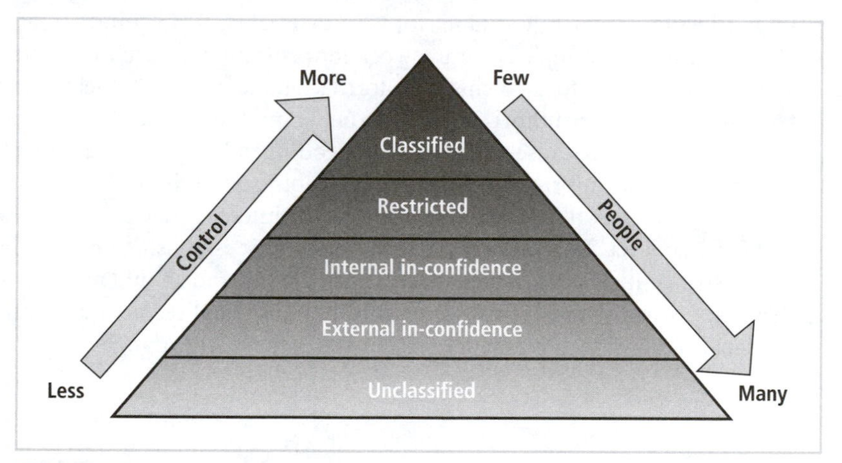

FIGURE 1-6 Confidentiality classification

Integrity

Integrity is the second principle of the C.I.A. triangle. For information integrity to exist, the data upon which it is based must be consistent and accurate throughout the system. You've probably heard the old expression "Garbage in, garbage out." For security, this means that consistent and valid data, if processed correctly, yields accurate information. The integrity aspect of information security is vital, because it focuses your attention on the most valuable asset, data, which in turn becomes information. Data is considered to have integrity if it is accurate and has not been tampered with intentionally or accidentally. Data must be protected at all levels to achieve full integrity.

Consider the following example. Employee A learns that his adversarial coworker in the next cubicle is earning a higher salary than he is. Somehow, employee A accesses an application program used by the Accounting department and manipulates the vacation hours and overtime hours of his colleague. Two security violations have occurred. First, sensitive data (salary), which is supposed to be confidential, was disclosed or obtained inappropriately. This is a violation of the confidentiality principle. Second, the disgruntled employee gained access to an application that allowed him to modify data. This is a violation of data integrity. These violations are connected and also interconnected with the third C.I.A. principle—availability. A security failure occurs when the application fails to detect this malicious act through an audit mechanism or other data controls that should be in place. An example of such a control in this case would be an application that cross-checks overtime hours against actual time cards, computes vacation hours, and verifies entered values. If the computed and entered values are different, the application should require an approval override from another person.

The integrity of the information system is measured by the integrity of its data. For the integrity of the data to be considered valid, it must avoid the pitfalls summarized in Table 1-2. One of the pitfalls is losing read consistency. When working with data that has **read consistency**, each user sees only his own changes and those that have been committed by other users.

TABLE 1-2 Degradation of data integrity

Type of Data Degradation	Description	Reasons for Data Losing Integrity
Invalid data	Indicates that not all the entered and stored data is valid without exception; checks and validation processes (known as database constraints) that prevent invalid data are missing.	▪ User enters invalid data mistakenly or intentionally. ▪ Application code does not validate inputted data.
Redundant data	Occurs when the same data is recorded and stored in several places; this can lead to data inconsistency and data anomalies.	▪ Faulty data design that does not conform to the data normalization process. (**Normalization** is a database design process used to reduce and prevent data anomalies and inconsistencies.)
Inconsistent data	Occurs when redundant data, which resides in several places, is not identical.	▪ Faulty database design that does not conform to the data normalization process.
Data anomalies	Exists when there is redundant data caused by unnormalized data design; in this case, data anomalies occur when one occurrence of the repeated data is changed and the other occurrences are not.	▪ Faulty data design that does not conform to the data normalization process.

TABLE 1-2 Degradation of data integrity (continued)

Type of Data Degradation	Description	Reasons for Data Losing Integrity
Data read inconsistency	Indicates that a user does not always read the last committed data, and data changes that are made by the user are visible to others before changes are committed.	▪ DBMS does not support or has weak implementation of the read consistency feature.
Data nonconcurrency	Means that multiple users can access and read data at the same time but they lose read consistency.	▪ DBMS does not support or has weak implementation of the read consistency feature.

Availability

Suppose you are asked to write a prescription for a corporation's success. You will probably prescribe three treatments: technology innovation and implementation, high-quality products, and excellent customer care and service. When a corporation skips any of these treatments, it probably loses the competitive edge and thus loses market share.

You may be asking yourself, "How is availability related to security?" To answer that question, consider this scenario. A prominent dot-com company sells a variety of products over the Web. You want to purchase a product, but when you try to visit the Web site, you receive an error message saying the site is unavailable. You call the company's customer service number to get more information about the product, but to your surprise, the customer service representative informs you that their system is not available and that you should call back.

If incidents such as these occur frequently, customers lose confidence in a company, the company loses customers, and eventually loses market share as well. Regardless of the reasons that led to system unavailability, the result is unsatisfied customers. Now, put system design and implementation aside and explore why a system becomes unavailable from a security point of view. An organization's information system can become unavailable because of the following security issues:

- External attacks and lack of system protection
- Occurrence of system failure with no disaster recovery strategy
- Overly stringent and obscure security procedures and policies
- Faulty implementation of authentication processes, which causes failure to authenticate customers properly

The **availability** principle with respect to information security means that the system should be available (accessible) to individuals who are authorized to access the information, and the system should determine what an individual can do with that information.

Information Security Architecture

An information system, as defined earlier, is a collection of components working together to solve a problem. Because data is processed into viable information by the information system, security becomes an important aspect of the system. This means that the information system must protect data and the information produced from the data from having its confidentiality, integrity, and availability violated on any layer. This section expands on the concept of information security by describing other aspects that make up the infrastructure required to build security procedures and policies.

Figure 1-7 shows that **information security architecture** is a model for protecting logical and physical assets. Information security architecture is the overall design of a company's implementation of the C.I.A. triangle. The architecture's components range from physical equipment to logical security tools and utilities. You can see in Figure 1-7 that if any of the principles of the C.I.A. triangle is violated, the information security model will fail to protect the company's logical or physical assets.

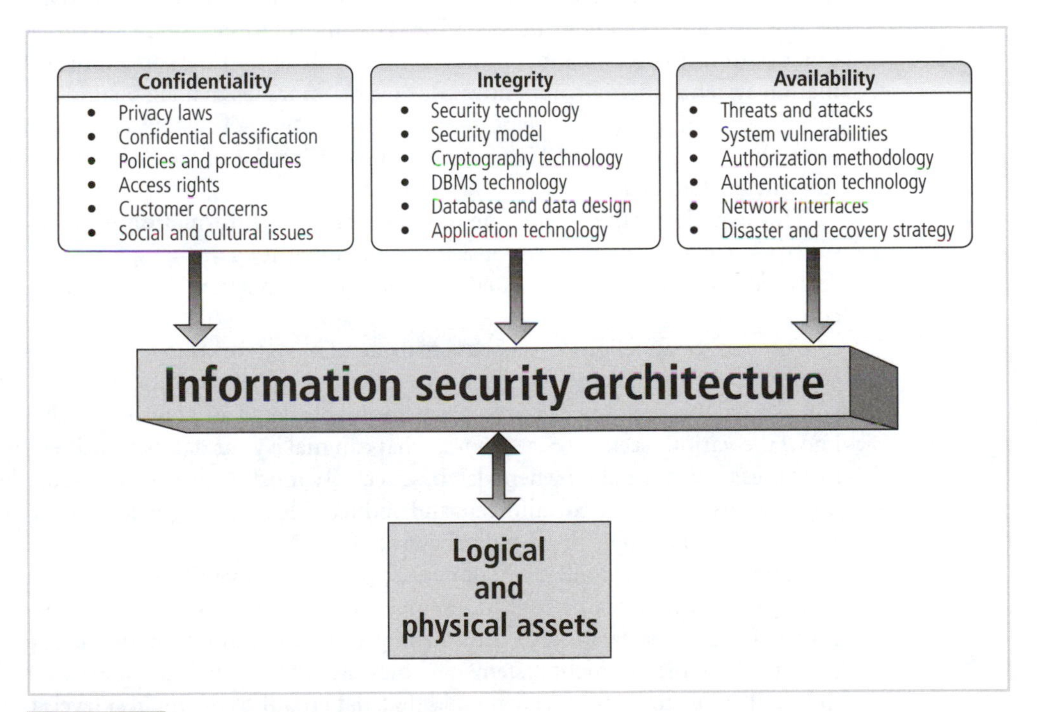

FIGURE 1-7 Information security architecture

The following list outlines the components of information security architecture:

- *Policies and procedures*—Documented procedures and company policies that elaborate on how security is to be carried out
- *Security personnel and administrators*—People who enforce and keep security in order

- *Detection equipment*—Devices that authenticate employees and detect equipment that is prohibited by the company
- *Security programs*—Tools that protect computer systems' servers from malicious code such as viruses
- *Monitoring equipment*—Devices that monitor physical properties, employees, and other important assets
- *Monitoring applications*—Utilities and applications used to monitor network traffic and Internet activities, downloads, uploads, and other network activities
- *Auditing procedures and tools*—Checks and controls put in place to ensure that security measures are working

Database Security

Business corporations and government institutions depend more and more on information technology as the sole tool for processing and storing data. This increased reliance on information technology in general, and on information systems specifically, allows organizations to become more productive and efficient. At the same time, use of information systems offers a competitive edge over companies that lag behind in technology. Reliance on information systems does not come without a cost. In fact, technology has not only introduced societal issues and problems, it has also created a vast range of security threats that could result in devastating situations.

Information is the foundation of knowledge, and information is not accurate if its source—data—does not have consistency and integrity. For this specific reason, most corporations employ sophisticated information systems that have a database component.

One of the functions of database management systems is to empower the database administrator to implement and enforce security at all levels of the database. In order for you as a database administrator to protect valuable data stored in the database, you must know the various security access points that can make your database vulnerable. A **security access point** is a place where database security must be protected and applied—in other words implemented, enforced, and audited. This section presents a list of security access points that apply to most databases.

Figure 1-8 presents all the major access points within a database environment where security measures must be applied, enforced, and audited. Figure 1-8 represents all the components of the database environment: people, applications, networks, operating system, database management system, data files, and data. Data is the most valuable asset of the database environment. Having said that, data requires the highest levels of protection, and therefore its data access point must be the smallest of all the components shown in Figure 1-8.

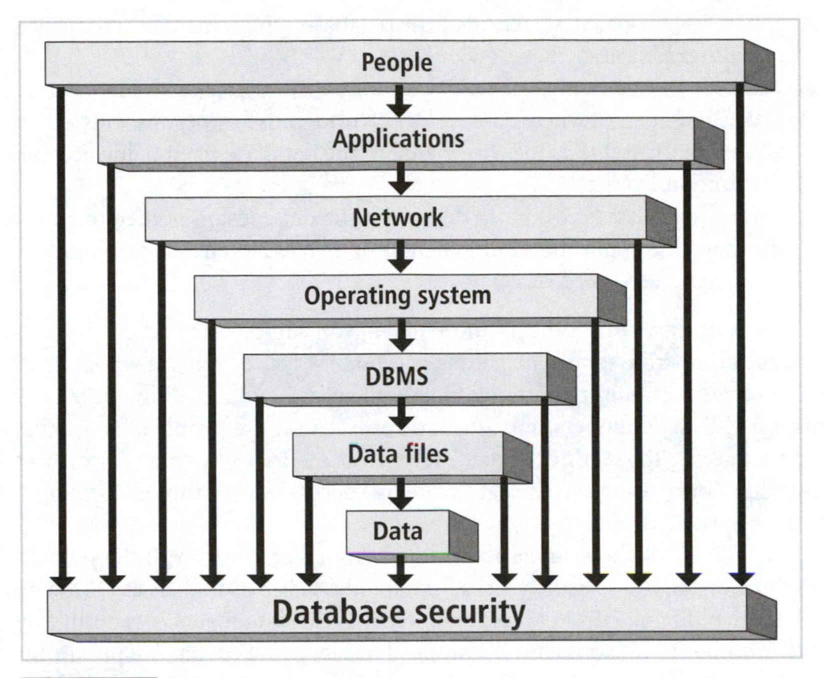

FIGURE 1-8 Database security access points

The security access points illustrated in Figure 1-8 are:

- *People*—Individuals who have been granted privileges and permissions to access applications, networks, workstations, servers, databases, data files, and data. This means that people represent a risk of database security violations. Therefore database security must entail all necessary measures to secure the data within the database against potential violations caused by people.
- *Applications*—Application design and implementation, which includes privileges and permissions granted to people. If these permissions are too loose, individuals can access and violate data. If these permissions are too restrictive, they do not allow users to perform their responsibilities. When granting security privileges to applications, be extremely cautious.
- *Network*—One of the most sensitive security access points. Be sure to use your best efforts to protect the network and provide network access only to applications, operating systems, and databases.
- *Operating system*—The operating system access point is defined as authentication to the system—the gateway to the data. For example, to access the data residing in a system, you must log on and your security credentials must be verified. The absence of good security measures at this access point is the cause of most security violations.

- *DBMS*—The logical structure of the database, which includes memory, executables, and other binaries.
- *Data files*—Another access point that influences database security enforcement is access to data files where data resides. Through use of permissions and encryption, you must protect data files belonging to the database from being accessed by unauthorized individuals.
- *Data*—This data access point deals with the data design needed to enforce data integrity, the application implementation needed to ensure data validity, and the privileges necessary to access data.

Examining access points in more detail, you can see each access point as a component of the entire system. The people component is the largest area because there is often a huge community of individuals who access data, including users, managers, visitors, outsiders, developers, and administrators. All these people increase the possibility of endangering the security of data. Therefore, security efforts and measures should be directed at decreasing the risks at the people access points, thus decreasing threats from people.

In Figure 1-8, the data file access point is smaller than any of the points above it, which means that the security risks for data files is not as high as at DBMS access points. Therefore reducing DBMS access points makes the data files access point even less accessible. Another point you may have noticed in Figure 1-8 is that the proximity of database security to the access point indicates how close you are to database security violations, and the area of the access point indicates the security risk. Having said that, Figure 1-8 indicates that you must start securing the database with people access points, followed by applications, and so on.

To see the other side of the coin, examine Figure 1-9, which shows that when the area size of the people access point is reduced, the only access to data is through all access points (layers or levels) above. Reducing access point size reduces security risks, which in turn increases database security.

As defined earlier, a security access point is a point at which security measures are needed to prevent access that can involve unauthorized actions. It is worth noting that security access points should not to be confused with security gaps or vulnerabilities. **Security gaps** are points at which security is missing, and thus the system is vulnerable. **Vulnerabilities** are kinks in the system that must be watched because they can become threats. In the world of information security, a **threat** is defined as a security risk that has a high possibility of becoming a system breach. The breach can be caused by either intentional or unintentional actions. Figure 1-10 shows the process of a security gap eventually resulting in a security breach. To complete this picture you need to know the formal definition of each security access point of the database environment, as defined earlier in this section and illustrated in Figure 1-8.

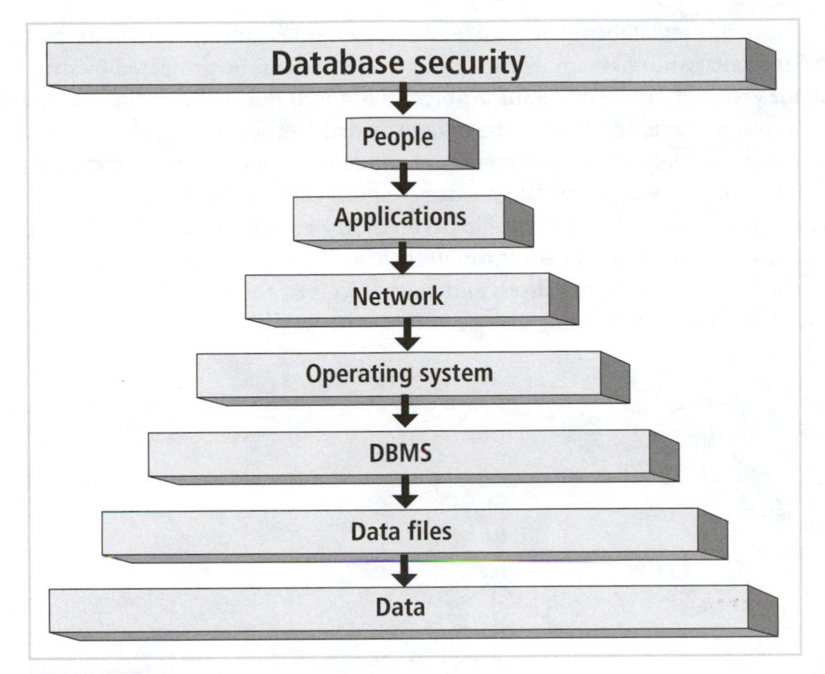

FIGURE 1-9 Database security enforcement

FIGURE 1-10 Data integrity violation process

Database Security Levels

As you know, a relational database is a collection of related data files; a data file is a collection of related tables; a table is a collection of related rows (records); and a row is collection of related columns (fields), as shown in Figure 1-11. As you have noticed, the structure of the database is organized in levels, and each level can be protected by a different security mechanism. For instance, a column can be protected by using a VIEW database object. A VIEW database object is a stored query that returns columns and rows from the selected tables. The data provided by the view object is protected by the database system functionality that allows schema owners to grant or revoke privileges. The data files in which the data resides are protected by the database and that protection is enforced by operating system file permissions. Finally, the database is secured by the database management system through the use of user accounts and password mechanisms as well as by the privileges and permissions of the main database functions—database shutdown, creating user accounts, and database backup and recovery, to name a few.

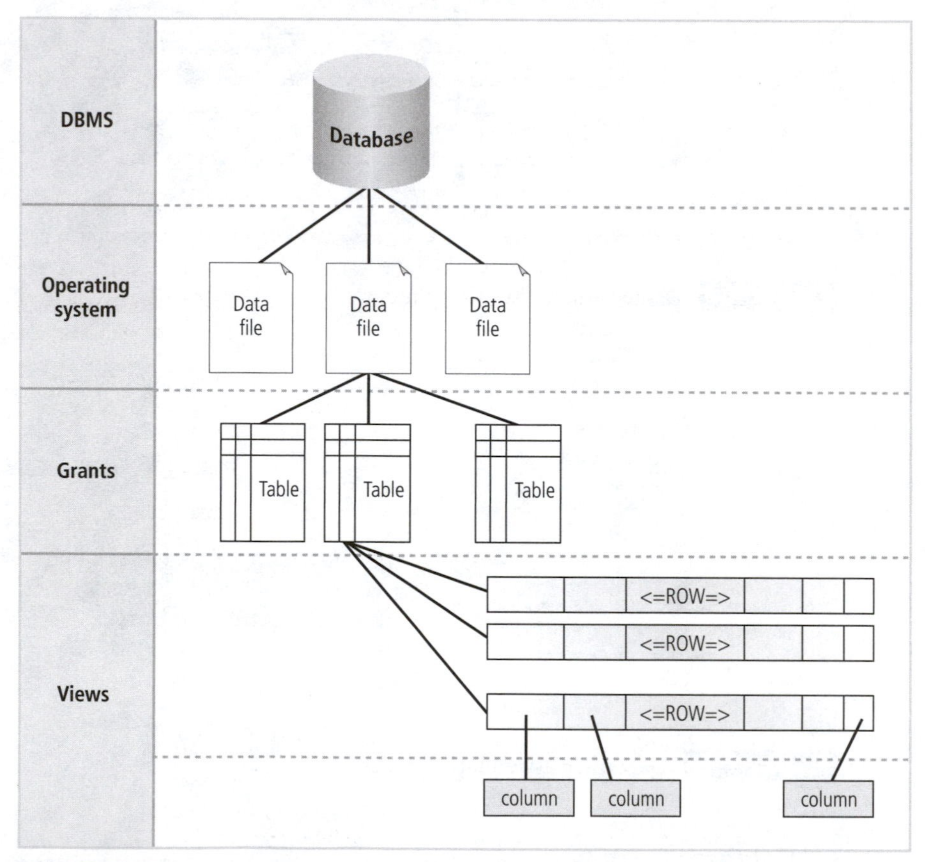

FIGURE 1-11 Levels of database security

Menaces to Databases

The following sections describe the kinds of menaces to database security that are commonly faced by today's organizations. The sections describe database vulnerabilities, threats, and risks. Although these terms have been used previously in this chapter, before proceeding with those descriptions, it is important that you understand the differences among these three terms, subtle as they may be:

- **Security vulnerability**—A weakness in any of the information system components that can be exploited to violate the integrity, confidentiality, or accessibility of the system
- **Security threat**—A security violation or attack that can happen any time because of a security vulnerability
- **Security risk**—A known security gap that a company intentionally leaves open

Types of Vulnerabilities

According to www.dictionary.com, vulnerability means "susceptible to attack." Why is this word important in security? The answer is simple—intruders, attackers, and assailers exploit vulnerabilities in your environment to prepare and start their attacks. From an information security perspective, hackers usually explore the weak points (design or code flaws) of a system until they gain entry through a gap in protection. Once an intrusion point is discovered, hackers unleash their array of attacks on the system, which could be viruses, worms, malicious code (code that could corrupt or adversely alter the state of your computer system), or other types of unlawful violations. To protect your system from these attacks, you must understand the types of vulnerabilities that may be found in your information security architecture. To conduct a review and examination of the different types of database security vulnerabilities, you need to understand how vulnerabilities are categorized. Vulnerability categorization is illustrated in Figure 1-12. A description of each category is presented in Table 1-3 with examples.

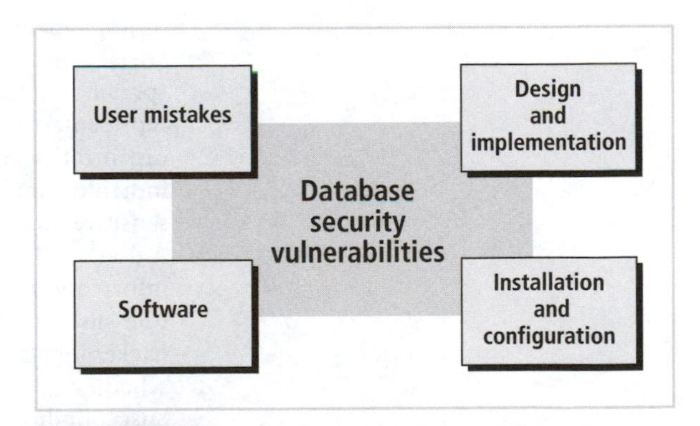

FIGURE 1-12 Categories of database security vulnerablilities

TABLE 1-3 Types of vulnerabilities with definitions and examples

Category	Description	Examples
Installation and configuration	This type of vulnerability results from using a default installation and configuration that is known publicly and usually does not enforce any security measures. Also, improper configuration or installation may result in security risks.	■ Incorrect application configuration that may result in application malfunction ■ Failure to change default passwords ■ Failure to change default permissions and privileges ■ Using default application configuration that leads to security vulnerability, as most applications do not enforce high-security measures for the default setup.
User mistakes	Although all security vulnerabilities are tied to humans, vulnerabilities listed in this category are mainly related to carelessness in implementing procedures, failure to follow through, or accidental errors.	■ Lack of auditing controls ■ Untested disaster recovery plan ■ Lack of activity monitoring ■ Lack of protection against malicious code ■ Lack of applying patches as they are released ■ Bad authentication process or implementation ■ Social engineering (pretending to be a representative of a legitimate organization to trick an individual into providing sensitive information) ■ A user's lack of technical information that leads to user susceptibility to various hacker intrusions and fraud schemes ■ Susceptibility to scams
Software	This category relates to vulnerabilities found in commercial software for all types of programs (applications, operating systems, database management systems, and other programs).	■ Software patches are not applied ■ Software contains bugs ■ System administrators do not keep track of patches

TABLE 1-3 Types of vulnerabilities with definitions and examples (continued)

Category	Description	Examples
Design and implementation	Vulnerabilities of this category are related to improper software analysis and design as well as coding problems and deficiencies.	■ System design errors ■ Exceptional conditions (special cases in which code fails to execute) and errors are not handled in program development ■ Input data is not validated

Type of Threats

Earlier in the chapter, you were shown that in the data integrity violation process, vulnerabilities can escalate into threats. As database administrator, database manager, or information security administrator, you need to be aware of these vulnerabilities and threats to protect your organization and its assets. As with the categorization of vulnerabilities, threats are categorized to ensure that everything that contributes to security risks is covered. Figure 1-13 presents threat categories.

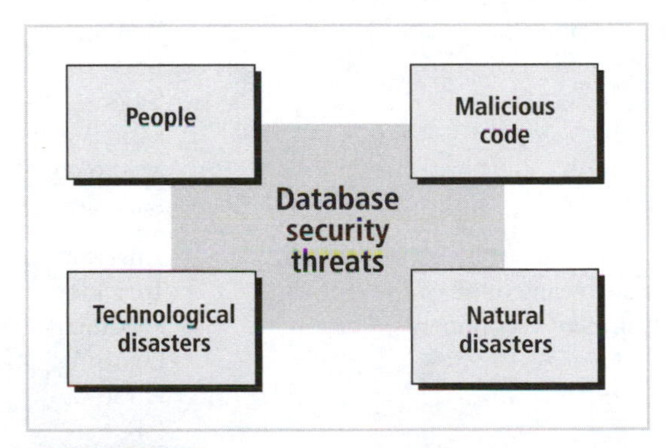

FIGURE 1-13 Categories of database security threats

As shown in Figure 1-13, four types of threats contribute to security risks. According to www.dictionary.com, a threat is defined as "An indication of impending danger or harm."

Table 1-4 defines and offers examples of each type of threat category shown in Figure 1-13.

TABLE 1-4 Threat types, definitions, and examples

Threat type	Definition	Examples
People	People intentionally or unintentionally inflict damage, violation, or destruction to all or any of the database environment components (people, applications, networks, operating systems, database management systems, data files, or data).	■ Employees ■ Government authorities or persons who are in charge ■ Contractors ■ Consultants ■ Visitors ■ Hackers ■ Organized criminals ■ Spies ■ Terrorists ■ Social engineers
Malicious code	Software code that in most cases is intentionally written to damage or violate one or more of the database environment components (applications, networks, operating systems, database management systems, data files, or data).	■ Viruses ■ Boot sector viruses ■ Worms ■ Trojan horses ■ Spoofing code ■ Denial-of-service flood ■ Rootkits ■ Bots ■ Bugs ■ E-mail spamming ■ Macro code ■ Back door
Natural disasters	Calamities caused by nature, which can destroy any or all of the database environment components.	■ Hurricanes ■ Tornados ■ Earthquakes ■ Lightning ■ Flood ■ Fire
Technological disasters	Often caused by some sort of malfunction in equipment or hardware, technological disasters can inflict damage to networks, operating systems, database management systems, data files, or data.	■ Power failure ■ Media failure ■ Hardware failure ■ Network failure

Terms used in the table:

- **Virus**—Code that compromises the integrity and state of a system
- **Boot sector virus**—Code that compromises the segment in the hard disk that contains the program used to start the computer
- **Worm**—Code that disrupts the operation of a system
- **Back door**—An intentional design element of some software that allows developers of a system to gain access to the application for maintenance or technical problems
- **Trojan horse**—Malicious code that penetrates a computer system or network by pretending to be legitimate code
- **Spoofing code**—Malicious code that looks like legitimate code
- **Denial-of-service-flood**—The act of flooding a Web site or network system with many requests with the intent of overloading the system and forcing it to deny service to legitimate requests
- **Rootkits and bots**—Malicious or legitimate software code that performs such functions as automatically retrieving and collecting information from computer systems
- **Bugs**—Software code that is faulty due to bad design, logic, or both
- **E-mail spamming**—E-mail that is sent to many recipients without their permission

A threat can result in a security risk that requires you to employ and execute security measures to prevent or foil security breaches or damage. In the next section you look at the security risks that can emerge from threats.

Types of Risks

Risks are simply a part of doing business. Managers at all levels are constantly working to assess and mitigate risks to ensure the continuity of departmental operations. As part of this game, you need not only to understand your system weaknesses and threats, but to walk the extra mile to diminish the probability of these threats actually occurring. So what are the risks to the security of the database environment? Simply put, the reliability of a database at all levels is at risk, and most importantly the integrity of the data. Figure 1-14 illustrates the categories of database security risks, and Table 1-5 defines those categories.

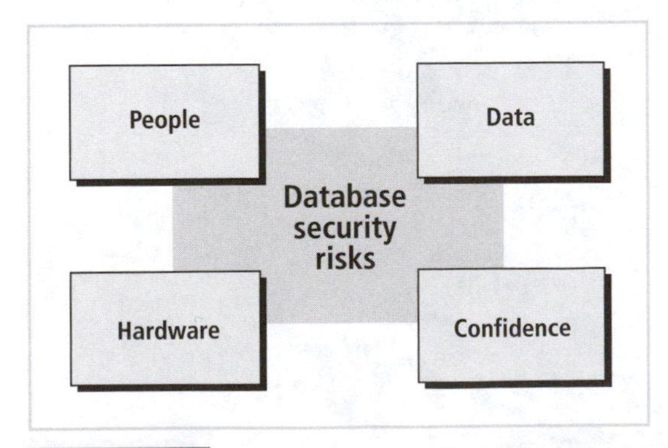

FIGURE 1-14 Categories of database security risks

TABLE 1-5 Definition and examples of risk types

Risk Type	Definition	Example
People	The loss of people who are vital components of the database environment and know critical information about the environment can create risks.	▪ Loss of key persons (resignation, migration, health problems) ▪ Key person downtime due to sickness, personal or family problems, or burnout
Hardware	A risk that mainly results in hardware unavailability or inoperability.	▪ Downtime due to hardware failure, malfunction, or inflicted damage ▪ Failure due to unreliable or poor quality equipment
Data	Data loss and data integrity loss is a major concern of the database administrators and management	▪ Data loss ▪ Data corruption ▪ Data privacy loss
Confidence	The loss of public confidence in the data produced by the company causes a loss of public confidence in the company itself.	▪ Loss of procedural and policy documents ▪ Database performance degradation ▪ Fraud ▪ Confusion and uncertainty about database information

If you were to rate vulnerabilities, threats, and risks according to most the common and important factors, you would list three factors: people, software, and data. The remaining factors act as amplifiers or supporters. Figure 1-15 represents this integration. Figure 1-15 shows that database security involves the protection of the main three components—people, software, and data—from vulnerabilities, which can become threats to the integrity of the system and consequently become a risk to the business operation.

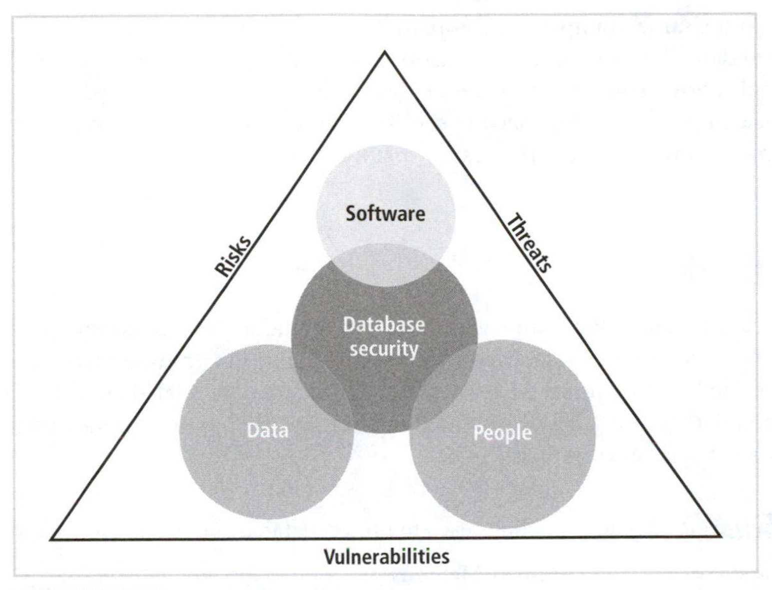

FIGURE 1-15 Integration of security vulnerabilities, threats, and risks in a database environment

Asset Types and Their Value

People always tend to protect assets regardless of what they are. For example, you may keep a memorable picture of your parents in a safe place. However, the degree of protection you provide is directly based on how much you value the assets. If you highly value the picture of your great-grandparents, you might take an extra measure of precaution by copying it and placing it in a fireproof safe where it is guarded from most natural disasters and from theft, or you may just put it in a frame because you have many similar pictures or because you can reproduce it.

Corporations treat their assets in the same way. Assets are the infrastructure of the company operation. Depending on the type of asset and how much the company values it, the company builds security policies and procedures and executes actions to protect these assets. In this section you explore the types of assets that business entities own in order to relate these concepts to database security. There are four main types of assets:

- *Physical assets*—Also known as tangible assets, these include buildings, cars, hardware, and so on
- *Logical assets*—Logical aspects of an information system, such as business applications, in-house programs, purchased software, operating systems, databases, and data
- *Intangible assets*—Business reputation, quality, and public confidence
- *Human assets*—Human skills, knowledge, and expertise

Security measures are implemented based on the value of each asset. For instance, if a company employs a scientist working on an important invention, the company may take extra measures to avoid losing the intellectual asset she represents. Similarly, every component in the database environment is protected according to its value. Continuing with the same example, the company may use no security measures to protect test-generated data that developers and quality assurance engineers use as part of the database application development phases. However, if the information is part of production data, the company probably executes specific security procedures and polices to protect that production data from all types of violations.

Security Methods

Security technology comprises a variety of methods that protect specific aspects of security architecture. In this section you explore the most common methods used to secure the database environment. Only methods that are data related are discussed in this book. Table 1-6 outlines the security methods that are used to protect the different components of a database environment.

TABLE 1-6 Security methods used to protect database environment components

Database Component Protected	Security Methods
People	■ Physical limits on access to hardware and documents ■ Through the processes of identification and authentication, make certain that the individual is who he or she claims to be through the use of devices, such as ID cards, eye scans, and passwords ■ Training courses on the importance of security and how to guard assets ■ Establishment of security policies and procedures
Applications	■ Authentication of users who access applications ■ Business rules ■ Single sign-on (a method for signing on once for different applications and Web sites)
Network	■ Firewalls to block network intruders ■ Virtual private network (VPN) (a remote computer securely connected to a corporate network) ■ Authentication
Operating system	■ Authentication ■ Intrusion detection ■ Password policy ■ User accounts
Database management system	■ Authentication ■ Audit mechanism ■ Database resource limits ■ Password policy
Data files	■ File permissions ■ Access monitoring

TABLE 1-6 Security methods used to protect database environment components (continued)

Database Component Protected	Security Methods
Data	■ Data validation ■ Data constraints ■ Data encryption ■ Data access

A business rule is the implementation of a business procedure or policy through code written in an application.

Database Security Methodology

By this point in this chapter, you have an overview of most of the essential aspects of security architecture. It is time to put the pieces of the database security jigsaw puzzle together to compose a process that will assist you in building your database security. This section presents an implementation process that can be used as a framework or methodology to outline the security tasks required for each stage. As shown in Figure 1-16, this process consists of phases similar to those of most software engineering methodologies, except the focus in each phase is security.

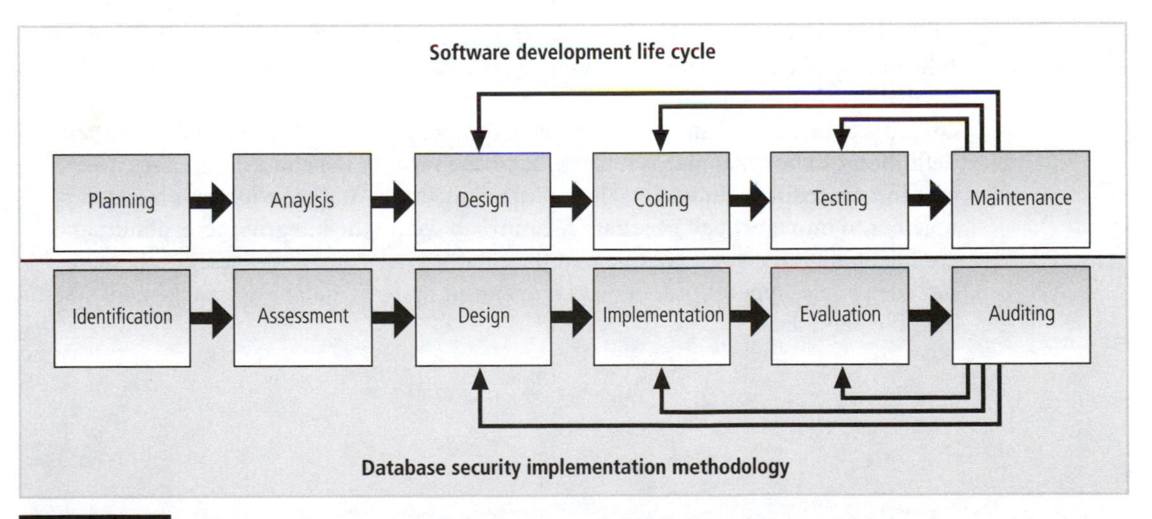

FIGURE 1-16 Database security methodology

Figure 1-16 presents database security methodology side by side with the software development life cycle (SDLC) methodology. Notice that phases in the database security methodology correspond to those of the SDLC. For example, suppose your company is carrying out a new inventory system project. Typically, your first phase in the SDLC is to plan for resources and devise a high-level project plan outlining major milestones. As a security architect or administrator, at the planning phase you are executing tasks in the identification phase. One of the tasks in this phase is identifying the security policy that will be adopted for this project. The following list presents the definition of each phase of the database security methodology.

- *Identification*—This phase entails the identification and investigation of resources required and policies to be adopted.
- *Assessment*—This phase includes analysis of vulnerabilities, threats, and risks for both aspects of database security: physical (data files and data) and logical (memory and code). You analyze system specifications and requirements to devise a security policy and procedures for all database modules and application data.
- *Design*—This phase results in a blueprint of the adopted security model that is used to enforce security. The blueprint shows how security measures are implemented to enforce data integrity and accessibility.
- *Implementation*—Code is developed or tools are purchased to implement the blueprint outlined in the previous phase.
- *Evaluation*—In this phase you evaluate the security implementation by testing your system against typical software attacks, hardware failures, natural disasters, and human errors. The result of this phase is a determination of the system's degree of security.
- *Auditing*—After the system goes into production, security audits should be performed periodically to ensure the security state of the system.

Database Security Definition Revisited

At the start of this chapter database security was defined as the degree to which all data is fully protected from tampering or unauthorized acts. You were warned, however, that you needed the chapter's quick tour of various information systems and information security concepts before confronting a complete definition. Now that you've had that tour, the definition can be expanded as follows: **Database security** is a collection of security policies and procedures, data constraints, security methods, and security tools blended together to implement all necessary measures to secure the integrity, accessibility, and confidentiality of every component of the database environment. These components include people, applications, networks, operating systems, database management systems, data files, and data.

Chapter Summary

- Security is defined as the level and degree of being free from danger and threats.
- Database security can be briefly defined as the degree to which data is fully protected from unauthorized tampering.
- Information systems are the backbone of the day-to-day company operations as well as the guide for long-term strategies.
- A typical information system consists of data, procedures, hardware, software, networks, and people.

- A client/server application is based on the business relationship in which the customer requests an order or service and the server responds to the request.
- A tier is a logical or physical platform in client/server architecture.
- DBMSs from different vendors vary in distinct implementation and architecture but they have close to the same functionality.
- The basic function of a DBMS is to enable developers and administrators to organize data; store, manipulate, and retrieve data efficiently; enforce data referential integrity; and provide a security mechanism to protect the data.
- Most companies employ an Information Security department to protect data and information.
- The concept of information security is based on the C.I.A. triangle in which "C" stands for confidentiality, "I" stands for integrity, and "A" stands for availability.
- There are two components to confidentiality: preventing unauthorized individuals from knowing or accessing secretive information, and keeping confidential information secret by not disclosing it to unauthorized individuals.
- Data is considered to have integrity if it is accurate and has not been intentionally or unintentionally tampered with.
- System availability is measured by how accessible the system is to individuals who are authorized to access information and how free individuals are to manipulate data.
- Database environment components are people, applications, networks, operating systems, database management systems, data files, and finally data.
- Data is the most valuable asset of the database environment.
- An access point is a gateway that requires measures to limit database security violations.
- A security access point is a point where security measures are needed to prevent access to unauthorized actions.
- Vulnerability is defined as being susceptible to attack.
- A threat is defined as an indication of impending danger or harm.
- A security risk is a result of a threat, which is a result of vulnerability.
- Information security architecture is a model for protecting logical and physical assets.
- Information security architecture is the overall design of a company's implementation of the C.I.A. triangle.
- Components of information security architecture include policies and procedures, security personnel and administrators, detection equipment, security programs, monitoring equipment, monitoring applications, and auditing procedures and tools.
- Database management systems empower the database administrator to implement and enforce security at all levels of the database.
- The security access points are people, applications, networks, operating systems, DBMS, data files, and data.

Review Questions

1. Security is best described as being totally free from danger. True or false?
2. Data is processed or transformed to become facts. True or false?
3. Data anomalies exist when there is redundant data caused by unnormalized data design. True or false?

4. Human error vulnerabilities are most often related to carelessness in implementing or following through on procedures. True or false?
5. Malicious code is software code written by hobbyists to test their capabilities. True or false?
6. A power failure is a type of natural disaster threat. True or false?
7. A system can become unavailable because of bad implementation of an authentication process. True or false?
8. Which one of the following is not a component of an information system?
   a. programmer
   b. report
   c. business procedure
   d. physical asset
9. Which one of the following is not a functionality of database management systems?
   a. allows users to validate data as it is entered
   b. allows developers and administrators to organize data
   c. enables developers and administrators to enforce data referential integrity and consistency
   d. allows administrators to enforce and implement data security
10. Which one of the following administrator functions is enabled by a database management system?
   a. Automatically back up data in case of a failure.
   b. Back up data in case of theft.
   c. Back up data in case of an intrusion.
   d. Back up data for auditing purposes.
11. Which one of the following is part of the information security triangle?
   a. intrusion
   b. integrity
   c. integral
   d. internal
12. Which one of the following is not part of a typical information security architecture?
   a. policies and procedures
   b. business rules
   c. detection equipment
   d. auditing procedures and tools
13. Data risk results in which of the following?
   a. data performance
   b. data access
   c. data privileges
   d. data corruption
14. Which of the following is not a logical asset?
   a. information system
   b. business application
   c. in-house programs
   d. purchased software
15. Outline the three components of the information security triangle and list one violation example for each.
16. Provide an example of how you can prevent physical access to an application database server.
17. Name three methods to enforce data integrity and provide an example for each method.

18. Provide three examples of people threats.
19. Explain how system vulnerabilities impact business.
20. Name three key measures that your business may adopt to protect data.

Hands-on Projects

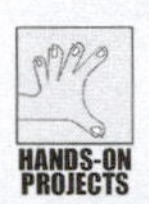

Hands-on Project 1-1

Using the Web as a resource, conduct a survey to compile a list of the top ten security vulnerabilities.

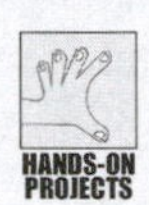

Hands-on Project 1-2

Why is the CI.A. triangle important?

Hands-on Project 1-3

You are a security officer working for a medium-sized research company. You have been assigned to guard a back entrance checkpoint. One day, a well-known manager walks out with a box of papers. A day later you are summoned to the security office by your manager and the security director for questioning about the manager who had been terminated the day before. The manager had walked out with highly confidential information.

1. Outline briefly what types of security measures were violated and how to avoid those violations.
2. Describe how this incident may result in security violations.

Hands-on Project 1-4

You are an employee of a company responsible for the administration of ten production databases. Lately, you have noticed that your manager is asking you frequent questions about the data used by one of the top researchers of the Engineering department. For two days, while conducting routine database tasks, you notice your manager exporting data from the database the top researchers are using.

1. What type of security threat is the exportation of data? How can you prevent it?
2. To what type of security risk could exporting data lead?
3. Explain briefly how you would react to this incident.

Hands-on Project 1-5

You were just informed by your manager that you are assigned to a new project that deals with financial data. Because you are the system analyst, your manager asked you to conduct a survey of users regarding what they require from the new project. After collecting all necessary data, you determine that this project requires high security measures. Outline the steps you should take to move forward.

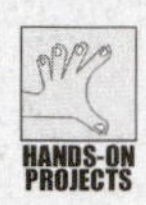

Hands-on Project 1-6

For each type of malicious code listed in Table 1-4, provide two examples of real-life code.

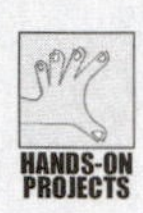

Hands-on Project 1-7

Describe a situation that illustrates each type of human threat listed in Table 1-4.

Case Project

You are a database administrator working for a national bank institution. One day, a lead developer sends you an e-mail requesting that you perform a data change. In the e-mail, he stresses the urgency and importance of this task. A minute later, you receive another e-mail but this is from the lead developer's manager to confirm the data change. This is the first time you have ever received this type of request. Usually, all requests go through the change management process.

1. List the security issues involved in this incident.
2. Describe the type of risks involved if you comply with the request and the types of risk involved if you do not.
3. Explain briefly how you would react to this incident, outlining your reasoning and whether you would comply or not.

2

Operating System Security Fundamentals

LEARNING OBJECTIVES:

On completion of this material, you should be able to:

- Explain the functions of an operating system
- Describe the operating system security environment from a database perspective
- List the components of an operating system security environment
- Explain the differences between authentication methods
- Outline useful user administration best practices
- List the criteria of strong password policies
- Describe operating system vulnerabilities
- Describe security risks posed by e-mail services

Introduction

As you already know, the operating system is the essence of a computer system—without it the computer hardware is not operable. The operating system is a collection of programs that manage the computer and allow programs and users to use its resources. No application, regardless of how simple and small, or complex and large, can be used without the operating system.

The operating system is one of the main access points to the database management system. Normally, when you want to access the database locally or remotely, you are authenticated by the operating system. Once you are authenticated to the system, you can operate or access the resources for which you have authorization, based on the set of privileges granted to you. Since the database resides on a machine operated by the operating system, the operating system becomes the first line of defense for any database security violations and infringements.

This chapter presents, from a database perspective, an overview of operating system security issues that help you gain an understanding of how security violations occur and where to focus your efforts to protect a database.

This chapter is not intended to provide technical details on configuring or running the operating system from an operational or security point of view. In addition, the information presented in the chapter is not tied to any specific operating system. In fact, the content of the chapter is applicable to any operating system. However, there are some instances in which technical details are presented. In these cases, UNIX and Windows 2000 or Windows XP are the operating systems referenced.

Operating System Overview

This section presents a quick overview of the operating system and its functions. If you are familiar with this topic, you can skip this section and move on to the discussion of the operating system security environment in the next section. Computer hardware consists of digital resources used to solve various computing problems at very high speed and with considerable accuracy. To take advantage of this technology, you need to communicate with the computer hardware in 0s and 1s, which is almost impossible for you to do. Several computer companies realized the need for a middleman between the user and the computer hardware to enable the user to operate the computer hardware. That was the birth of the operating system. An **operating system** is a collection of programs that allows the user to operate the computer hardware. But this is not the only thing that the operating system does. In fact, the operating system performs many tasks that vary from managing resources to scheduling jobs.

The following describes the three layers of a computer system, as shown in Figure 2-1:

- The inner layer represents the computer hardware, which is managed and controlled by the middle layer.
- The middle layer is the operating system.
- The outer layer represents all the different software used by users to solve a problem or perform a specific task.

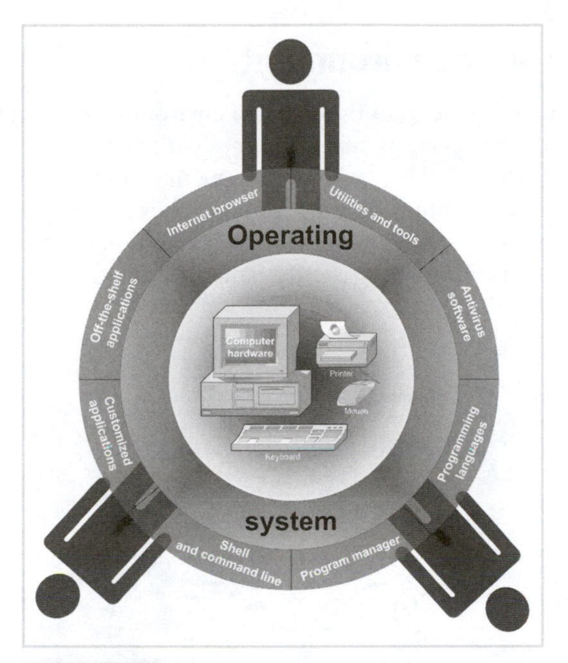

FIGURE 2-1 Three layers of a computer system

An operating system has a number of key functions and capabilities as outlined in the following list:

- Performs multitasking, that is, runs multiple jobs (tasks) at the same time
- Performs multisharing, that is, allows multiple users to use the computer hardware at the same time
- Manages computer resources such as CPU (central processing unit), memory, input and output devices, and disk storage
- Controls the flow of activities
- Provides a user interface to operate the computer
- Administers user actions and accounts
- Runs software utilities and programs
- Provides functionality to enforce security measures
- Schedules jobs and tasks to be run
- Provides tools to configure the operating system and hardware

There are many different vendors of operating systems including Windows by Microsoft; UNIX by companies such as Sun Microsystems, HP, and IBM; Linux "flavors" from various vendors such as Red Hat; and Macintosh OS by Apple.

This quick tour of operating system basics is designed to prepare you for the next section, in which you learn about the operating system security environment to gain an understanding of the security risks posed by operating system security access points.

The Operating System Security Environment

Figure 2-2 illustrates the components of the database environment, as explained in Chapter 1. If it is exposed, the operating system component of the database environment can open the door to unlawful individuals who contravene rules imposed to protect the database and its data. The door of the room that contains the computer that runs the operating system must be protected as securely as an organization can manage through padlocks, chain locks, guards, peep holes, security cameras, and other detection and authentication measures.

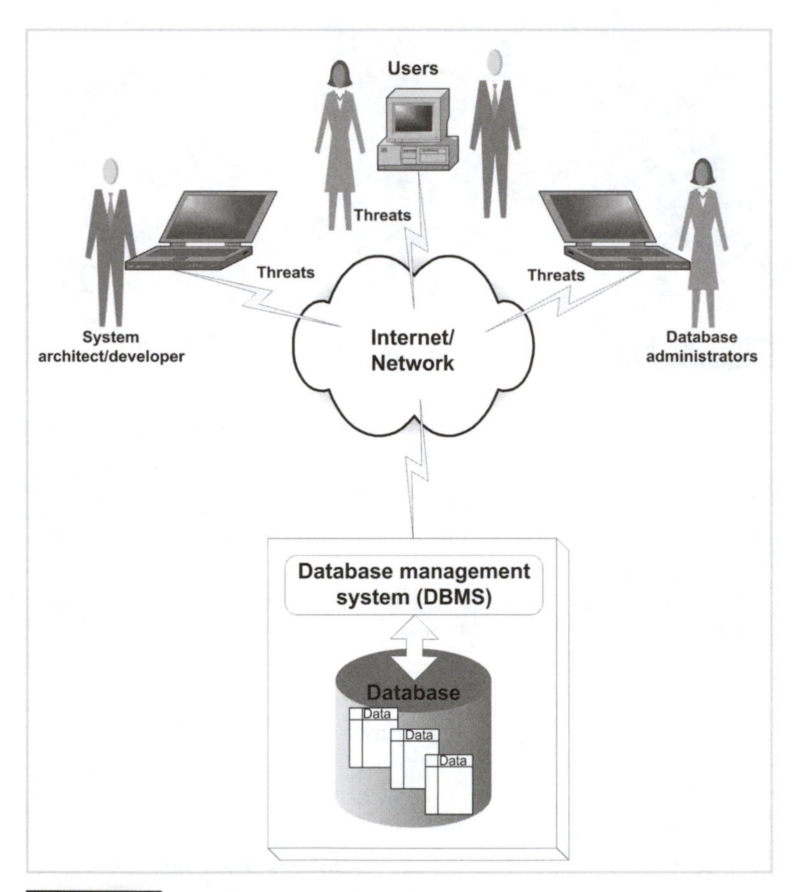

FIGURE 2-2 Database security environment

When thinking of how to guard an operating system, it may help you to think of the physical building of a bank (the operating system), a safe (the database), and money (the data). To rob the bank, thieves must get inside the bank property and then break into the safe. Over the years, bank administrators have learned from a history of robberies how to protect their institutions. An increasing number of security measures are installed and enforced, including the architectural structure of the building, mechanical equipment, and electronic and digital devices. All of these measures are costly, but banks invest the necessary money and resources for two main reasons: to protect their clients' money and to maintain their clients' confidence. If you are the system administrator protecting the access door to the database where data resides, you must work tirelessly to secure that door. In this section you explore the operating system security environment components that can be exploited and thus lead to data violations.

The Components of an Operating System Security Environment

The components that make up the operating system security environment are used as access points to the database. These access points are weak and insecure links to data within the database. The three components (layers) of the operating system are represented in Figure 2-3: memory, services, and files. As shown in the illustration, the services layer is an entry point and a gateway to the operating system as well as to the other components—memory and files. The services component comprises such operating system features and functions as network services, file management, and Web services.

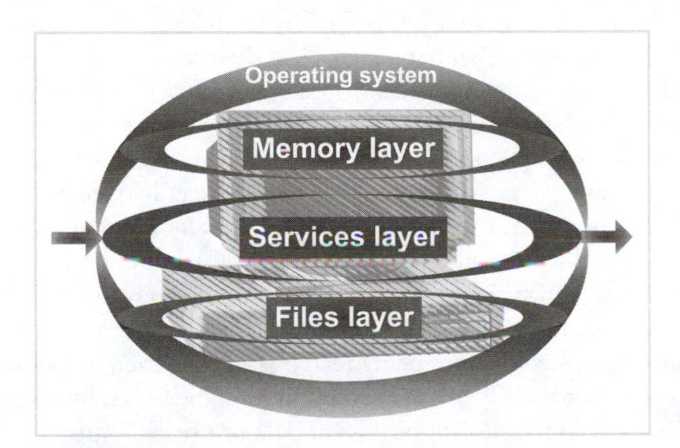

FIGURE 2-3 Operating system security environment

The memory component is the hardware memory available on the system, whereas the files component consists of the files stored on the disk. Why are these components important? Improper protection of these components can jeopardize the security of the database system. The following sections explore each component of the operating system from a database security perspective.

Services

The main component of the operating system security environment is services. The services component consists of functionality that the operating system offers as part of its core utilities. Users employ these utilities to gain access to the operating system and to all the features the users are authorized to use. If not secured and configured properly, each service becomes a vulnerability and access point and can lead to a security threat. These services vary a good deal and include the following: fundamental functionality such as authentication of the identity of users; core features such as remote copy (a program that allows you to copy a file from or to a remote computer system); and common practices such as user administration tasks and password policies.

Files

Here is an incident in which data was compromised. A system engineer was hired as a contractor by a small telecommunication company to upgrade the operating system and the main application to a higher software version. One of this engineer's tasks was to set up UNIX scripts to monitor the database. After long hours of implementation and testing, the system went into production. A few months later while the system administrator was browsing through the system logs, he noticed some unusual activities that shocked him. The system had been violated! There had been an intrusion from one specific IP address, and for the last three months it had been causing a huge amount of traffic every night at different hours.

The system administrators and other engineers in the company spent hours analyzing all logs and finally, with the help of the database administrator, they pinpointed what happened. A hacker broke into the system and somehow got the password to a database account and transferred customer data from two tables in the database, as determined through a trace of spool files that were left behind by the hacker. It was not enough for the operations manager to find out what happened. How did the hacker get the password? Another audit was performed on the machine looking at every single file, examining file permission, date and time stamp, contents, and so forth. The audit was in its sixth straight day when one of the system operators located a file that contained the account name and password for the database. It was a file that the system engineer had created for monitoring the scripts that he had installed. The scripts used this file to look up the user name and password. Not only was the file in plaintext (not encrypted), but also the file permission was set to `-rw-r--r--`, which meant that everyone was able to read the file.

File permission and sharing files are common threats to system security when not set properly. Files must be protected from being read by unauthorized individuals and kept safe from being written or modified. Data resides in files; therefore, improper file permissions on the file could lead to catastrophic loss of data or breach of privacy. Most operating systems have an elaborative function to implement any desired method of file permission. File sharing is another phenomenon in which individuals are using different types of peer-to-peer software, which may impose a high security risk.

File Permissions

Every operating system has a method of implementing **file permission** to grant read, write, or execute privileges to different users. In the sections that follow, first you look at the Windows 2000 implementation of file permission, followed by UNIX implementation.

Windows 2000

In Windows 2000 you can change file permissions by clicking a file's Properties to open it and clicking on the Security tab as shown in Figure 2-4. The security tab shows all permissions that have been assigned for each user. In this screen you may grant and revoke privileges to and from users. Note that Allow indicates grant, and Deny indicates revoke.

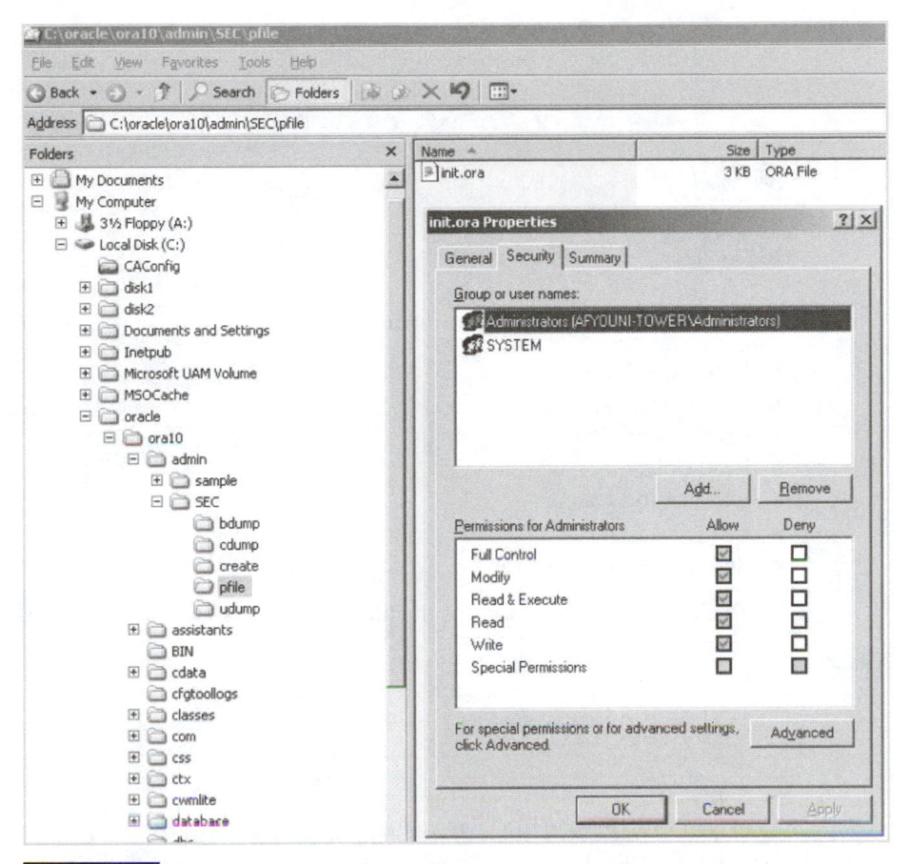

FIGURE 2-4 File properties function showing Security tab

UNIX

In UNIX, file permissions work differently than in Windows. For each file there are three permission settings: one for the owner of the file, one for the group to which the owner belongs, and finally one for all other users. Each setting consists of `rwx` as illustrated in Figure 2-5, in which `r` stands for read permission, w stands for write permission, and `x` stands for execute permission. In Figure 2-5, the `initSAM.ora` file permissions indicates the following: read and write permission for owner of file, read permission for the `oinstall` group to which the file owner belongs, and finally read permission to all other users.

```
-rw-r--r--    1 oracle   oinstall    4568 Mar 27 11:20
initSAM.ora
```

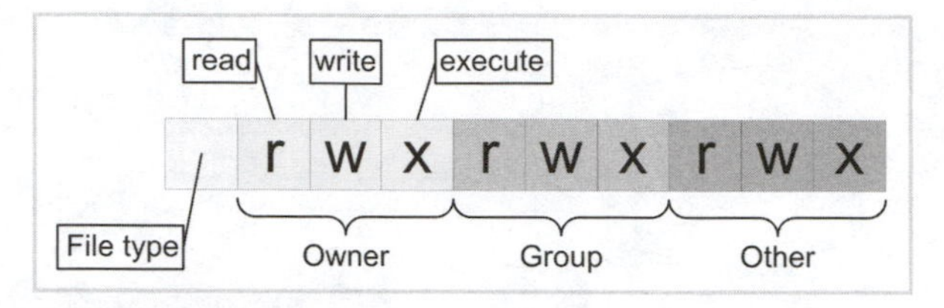

FIGURE 2-5 UNIX file permissions

You can use the CHMOD command to change the following file permissions:

- Execute only
- Write only
- Execute and write
- Read only
- Execute and read
- Read and write
- Read, write, and execute

To change the permissions of the file mail_list to `-rw-rw-r--` you issue the following command:

```
$ chmod 664 mail_list
```

Consult your UNIX operating system documentation for more information on the CHMOD command.

File Transfer

Steve is a production Oracle DBA for a regional chain of retail stores. He is responsible for administering over 20 database applications. He is part of a team of three database administrators and a database manager. One of his weekly tasks is to refresh the QA (quality assurance) database. Because the database is not large, Steve uses the Oracle export and import utilities. The process goes like this—On Sunday a scheduled job performs a full export of the database. The generated file from the export utility is transferred to the host machine where the QA database resides, and then it is imported.

One day Steve was summoned to an urgent meeting held by the chief technology officer (CTO). To Steve's surprise, the only people attending the meeting were his manager, the system manager, and the security director. Steve's anxiety and curiosity about the meeting made him uncomfortable. The CTO opened the meeting by talking about the refresh process that takes place every week and explained that it was necessary to change it because a violation was detected and fortunately prevented. One of the developers who had access to the machine and to the file was able to copy the file and transfer it to his home computer. The developer's intention was to be able to work from home, not to compromise the privacy of the company data.

This sort of "innocent" security breach happens all too frequently in organizations of all sizes and types. Who is responsible for protecting this process? Everyone is responsible.

What could've happened if this violation had not been detected? Data integrity and confidentiality could have been compromised, which is a major security violation. How can you protect this process from being violated again? This section presents best practices that you can adopt to secure file-transferring tasks.

First, you must know the following facts about FTP (File Transfer Protocol). FTP is an Internet service that allows transferring files from one machine to another.

- **File transfer** is a tool to send files from one computer to another.
- FTP clients and servers transmit user names and passwords in plaintext format (not encrypted). This means any hacker can sniff network traffic and be able to get the logon information quite easily.
- Logon information is not the only data that is transmitted in plaintext. Files are also transmitted unencrypted.
- A root account cannot be used to transfer files using FTP.
- Anonymous FTP is the ability to log on to the FTP server without being authenticated. This method is usually used to provide access to files in the public domain.

Here are some best practices for transferring files:

- Never use the normal FTP utility. Instead, use the Secure FTP utility, if possible.
- Make two FTP directories: one for file uploads with write permission only, and another one for file downloads with read permission only.
- Use specific accounts for FTP that do not have access to any files or directories outside the UPLOAD and DOWNLOAD directories.
- Turn on logging, and scan the FTP logs for unusual activities on a regular basis.
- Allow only authorized operators to have FTP privileges.

Sharing Files

No matter what the circumstances, sharing files naturally leads to security risks and threats. The peer-to-peer technology phenomenon is on the rise. **Peer-to-peer programs** allow users to share files with other users over the Internet. These types of programs introduce a whole new can of worms—in this case, worms and viruses that can infect your system. If you were to conduct a survey of users that use peer-to-peer programs, you would most likely find that the majority of the users' machines are infected with some sort of virus, spyware, or worm.

Most companies prohibit the use of such programs, and rightfully so. They should prosecute users who don't adhere to this policy. The main reasons for blocking these programs are:

- **Malicious code**—Peer-to-peer programs are notorious for malicious code, and most of the files that are being shared are infected with code that could harm your computer. Malicious code is a program, macro, or script that deliberately damages files or disrupts computer operations. There are several types of malicious code, such as viruses, worms, Trojan horses, and time bombs.
- **Adware and spyware**—Similar to malicious code, these types of programs are used to generate pop-up advertisements as well as capture key strokes, Web sites visited, clicks on pages, and more. Peer-to-peer programs are sponsored by Web sites and individuals.

- **Privacy and confidentiality**—If peer-to-peer programs are configured improperly, users can see and obtain all sorts of private and confidential data residing on the computer system.
- **Pornography**—A huge community of individuals uses peer-to-peer programs to exchange pornographic images or clips that may be offensive or inappropriate to other users.
- **Copyright issues**—Peer-to-peer programs make sharing music files, video clips, software applications, gaming software, or electronic books very easy. This medium encourages individuals to infringe on other people properties and promotes piracy as an acceptable action.

Now that you are aware of the risks of peer-to-peer programs, you should monitor all network activities to determine who is using this type of program. Employees who are caught using these programs should be disciplined to educate them about the possible risks that peer-to-peer programs bring to the company.

Memory

The last component of the operating system to be discussed in this chapter is memory. You may wonder how memory is an access point to security violations. There are many badly written programs and utilities that could damage the contents of memory. Although these programs do not perform deliberate destructive acts, you need to be aware of them. In most cases, when you use a program that violates the integrity of your data, you have two options: either stop using it or apply a patch (service pack) to fix it. On the other hand, programs that intentionally damage or scan data in memory are the type that not only can harm data integrity, but may also exploit data for illegal use.

Authentication Methods

Authentication is a fundamental service of the operating system. It is a process that verifies the identity of the user to permit access to the operating system. A weak authentication method exposes the system to security risks and threats. Most security administrators implement two types of authentication methods, physical and digital.

The **physical authentication** method allows physical entrance to the company property. Most companies use magnetic cards and card readers to control entry to a building, office, laboratory, or data center. For mission-critical or sensitive operations, personnel are physically authenticated using biometric or biomagnetic technologies. Examples of these technologies are eye retina scans, fingerprint scans, handprint scans, voice recognition, signature recognition, thermal sensing, and others.

The **digital authentication** method is the process of verifying the identity of the user by means of a digital mechanism or software. The following sections list digital authentication mechanisms used by many operating systems and implemented by many organizations.

Digital Certificate

A **digital certificate** is a type of authentication that is widely used in e-commerce (conducting business over the Internet). A digital certificate is a digital passport that identifies and verifies the holder of the certificate. The holder can be a person, a computer, a Web site, or a network system. This digital certificate is an electronic file issued by a trusted party (known as certificate authority) and cannot be forged or tampered with.

Digital Token (Security Token)

A **digital token** is a small electronic device that users keep with them to be used for authentication to a computer or network system. Usually, this device displays a number unique to the token holder, which is used with the user's PIN (personal identification number) as the password. This token constantly displays a new number, which means that each time a user is authenticated, a different password is used, making it very hard for hackers to pass authentication. Many companies commonly use this method. One note worth mentioning: if a user loses this token, it should be reported immediately, and the user account should be locked until the situation is rectified.

Digital Card

A **digital card** is also known as a security card or smart card. This card is similar to a credit card in dimensions but instead of a magnetic strip, it has an electronic circuit that stores user identification information such as name, ID, password, and other related data. The card is used to authenticate the holder to a computer system by having a card reader device read the card.

Kerberos

Kerberos was developed by the Massachusetts Institute of Technology (MIT) to enable two parties to exchange information over an open network by assigning a unique key, called a ticket, to each user. This ticket is used to encrypt communicated messages.

Lightweight Directory Access Protocol (LDAP)

LDAP is an authentication method developed by the University of Michigan that uses a centralized directory database storing information about people, offices, and machines in a hierarchical manner. An LDAP directory can be easily distributed to many network servers. You can use LDAP to store information about:

- Users (user name and user ID)
- Passwords
- Internal telephone directory
- Security keys

LDAP servers are efficient for reading information from the directory but are not suited for data that is frequently changing. Many operating systems and applications use this method of authentication because it is simple to implement. Use LDAP for these reasons:

- LDAP can be used across all platforms (operating system independent).
- It is easy to maintain.
- It can be employed for multiple purposes.

LDAP architecture is client/server based, in which the client requests information from the directory, and the LDAP server supplies a response.

NTLM

NT LAN Manager, developed and used by Microsoft, employs a challenge/response authentication protocol that uses an encryption and decryption mechanism to send and receive passwords over the network. This method is no longer used or supported by new versions of the Windows operating system.

Public Key Infrastructure (PKI)

PKI, also known as public key encryption, is an authentication method in which a user keeps a private key and the authentication firm holds a public key. These two keys are used to encrypt and decrypt communication messages between the two parties. The private key is usually kept as a digital certificate on the user's system.

RADIUS

Remote Authentication Dial-In User Services (RADIUS) is an authentication method commonly used by network devices to provide a centralized authentication mechanism. RADIUS is client/server based, and uses a dial-up server, a virtual private network (VPN), or a wireless access point communicating to a RADIUS server.

Secure Sockets Layers

Secure Sockets Layers (SSL) is a method in which authentication information is transmitted over the network in an encrypted form. This method is commonly used by Web sites to secure client communications. This protocol was developed by Netscape Communications to provide secure communication between client and server.

Secure Remote Password (SRP)

SRP was developed by Stanford University. It is a protocol in which the password is not stored locally in encrypted or plaintext form. This method is very easy to install and does not require client or server configuration. Also, this method is invulnerable to brute force or dictionary attacks.

Authorization

Authentication is the process of proving that users really are who they claim to be. **Authorization**, on the other hand, is a process that decides whether users are permitted to perform the functions they request. Authorization is not performed until the user is authenticated. Authorization deals with privileges and rights that have been granted to the user. For example, suppose you have created a user account to perform file transferring only. This user is provided with a user name and password to allow the user to be authenticated. You may provide this user read permission on the DOWNLOAD directory and write permission to the UPLOAD directory. This means that this user is only authorized to read and write on these two directories and is not permitted to perform other tasks.

User Administration

Authentication and authorization are essential services that every operating system provides in order to secure access to the computer's logical and physical resources. Another related service is **user administration**. Administrators use this functionality to create user accounts, set password policies, and grant privileges to users. Although hackers do not often tap into user administration, improper use of this feature can lead to security risks and threats. The following is a compilation of best practices for user administration, in no specific order.

- Use a consistent naming convention by adopting a combination of first name and last name for the user account.
- Always provide a password to an account and force the user to change it at the first logon.
- Make sure that all passwords are encrypted in a well-protected file.
- Do not use default passwords for any account.
- If a machine is compromised or you suspect it is compromised, change all passwords for all existing accounts.
- Use different accounts for different applications and users.
- Create a specific file system for users, separate from applications and data.
- Educate users on how to select a password.
- Lock a user account when a user's employment is terminated or ended.
- Lock accounts that are not used for a specific period of time.
- If possible, grant privileges on a per host basis.
- Do not grant privileges to all machines, but only to those users who are absolutely in need.
- If connected remotely, use Secure Shell (ssh), Secure Copy (scp), and Secure FTP for telneting, copying files, and transferring files, respectively.
- When a computer system is compromised, isolate the system from other systems to prevent further intrusion.
- When a system is compromised, work with management and the security office to determine the cause of the infringement.
- Perform random auditing procedures on a regular basis.

Password Policies

A good **password policy** is the first line of defense against the unwanted accessing of an operating system. Usually, hackers try to access the system through the front door using an account and password. If this method fails, they try other methods. In fact, most hackers utilize tools that use the dictionary method to crack passwords. These tools use the permutation of words in the dictionary to guess the password. As the system administrator, you should work with the security manager to establish a password policy to make it difficult for hackers to enter your system.

There are many different practices and policies that you can adopt for your company. However, the best password policy is the one that matches your company missions and is enforced at all levels of the organization. The following password practices—all or a combination of them—can be employed to devise a policy plan that suits your company.

- **Password aging**—Tells the system how many days a password can be in effect before it must be changed. Most companies practice a three-month policy, but you should determine the number of days based on your business and security requirements.

- **Password reuse**—This practice can be interpreted and applied in three different ways:
  - Tells the system how many times you can reuse a password
  - Indicates the number of days that must pass before you can reuse a password
  - Determines whether the system allows passwords to be reused
- **Password history**—This practice is related to password reuse, and it tells the system how many passwords it should maintain for an account. The password history can be used to determine if a password can be reused or not.
- **Password encryption**—A method that encrypts (scrambles) the password and stores it in a way that it cannot be read directly.
- **Password storage**—The place where the password is stored and kept hidden from the public.
- **Password complexity**—This is one of the most important password practices that should be implemented for any password policy. Complex passwords are those that are made up of a combination of upper- and lowercase letters, digits, and symbols. Having a password complexity requirement forces users to choose a password that is not easily cracked. The following is a list of standards that can be used when creating complex passwords:
  - The password must contain digits, symbols, and alphabetic characters (a-z, A-Z, 0-9, !@#$%^&*()_+}{":><?).
  - The password must have a minimum length which is usually six characters, but eight characters are recommended.
  - The alphabetical characters must use mixed letter cases (uppercase and lowercase).
  - The password must not contain any part of your account, first name, last name, birthday, telephone number, license number, registration number, employee number, spouse's name, child's name, parent's name, sibling's name, city you live in, or country in which you reside.
- **Logon retries**—A good practice is to allow a user to unsuccessfully try to log on up to three times before the account is locked and an administrator is contacted.
- **Password protection**—Although this practice is very hard to enforce, you, the manager, system administrator, security manager, or human resources manager, must train your employees and make them aware of the danger of concealing a password in a place from which it can be retrieved in case it is forgotten. It is bad practice to record a password on paper even if the paper is stored in a locked place. If you must record a password, use an encrypted file that can be accessed only by you.
- **Single sign-on**—Single sign-on allows you to sign on once to a server (host machine) and then not have to sign on again if you go to another server where you have an account. Although a single sign-on provides great convenience, it should not be practiced for mission-critical operations, financial institutions, government agencies, or other similar organizations.

Vulnerabilities of Operating Systems

In this section you are presented with a list of the top ten vulnerabilities of Windows and UNIX. The list identifies the tools that hackers use as a gateway to break into the system, and in most cases these intrusions lead to loss of service, loss of data, invasion of privacy, data corruption, or a combination of all these. This list was released by the U.S. Department of Homeland Security, along with its Canadian and British counterparts and the SANS Institute, on October 28, 2003.[1]

The top vulnerabilities to Windows systems are:

- Internet Information Services (IIS)
- Microsoft SQL Server (MSSQL)
- Windows Authentication
- Internet Explorer (IE)
- Windows Remote Access Services
- Microsoft Data Access Components (MDAC)
- Windows Scripting Host (WSH)
- Microsoft Outlook and Outlook Express
- Windows Peer-to-Peer File Sharing (P2P)
- Simple Network Management Protocol (SNMP)

The top vulnerabilities to UNIX systems are:

- BIND Domain Name System
- Remote Procedure Calls (RPC)
- Apache Web Server
- General UNIX authentication accounts with no passwords or weak passwords
- Clear text services
- Sendmail
- Simple Network Management Protocol (SNMP)
- Secure Shell (SSH)
- Misconfiguration of Enterprise Services NIS/NFS
- Open Secure Sockets Layer (SSL)

E-mail Security

E-mail may be the tool most frequently used by hackers to exploit viruses, worms, and other computer system invaders. This is true no doubt because e-mail is the tool most widely used by public and private organizations as a means of communication. If you were to research the number of incidents that have occurred in the last five years, you would find that e-mail was the medium used in many of the most famous worm and virus attacks[2]; for example, the Love Bug worm, the ILOVEYOU worm, the Mydoom worm, and the Melissa virus were all spread through e-mail.

More worrisome and threatening is that e-mail is not only used to send viruses and worms, but to send spam e-mail, private and confidential data, as well as offensive messages. Here is another incident that actually occurred.

An Oracle database developer was hired to work on back office modules for a well-known department store. This developer was responsible for writing PL/SQL code to implement business rules and other processing logic. Several weeks later, the application went into production without any issues or hiccups. A few months after that, this developer was laid off because of the downturn of the economy. Although the developer had left the scene, his presence would be felt for months to come.

Before long, the department store started to get complaints from customers about credit card charges that were incurred without their knowledge. The fraud office of the department store investigated these complaints and soon verified that the customers' complaints were valid. A memo was issued to all employees in every department to be on the watch for any suspicious activity. In addition, a special meeting was held by the information technology group to talk about ways to monitor and audit all database activities. The meeting resulted in the creation of an internal audit group made up of three database engineers.

After a thorough investigation, the group came across a module within a PL/SQL package that sent e-mail to an ambiguous address. This module created and sent a report listing all customer and credit data for 20 customers. Also, the module was scheduled to run every week via the Oracle job scheduler, DBMS_JOB.

This did happen, and it could happen to other companies that do not follow stringent security procedures to secure data. In the case above, system operations failed to monitor the activities of the e-mail server where the database resided. In addition, the database administration team failed to examine the PL/SQL code to get an idea of what the code was doing. The developer team failed to review the code submitted by the database developer. The whole system development process failed because it did not have security checks and controls to catch this mishap before it occurred.

E-mail is used by many employees to communicate with clients, colleagues, and friends, and some of these employees may violate the security policies of the company by sending confidential data. Many reports and research studies claim that e-mail is being used more frequently by unhappy and disgruntled employees to expose sensitive and confidential data inside and outside the company. What does this mean to you? Regardless of your position, you should have the integrity to comply and adhere to the company policies and respect others' privacy and confidentiality. For those individuals who do not understand what this means, you need to install auditing and monitoring controls to detect any suspicious activities and report them immediately to management.

To prevent incidents similar to the scenario just described, do not configure the e-mail server on a machine in which sensitive data resides, and do not disclose technical details about the e-mail server without a formal written request from the technology group manager explaining the reasons the e-mail server information is needed.

Chapter Summary

- An operating system is a collection of programs that allows the user to interact with the computer hardware.
- An operating system is one of the main access points to the database management system.
- If the operating system component of the database security environment is exposed, it can open the door for unlawful individuals to contravene all rules imposed to protect the database.

- Authentication is a process that validates the identity of the user in order to permit access to the operating system.
- Physical authentication methods allow physical entrance to the company property.
- Digital authentication methods are the processes of verifying the identity of the user by means of a digital mechanism or software.
- A digital certificate is a digital passport that identifies and verifies the holder of the certificate.
- A digital token is a small electronic device that users keep with them to be used for authentication to a computer or network system.
- A digital card is similar to a credit card; it holds user identification information such as name, ID, and password.
- Kerberos enables two parties to exchange information over an open network by assigning a unique key to each user.
- LDAP is an authentication method that uses a centralized directory database to store information about people, offices, and machines in a hierarchical manner.
- PKI is an authentication method whereby a user keeps a private key and the authentication firm holds a public key.
- Remote Authentication Dial-In User Services (RADIUS) is an authentication method commonly used by network devices to provide a centralized authentication mechanism.
- Secure Sockets Layers is a method whereby authentication information is transmitted over the network in an encrypted form.
- SRP is a protocol in which the password is not stored locally in either encrypted or plaintext form.
- Authorization is a process that determines whether the user is permitted to perform the function he or she requests.
- Authorization is not performed until the user is authenticated.
- Authorization deals with privileges and rights that have been granted to the user.
- A good password policy is the first line of defense for protecting access to an operating system.
- The best password policy is the one that matches your company missions and is enforced at all levels of the organization.
- When set improperly, file permission and file sharing are common threats to system security.
- Sharing files naturally leads to security risks and threats.
- E-mail may be the tool most frequently used by hackers to exploit viruses, worms, and other computer system invaders.

Review Questions

1. The graphical user interface program found on the desktop of most machines is called an operating system. True or false?
2. Authorization is a process that validates the identity of the user in order to permit access to the operating system. True or false?
3. Digital authentication is a digital passport that identifies and verifies the holder of the certificate. True or false?
4. FTP clients and servers encrypt all transmitted data. True or false?

5. It is acceptable to use peer-to-peer programs to download files as long as these files are public domain and your system is protected by an antivirus program. True or false?
6. LDAP can be used to store information not related to authentication. True or false?
7. Which of the following is not a valid authentication method?
    a. Lightweight Directory Access Protocol
    b. NLM
    c. Kerberos
    d. RADIUS
8. Which of the following is a malicious code?
    a. bug
    b. patch
    c. service pack
    d. time bomb
9. Which of the following is *not* true about operating system security environment?
    a. An operating system is a collection of programs that allows the user to operate the computer hardware.
    b. The operating system component of the database environment can be used as a gateway to violate database integrity.
    c. The components that make up the operating system security environment are used as access points to the database and can be weak or insecure links to connect to data within the database.
    d. Files, services, and memory are the three components of the operating system security environment.
10. Why is it important to protect the operating system?
11. Name three methods of protecting your operating system.
12. What is the difference between authentication and authorization? Provide an example.
13. Name two best practices for user administration, and provide an example of how each practice enhances operating system security.
14. Name three sources of detailed information about viruses.
15. What should you do if a developer needs one of the files that you own?

Hands-on Projects

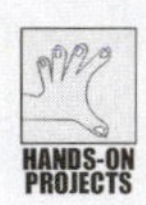

Hands-on Project 2-1

Find owner, group, and others file permissions for the Oracle file called orapwSID_NAME found in:

- ORACLE_HOME/dbs (UNIX)
- ORACLE_HOME/database (WINDOWS)

If the permissions for this file were accessible to all, explain what the implications would be.

Hands-on Project 2-2

Suppose your system were attacked by a worm. Use information found on the Web to outline steps to rid your system of the worm.

Hands-on Project 2-3

Compile a list of five system administration best practices for any two operating systems.

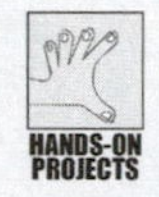

Hands-on Project 2-4

Using the Internet as a source, compile a list of three known vulnerabilities for Windows 2000. What would you do to protect your system if your operating system were Windows 2000?

Hands-on Project 2-5

Suppose you were hired as a system administrator for a small company. On your first day of work, you were asked by a developer to configure an e-mail server on one of the systems because the e-mail server was needed as part of the new system being developed. Outline the steps that you would take in response to this request.

Case Project

Suppose you are the security manager for a small high-tech company. Outline security measures that you would implement to protect the operating system containing code for a new product innovation.

Endnotes

1 This list is adapted from *www.sans.org/top20*.

2 For a full list and details of viruses and worms, visit *http://securityresponse.symantec.com/avcenter/vinfodb.html*.

3

Administration of Users

LEARNING OBJECTIVES

Upon completion of this material, you should be able to:

- Explain the importance of administration documentation
- Outline the concept of operating system authentication
- Create users and logins using both Oracle10*g* and SQL Server
- Remove a user from Oracle10*g* and SQL servers
- Modify an existing user using both Oracle10*g* and SQL servers
- List all default users on Oracle10*g* and SQL servers
- Explain the concept of a remote user
- List the risks of database links
- List the security risks of linked servers
- List the security risks of remote servers
- Describe best practices for user administration

Introduction

A small financial company hired a new information security director to assume responsibility for all security issue policies. A team of two security administration officers reported to the director. One of the first projects that the director initiated was a full review and analysis of the company's system and database administration practices. When the review was finished, the director implemented policies and procedures to be carried out company-wide. The director then hired a hacker to perform a single task—to try breaking into the system and database application. A day later the hacker was able to break into the system, and within a few minutes he broke into the database and obtained a listing of all users in the database. Of course, the hacker documented all the hacking steps he took in a report, which outlined the main gaps of security. One of the security gaps uncovered was that some of the database users were not authenticated by the database.

What does it mean to have database users who are not authenticated by the database? This chapter presents the answer and introduces you to many essential concepts of user administration that are useful in buttoning up database security.

Chapters 1 and 2 provided you with building blocks upon which to base further exploration of security concepts and fundamentals. This chapter covers user administration, which is the core of database security. This chapter presents an overview of database practices from the point of view of both management and operations. It also supplies scenarios that broaden your understanding of user administration. The chapter begins with a discussion of the importance of documentation with respect to user administration and then continues with a presentation of the user account and login concepts from two database perspectives (Oracle10*g* and Microsoft SQL Server 2000) followed by step-by-step instructions on how to put these concepts into practice. In addition, this chapter provides an explanation and implementation of profiles and roles for professionals in security. The chapter ends with a summary of best practices for user administration.

Because this book covers implementation of concepts in Oracle10*g* and SQL Server databases, it is worth noting that SQL Server DBAs tend to use a graphical user interface tool called Microsoft Enterprise Manager more than the command line or SQL statements for their database administration tasks. This tendency evolved from the promotion of graphical user interfaces by Microsoft. On the other hand, Oracle DBAs, especially those who are experienced, tend to use the traditional SQL*Plus tool to perform their tasks. However, this tradition is slowly changing. Nowadays, new Oracle DBAs use mainly Oracle Enterprise Manager, especially for manual and unscripted tasks. This book covers many commonly used graphical tools and command-line interfaces.

Documentation of User Administration

At organizations of every type, many security violations are caused by negligence and ignorance, and in particular by failing to consider documentation of practices to be a part of the process of administering users. Here are the top three excuses for failing to incorporate documentation as part of the administration process:

- Lack of time
- Belief that the administration process is already documented in the system
- Reluctance to complicate a process that is simple

Everything should be documented for two reasons: to provide a paper trail to retrace exactly what happened when a breach of security occurs; and to ensure administration consistency.

It cannot be overemphasized that you introduce risks when you fail to document your work, which includes every process of database administration. Documentation in this context includes the following:

- *Administration policies*—This type of documentation includes all policies for handling new and terminated employees, managers, system and database administrators, database managers, operation managers, and human resources. A detailed document should describe guidelines for every task that is required for all common administrative situations.
- *Security procedures*—This is an outline of a step-by-step process for performing an administrative task according to company policies.
- *Procedure implementation scripts or programs*—This is documentation of any script or program used to perform an administrative task. This includes a user's manual and operational manual.
- *Predefined roles description*—This provides a full description of all predefined roles, outlining all tasks for which the role is responsible and the role's relationship to other roles.
- *Administration staff and management*—This is usually a detailed description of each administration staff and management position. The document includes an organizational chart.

Many companies develop procedures and forms used to perform any security-related process. In this case, the process of creating a user account in the database is documented and provided to employees responsible for carrying out this task. Figure 3-1 presents a sample process for creating a database user account that you can customize per your business requirements and company policies. This process illustration is followed by Figure 3-2, which presents a sample form for a user account application.

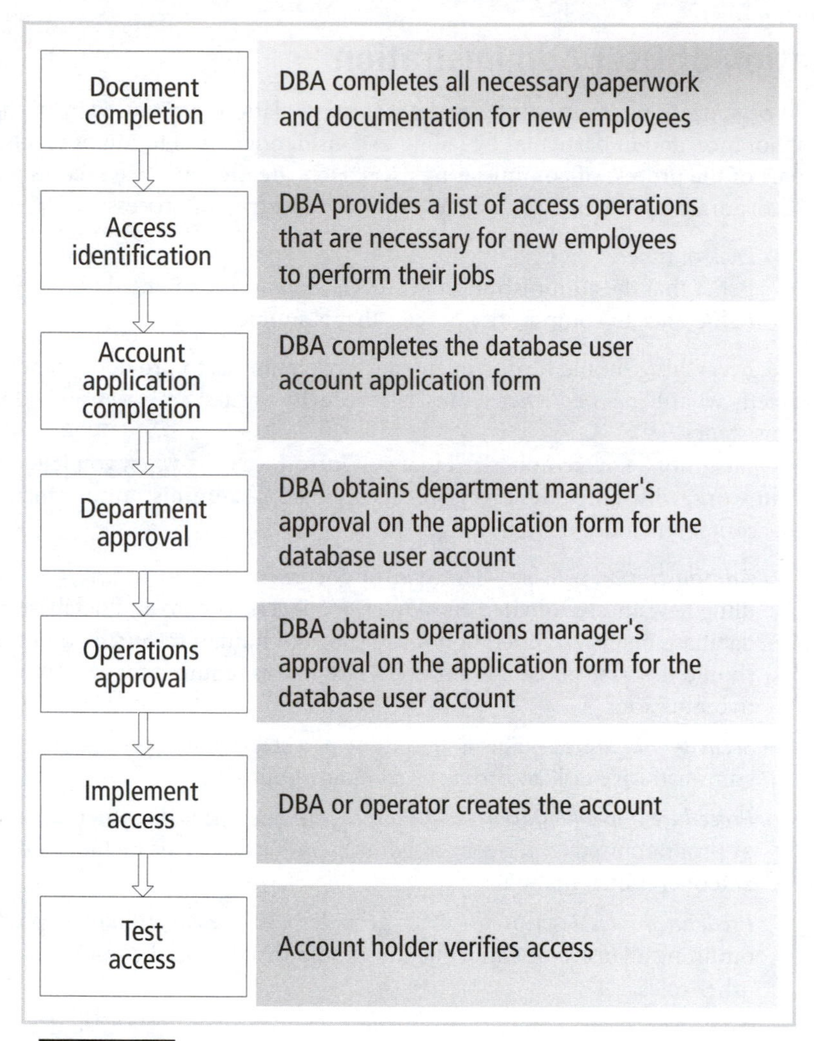

FIGURE 3-1 Database account access procedure

Acme Pharmaceutical Company
Database User Account Form

Requested For	
Name (First, MI, Last)	
Employee Type	☐ Employee ☐ Contractor ☐ Temporary ☐ Intern
Title	
Employee# (if available)	

Requested By			
Name (First, MI, Last)			
E-mail		Telephone Ext.	

Date			
Requested		Expected	

Action
☐ Add ☐ Modify ☐ Password Change ☐ Lock ☐ Unlock ☐ Remove

Location & Department	
Location	
Department	

Database Application

Database Role
☐ Operations Manager ☐ Business Manager ☐ Analyst ☐ Administrator
☐ Developer ☐ Operator ☐ Clerk ☐ QA
☐ Other:

Reason for the request

Approved by	
Requester Manager:	
Operation Manager:	

Comments

Completed by			
Administrator		Date	

FIGURE 3-2 Database user account application form

Operating System Authentication

As you recall from Chapter 2, the operating system is the gateway to database access. If a hacker breaks into the operating system through the host machine (server), the possibility of that hacker accessing the database residing on the server is high. Once again, a tight grip on the operating system is warranted and essential. Many database management systems, including Microsoft SQL Server 2000, depend on the operating system to authenticate users.

The only type of authentication the default installation of Microsoft SQL Server 2000 supports is Windows authentication. You and your users authenticate to the operating system and authorization is handled within SQL Server. You'll learn more about how this works later in this chapter and in Chapter 5. Oracle10*g* supports three types of database authentication, which are described in the next section.

The main question is why some database vendors rely on the operating system for user authentication and bypass database authentication. The primary reason for such a method is the assumption mentioned previously—once an intruder gets control of the operating system, it takes very little knowledge, or effort, to access the database. In addition, database architects and administrators may prefer to centralize the management of all user administration in one location by using operating system authentication only.

Figure 3-3 presents the ideal authentication enforcement for database applications. As you can see, the user must be authenticated at each level: the network, the operating system, and the database system.

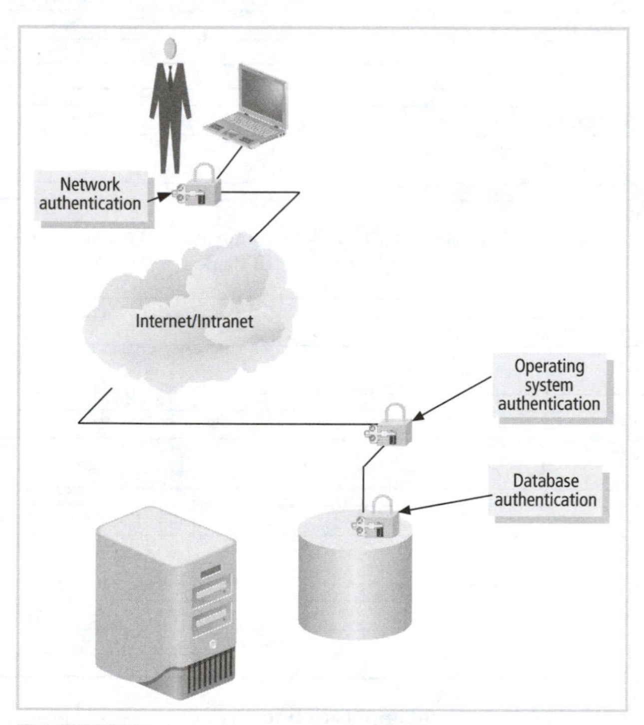

FIGURE 3-3 Ideal authentication levels for database applications

Creating Users

Creating users is one of the main tasks you will perform as a database operator or database administrator. In most organizations, this process is standardized (standards and user administration conventions are in place), well-documented, and securely managed. However, if this task is performed arbitrarily without following specific standards and policies, the database is probably exposed to many types of security risks and violations. Consider the following scenario.

A medium-sized telecommunication company was in the midst of a large, lengthy, and critical project, which involved a database application. A team of three administrators administered the database portion of the application. The development portion of the application included a team of 30 full-time developers and several contractors. To simplify their tasks, the administrators created users with a default password that was known to everyone. After 20 months, the project was completed and commissioned to production.

Soon afterwards, a sizable telecommunication corporation presented a lucrative offer to buy this company. The offer was accepted immediately, and the process to transition and transfer resources, assets, and all other business logistics started. As part of the transition, a full audit was performed on all assets, financial accounts and transactions, hardware, software, and business processes to verify that everything complied with the corporate policies of the purchaser. To the surprise of the database manager, it was discovered that there could be an issue with one of the database applications.

The audit team drafted a quick memo about a database security leak that could jeopardize the buyout deal, and it was sent to all managers at all levels in the company. The president of the company convened a meeting to discuss this issue and and resolve it. The consensus was to establish database auditing measures to monitor suspicious activities and to continue with the transition process. If unlawful activities were discovered, they would be dealt with immediately by assessing security risks and threats.

Two weeks into monitoring activities, investigation, and log tracing, an auditing report showed that a user account was spooling sensitive data to a file, which was then transferred to another internal machine, which was in turn transferred to an external IP address. The issue was resolved and the individual who violated this company policy was caught, prosecuted, and jailed. The database manager was terminated on the spot for her negligence in failing to establish and enforce user account policies. The database administration team was required to attend on-site training on security best practices by one of the database security experts. The buyout deal was completed.

What was the source of the problem? Consider the issues that could have led to this incident. Remember that it is related to policies and documentation. Here is what happened.

The database administrators had written a script to create a user for every developer working on the project. The script granted privileges to read and write data to the database schema. Once the database went into production, the database administrators took the development database configuration, including all existing accounts, and applied it to production. A disgruntled developer who worked on the project left the company. This developer recorded all passwords and soon was able to hack his way into the hosting machine where the database resided. He was then able to log on to the database with his existing account and password.

This scenario demonstrates that creating a user without following specific standards or policies is not as trivial as you might think. In fact, many companies have created documentation on how to create database user accounts that entails the creation of the users and all specific options that the database offers. In this section you are shown how to create a user using Oracle10*g* and MS SQL Server. Regardless of the database you use, creating the user is generally an easy task once a policy is documented and followed.

Creating an Oracle10*g* User

This section outlines the steps required to create a user using Oracle10*g*. The full syntax for creating a user is outlined in the boxed code. Before you delve into this function, you need to understand conceptually each option that the CREATE USER statement offers.

```
CREATE USER username IDENTIFIED { BY password | EXTERNALLY |
GLOBALLY AS 'external_name' }
   [ DEFAULT TABLESPACE tablespace]
   [ TEMPORARY TABLESPACE  { tablespace |
tablespace_group_name } ]
   [ QUOTA { integer { K | M } ON tablespace | UNLIMITED ]
   [ PROFILE profile ]
   [ PASSWORD EXPIRE ]
   [ ACCOUNT { LOCK | UNLOCK } ]
```

The syntax presented here is derived from the online documentation that Oracle provides at the Oracle Technology Network site: *www.otn.oracle.com*.

First of all, the CREATE USER statement is a SQL statement that is supported by all relational databases. The CREATE USER statement is part of a Data Definition Language (DLL), which is a component of the SQL language. This statement enables database administrators to create a database user account. When granted the proper privileges, this user account can perform various tasks and own different database objects. The various Oracle10*g* options are each examined in the sections that follow.

IDENTIFIED Clause

This clause tells Oracle10*g* how to authenticate a user account. Oracle10*g* has three authentication methods. Figure 3-4 presents the architecture of authentication provided by Oracle10*g*.

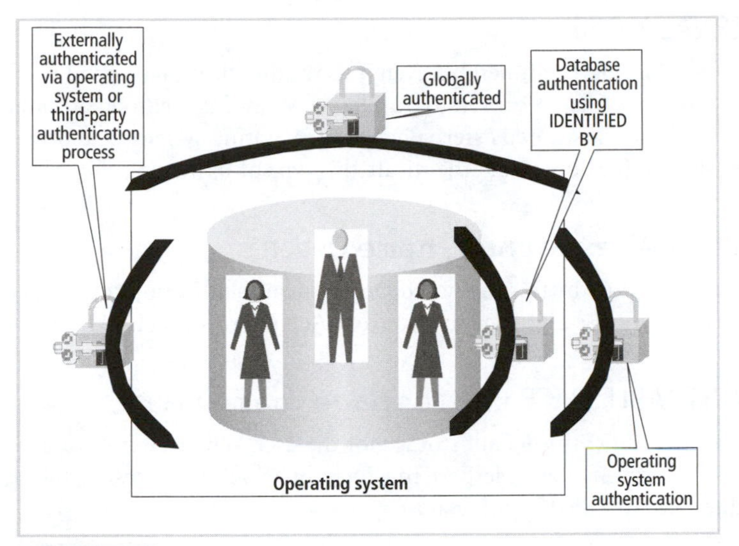

FIGURE 3-4 Architecture of Oracle authentication methods

BY Password Option

This is one of the options for authentication of database users. In addition, it is Oracle10*g*'s most commonly used method of authentication. Oracle10*g* encrypts and stores an assigned password in the database. Because the password is encrypted, the password is not visually readable. The encrypted password can be retrieved from the PASSWORD column in the DBA_USERS data dictionary view as shown in the following example.

```
SQL> SELECT USERNAME, PASSWORD
  2     FROM DBA_USERS
  3    WHERE USERNAME = 'HR'
  4  /

USERNAME                        PASSWORD
------------------------------ -----------------
HR                              6399F3B38EDF3288
```

Note that the results of the query show that the user HR was created by Oracle10*g* using the plain vanilla (typical) installation, and it shows that the password is encrypted.

EXTERNALLY Option

This method is not recommended even if the authentication method used outside the database is stringent and proven to be reliable. With this method you can create a user account to be authenticated externally by the operating system or another authentication process. Oracle10*g* does not authenticate this type of user.

GLOBALLY AS `external_name` Option

User authentication by this method depends on authentication through centralized user management such as Lightweight Directory Access Protocol (LDAP) or a similar method.

DEFAULT TABLESPACE `tablespace_name` Clause

This clause specifies the default storage for the user. When a user creates a database table and does not specify the tablespace in which the table should reside, the table is created in the tablespace specified by this clause.

TEMPORARY TABLESPACE `tablespace_identifier` Clause

In Oracle10*g* tablespace identifier can be:

- `tablespace`: The name of an existing temporary tablespace (temporary in this context is a classification of tablespace type). For further information on how to create a temporary tablespace, consult Oracle10*g* documentation.
- `tablespace_group_name`: The name of a tablespace group. A tablespace group is a collection of temporary tablespaces.

QUOTA Clause

This clause tells Oracle10*g* how much storage space a user is allowed for a specified tablespace. This option is used when storage space is scarce or to prevent users from wasting space by retaining unnecessary data. Although most DBAs use UNLIMITED to indicate no restriction on space for a tablespace, it is a good practice to enforce a storage quota. You can view each user's quota using the data dictionary view DBA_TS_QUOTAS as shown in Figure 3-5. Notice the value of the MAX_BYTES column. Minus one (-1) indicates unlimited, which means that there is no quota on the specified tablespace and therefore the user can utilize all allocated storage to the tablespace. Zero (0) indicates no quota, which means that a user is not allowed to use any storage on the specified tablespace, and therefore the user cannot create any objects on the tablespace. Other values indicate the total number of bytes of storage allowed on that tablespace.

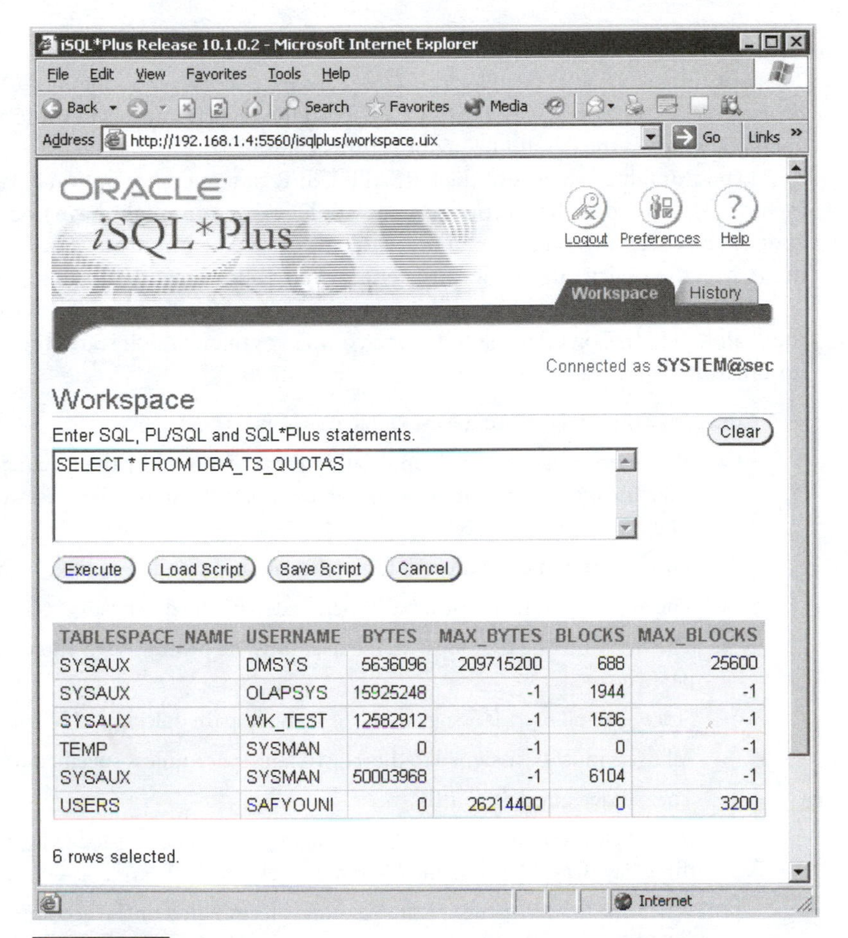

TABLESPACE_NAME	USERNAME	BYTES	MAX_BYTES	BLOCKS	MAX_BLOCKS
SYSAUX	DMSYS	5636096	209715200	688	25600
SYSAUX	OLAPSYS	15925248	-1	1944	-1
SYSAUX	WK_TEST	12582912	-1	1536	-1
TEMP	SYSMAN	0	-1	0	-1
SYSAUX	SYSMAN	50003968	-1	6104	-1
USERS	SAFYOUNI	0	26214400	0	3200

FIGURE 3-5 Contents of data dictionary view DBA_TS_QUOTAS

PROFILE Clause

This clause indicates the profile used for limiting database resources and enforcing password policies. The default profile is DEFAULT; this profile has no resource limitations or password restrictions. See the next section for more details on profiles.

PASSWORD EXPIRE Clause

This clause tells Oracle10*g* to expire the user password and prompts the user to enter a new password. Consider the following statement:

```
SQL> ALTER USER sam PASSWORD EXPIRE;
```

This statement will flag user account sam that the password is expired, which means that next time sam tries to log on he will be prompted to enter a new password. Typically, this clause is used when a new account is created with a default password.

ACCOUNT Clause

This option tells Oracle10*g* to lock (disable) account access when the ACCOUNT LOCK clause is selected. The selection of ACCOUNT UNLOCK indicates that the account is enabled, which is the default behavior for this clause.

Even though all clauses of the CREATE USER statement are optional, you should include clauses according to your company policies. As you might have noticed, the word "policy" has been repeated several times to emphasize the importance of such a document, because policy outlines a consistent procedure for creating database users.

At this point you have been exposed to a number of options, and it is time to pull them all together. Consider the following company policy, which states the requirements for creating a new user:

Company Policy for Creating New Database Users

- New users must have their application form approved by their manager and database manager; then the form can be submitted to the database administration group.
- The application form must describe the role and purpose of the user account.
- The application form must be authenticated by the database.
- The first time the user logs on, the user must be prompted to enter a new password.
- Storage for all users is assigned to USERS tablespace unless specified otherwise.
- All users must be assigned to the TEMP tablespace unless specified otherwise.
- The storage quota is 25 MB.
- If the new account is not used immediately, the account is locked within three days from the time the account was established.
- When the holder/owner of the account is terminated or the account is not used for a period of 30 days, the account is locked.

The following CREATE USER statement implements the creation of a user called safyouni (for a user named Sam Afyouni).

```
SQL> CREATE USER safyouni IDENTIFIED BY safyouni01
  2         DEFAULT TABLESPACE users
  3         TEMPORARY TABLESPACE temp
  4         QUOTA 25M ON users
  5         PROFILE default
  6         PASSWORD EXPIRE
  7         ACCOUNT UNLOCK
  8  /

User created.
```

Once a user is created, you can modify a user account with an ALTER USER statement using the same clauses listed in the previous sample.

When an Oracle10*g* user account is created, the new user cannot log in to the account until the database administrator provides the CREATE SESSION system privilege to allow the account to connect to the database.

Creating an Oracle10*g* User Using Database Authentication

The previous example showed how to to issue a SQL statement to create an Oracle10*g* user using database authentication. The graphical interface of Oracle Enterprise Manager also provides a function for creating a user. This function is called Security Management and is shown in Figure 3-6.

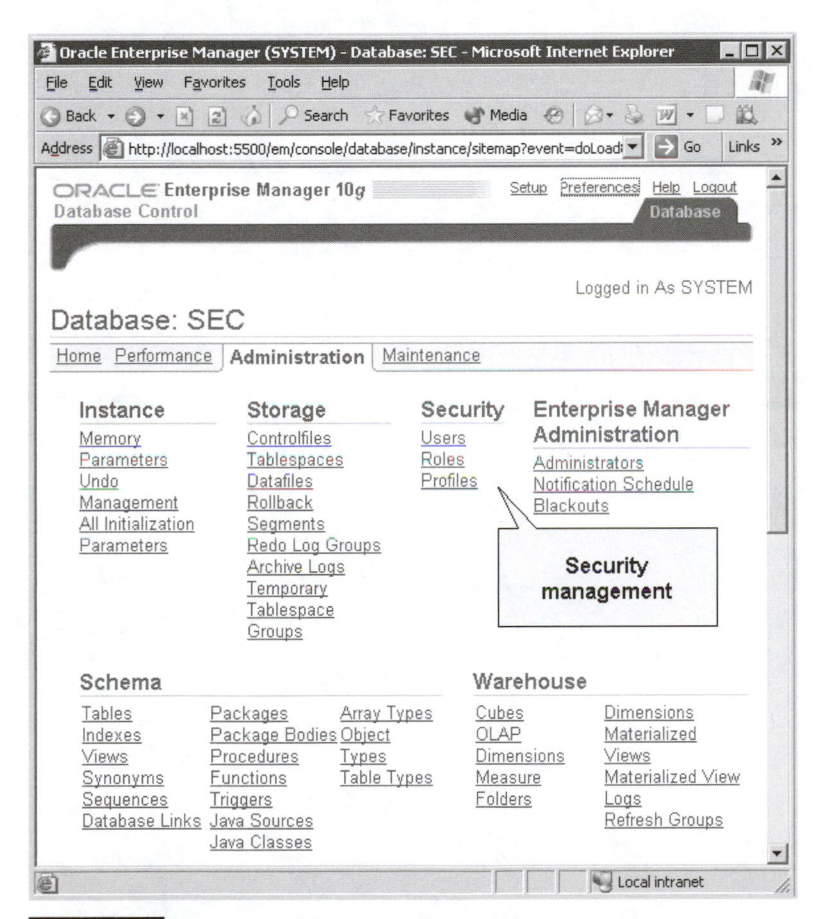

FIGURE 3-6 Enterprise Manager Console showing the Security Management function

Use the following steps to create a new user:

1. Select the **Users** node under the Security function and click the **Create** button on the toolbar on the right as shown in Figure 3-7. When this button is clicked, a dialog box appears offering a choice of objects that can be created.

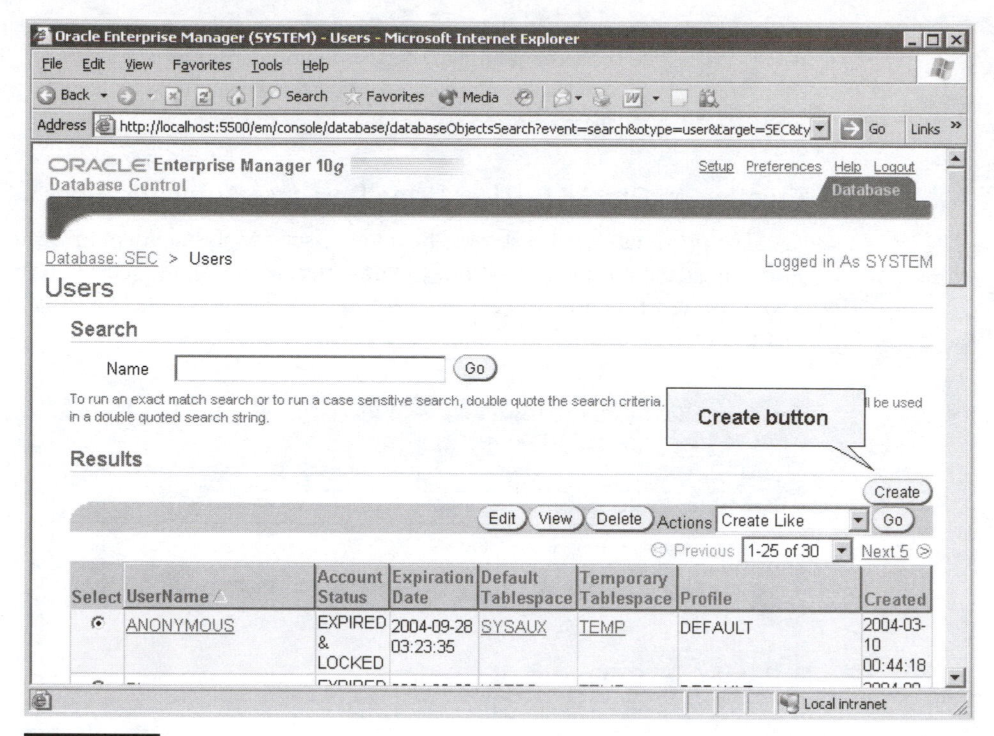

FIGURE 3-7 Oracle Enterprise Manager Console showing the Create Objects button

2. The Create User dialog box appears, as shown in Figure 3-8.

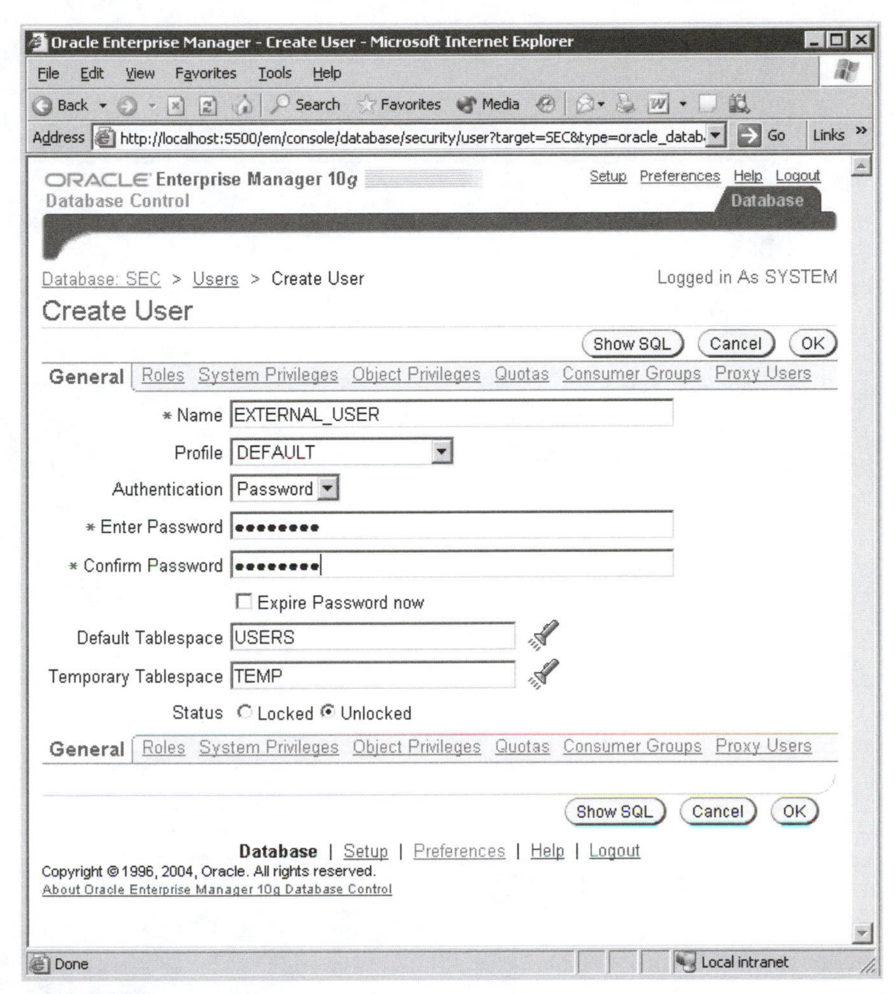

FIGURE 3-8 Creating a new user

3. Once you click **Create**, the Create User dialog box appears. Here you enter the user account details as shown in Figure 3-9. Note that in the example, only the fields on the General tab page are filled in; the other tabs are demonstrated in upcoming sections.

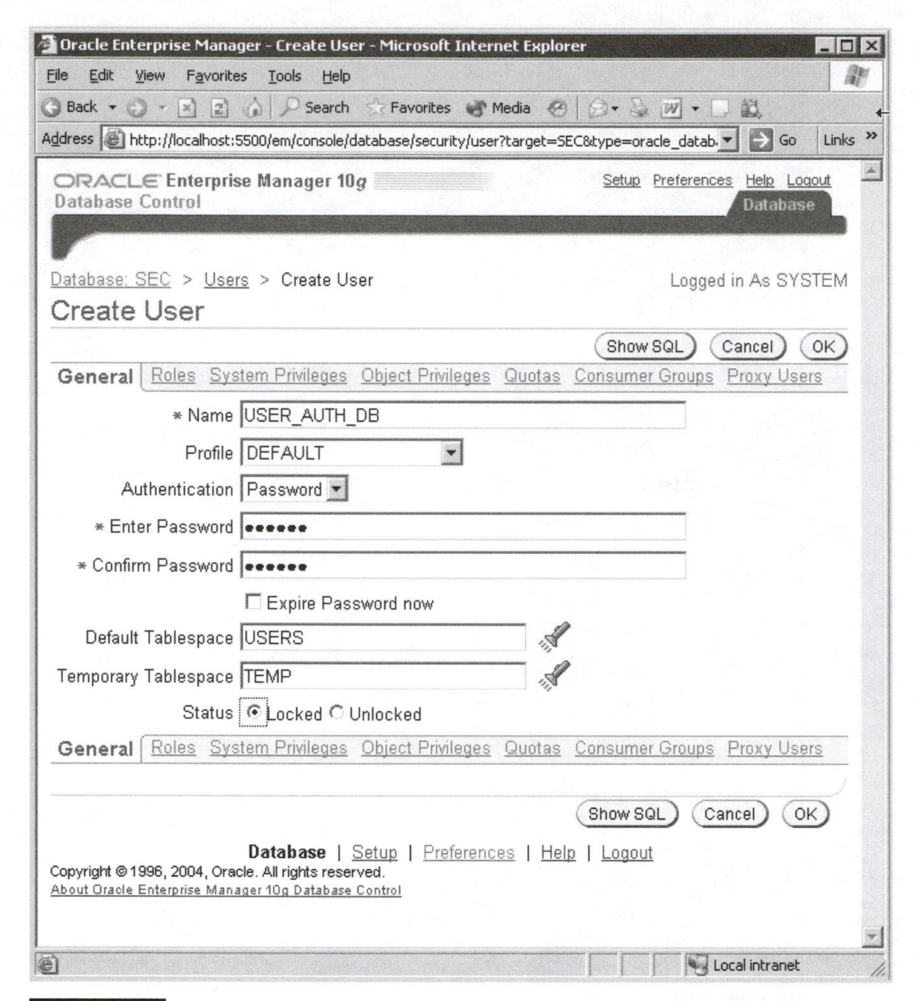

FIGURE 3-9 Create User screen showing the General tab

4. Click the **OK** button. The user is created and a message is displayed confirming successful creation of a user, as shown in Figure 3-10.

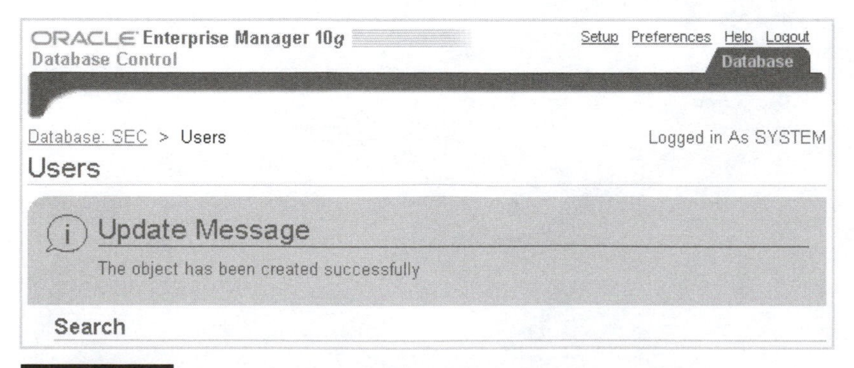

FIGURE 3-10 Create user success message

Creating an Oracle10*g* User Using External (Operating System) Authentication

For this type of authentication, Oracle10*g* depends on an external party to authenticate the user. Once the user is authenticated, Oracle10*g* allows the user to log in without a password if the user has the privilege to connect to the database. The following steps demonstrate how to create an Oracle10*g* user using external (operating system) authentication.

1. The Windows operating system account that you want Oracle10*g* to use for external authentication must belong to the ORA_DBA group. You can verify this by using the Computer Management tool found under Administrative Tools in Control Panel (see Figure 3-11 for an illustration). In this case, you are using an operating system account called EXTERNAL_USER.

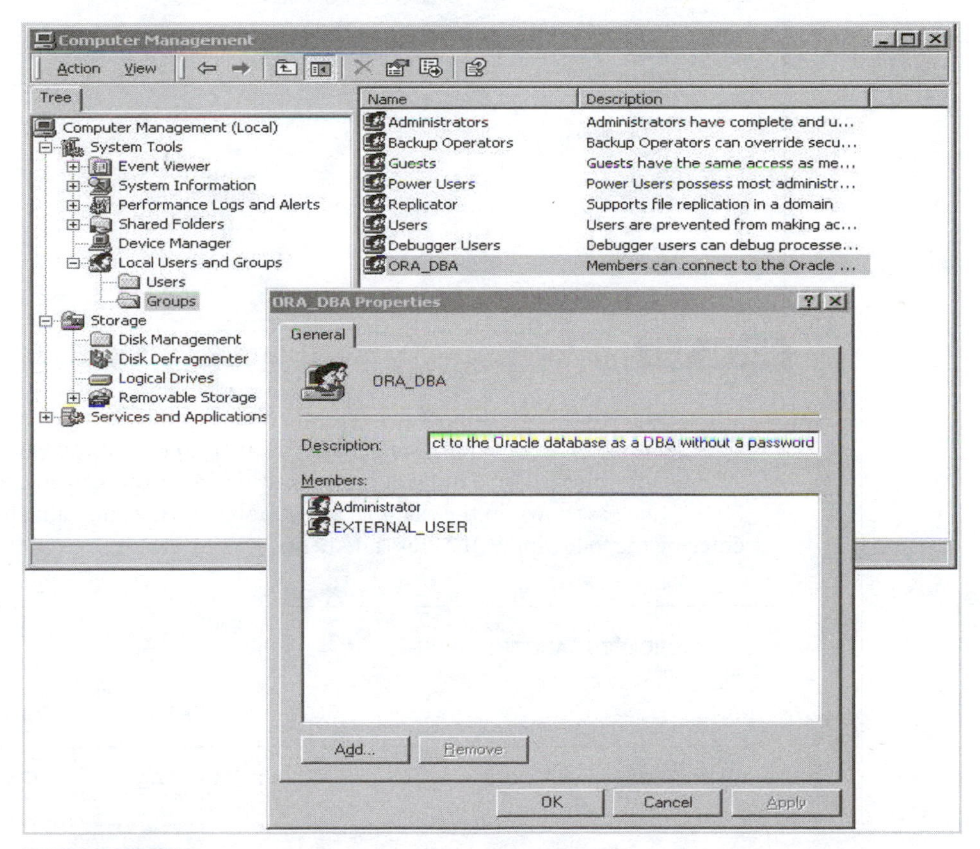

FIGURE 3-11 Computer management tool showing the ORA_DBA group properties

2. You must set the Windows registry string OSAUTH_PREFIX_DOMAIN to FALSE. To locate this parameter, navigate to **HKEY_LOCAL_MACHINE**, **SOFTWARE**, **ORACLE**, **HOME1** where HOME1 is the home of Oracle10*g* (see Figure 3-12). In your case, it could be different. If this parameter does not exist, create one. To create the parameter, right-click in the details pane, click **New**, then click **String Value** on the context menu, and enter the new parameter.

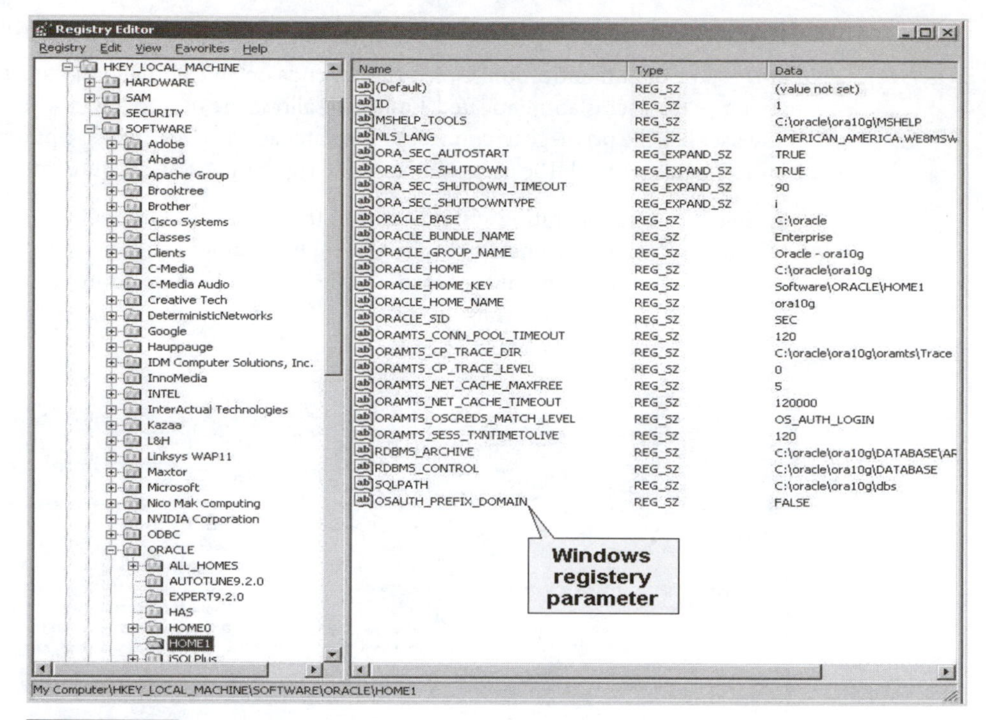

FIGURE 3-12 Windows registry showing the OSAUTH_PREFIX_DOMAIN parameter

3. You need to view the setting of the OS_AUTHENT_PREFIX initialization parameter. This parameter tells Oracle10*g* which prefix is used to identify an external account. The default value is OPS$ but you can use any desired value or you can use the NULL value as shown in the example below. You can view this value from SQL by entering the following SQL*Plus statement.

```
SQL> SHOW PARAMETER PREFIX

NAME                                 TYPE        VALUE
------------------------------------ ----------- ------
os_authent_prefix                    string      OPS$
```

4. Change the OS_AUTHENT_PREFIX initialization parameter value to **NULL** in the spfile or parameter file (pfile). The entry in this file should look like this:

```
os_authent_prefix = ''
```

NOTE

If the os_authent_prefix parameter does not exist in your intialization parameter files, you must add it.

5. Now create an Oracle user, EXTERNAL_USER, with the same user name as the Windows user name that is used for external authentication. See the following code for an example.

```
SQL> CREATE USER EXTERNAL_USER IDENTIFIED EXTERNALLY
  2  /

User created.
```

6. Provide the new user with the CREATE SESSION privilege, which enables the user to connect to the database. As explained earlier, no Oracle10*g* account can connect to the database without this privilege.

```
SQL> GRANT CREATE SESSION TO EXTERNAL_USER;

Grant succeeded.
```

7. Log off the Oracle SYS or SYSTEM account, and log off the current Windows account.
8. Log on as the EXTERNAL_USER windows account.
9. From a command prompt, log on to Oracle as EXTERNAL_USER as illustrated in Figure 3-13. Notice that the user logged on to Oracle by merely typing this command: `sqlplus /`. The (/) tells Oracle10*g* to verify that the Windows user has an Oracle10*g* account with the same name.

The user must be part of the DBA group and Administrator group.

Rather handy, isn't it, to log on to the database without a password. On the other hand, imagine yourself as a hacker who just broke into the database. He or she would be elated because the lack of a password made it so easy. Of course, Oracle10*g* external authentication offers one excellent advantage—it allows administrators to use one generic user to run maintenance scripts without the need to hard-code a password or spend extra effort to hide the password or encrypt it.

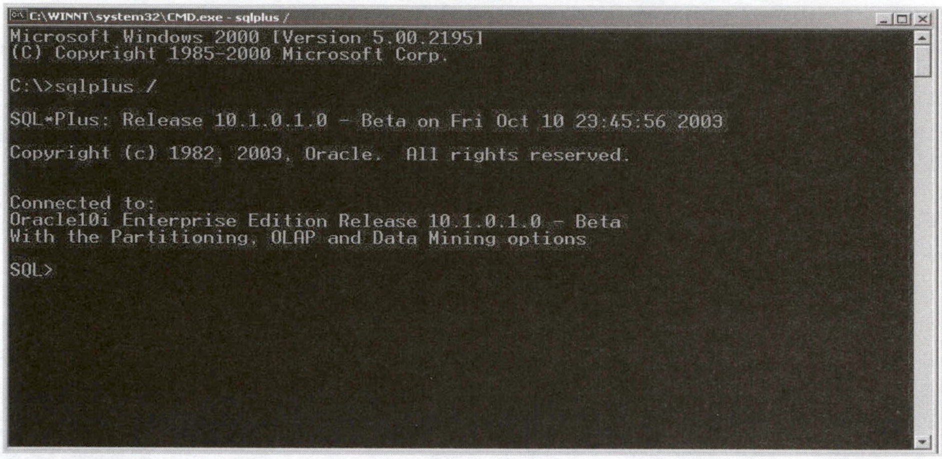

```
C:\WINNT\system32\CMD.exe - sqlplus /
Microsoft Windows 2000 [Version 5.00.2195]
(C) Copyright 1985-2000 Microsoft Corp.

C:\>sqlplus /

SQL*Plus: Release 10.1.0.1.0 - Beta on Fri Oct 10 23:45:56 2003

Copyright (c) 1982, 2003, Oracle.  All rights reserved.

Connected to:
Oracle10i Enterprise Edition Release 10.1.0.1.0 - Beta
With the Partitioning, OLAP and Data Mining options

SQL>
```

FIGURE 3-13 Example of external authentication

Creating an Oracle User Using Global Authentication

Oracle10*g* does provide one more method of authentication called GLOBAL. This method is used as an enterprise-level authentication solution. To create a user you still need to issue the following statement:

```
CREATE USER GLOBAL_USER IDENITFIED GLOBALLY AS 'global_name_propoerties'
...
```

Normally after you create any object or user, you should verify what you did. In the case of users, you can view all users by querying the data dictionary view DBA_USERS. Table 3-1 presents all the columns contained in this view.

Table 3-1 Columns of DBA_USERS view

Column Name	Description
USERNAME	Name of the database user; name is used to log on to the database; Oracle10*g* does not allow duplicate values of this column
USER_ID	Unique identification number to identify a user; this column is not used as frequently as USERNAME; used most commonly for auditing
PASSWORD	Keeps encrypted password for a database-authenticated user; use EXTERNAL for externally authenticated users and GLOBAL for globally authenticated users
ACCOUNT_STATUS	Indicates whether the user account is EXPIRED, LOCKED, EXPIRED and LOCKED, or OPEN

Table 3-1 Columns of DBA_USERS view (continued)

Column Name	Description
LOCK_DATE	Date and time the account was locked
EXPIRY_DATE	Date and time the account password expired
DEFAULT_TABLESPACE	Name of the tablespace assigned to this user account
TEMPORARY_TABLESPACE	Name of the temporary tablespace assigned to this user account
CREATED	Date and time this user account was created
PROFILE	Name of the profile assigned to this user account (for more details on profiles see Chapter 4)
INITIAL_RSRC_CONSUMER_GROUP	Name of the resources group to which this account belongs
EXTERNAL_NAME	External name for a GLOBAL authenticated user

The next section covers Microsoft SQL Server 2000 and how to create a SQL Server user.

Creating a SQL Server User

Before you can create a database user on a SQL server, you must first create a login ID for that user. A login ID controls access to the SQL Server system. This login ID is then associated with a database user in each database to which you want to grant the user access. You cannot log in to a SQL Server database without first supplying a valid login ID and password. To create a login ID, you must be a member of the SYSADMIN or SECURITYADMIN fixed server roles. You'll learn more about roles in Chapter 5. There are two types of login IDs: Windows Integrated (trusted) login, and SQL Server login.

Windows Integrated (Trusted) Logins

As you learned in an earlier section, you can associate a Microsoft Windows account or group with either the server in which SQL Server is installed or the domain in which the server is a member.

Creating Windows Integrated Logins

From the Command Line

To create a new login associated with a Windows account (Windows Integrated), in the Query Analyzer tool (see Figure 3-14) use the SP_GRANTLOGIN system stored procedure. The syntax is as follows:

```
sp_grantlogin [@login =] 'login'
```

In the above code, `login` is the fully qualified name of the Windows user account in the form of *`machine_name\username`* for local Windows users and *`domain\username`* for Windows domain accounts. Windows Integrated logins can also be associated with Windows groups on either the local server or domain. In this case, `login` is in the form of `machine_name\group_name` or `domain\group_name`.

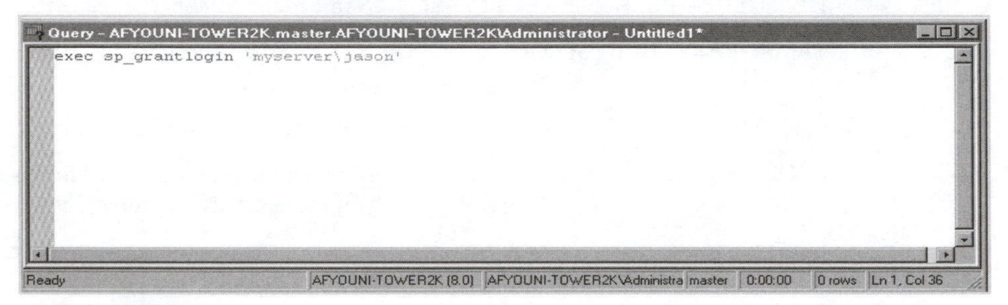

FIGURE 3-14 **Microsoft SQL Server 2000 Query Analyzer tool used as a command line interface**

For example, if you have a local Windows account named Jason on the SQL Server itself where the server name is myserver, you enter the following:

```
exec sp_grantlogin 'myserver\jason'
```

For a Windows domain account named sam in the domain mydomain, you enter the following:

```
exec sp_grantlogin 'mydomain\sam'
```

To associate a local Windows group called SQL_DBA, you enter the following:

```
exec sp_grantlogin 'myserver\sql_dba'
```

A login must be between 1 and 128 characters in length and cannot contain any spaces.

From Enterprise Manager

To create a new login associated with a Windows account (Windows Integrated) in Enterprise Manager, you take the following steps:

1. Open Enterprise Manager (see Figure 3-15).

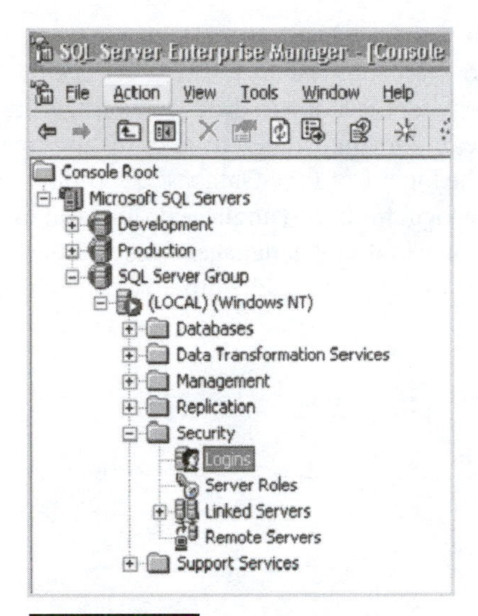

FIGURE 3-15 Opening Microsoft SQL Server Enterprise Manager

2. Expand the server group in which your server is functioning.
3. Expand the server you want to create the login for.
4. Expand the **Security** container.
5. Click **Logins.**

Typical of Microsoft Windows applications, you can reach the new login dialog box in several ways, but this series of instructions sticks with the menus to keep things simple.

6. On the menu bar, click **Action**, then click **New Login** (see Figure 3-16).

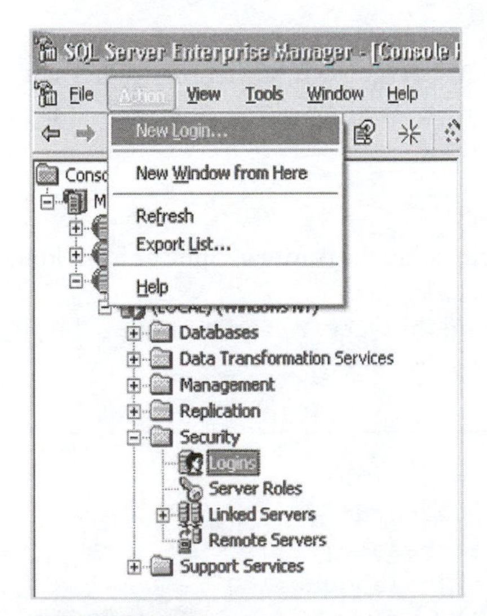

FIGURE 3-16 SQL Server Enterprise Manager showing action menu

7. Type the name of the user; in this case use **jason** as the user name.
8. Depending on the type of Windows account you are creating, select either the local server name or the domain name from the Domain drop-down list. Enterprise Manager automatically fills in the machine or domain name in front of the user name.
9. Select the default database for the login from the Database drop-down list.
10. Select the default language for the login from the Language drop-down list, or accept the default to set the language to the current default language of the SQL Server Instance.
11. Click **OK**. See Figure 3-17. (You will learn about the other two tabs in the dialog box later in this chapter.)

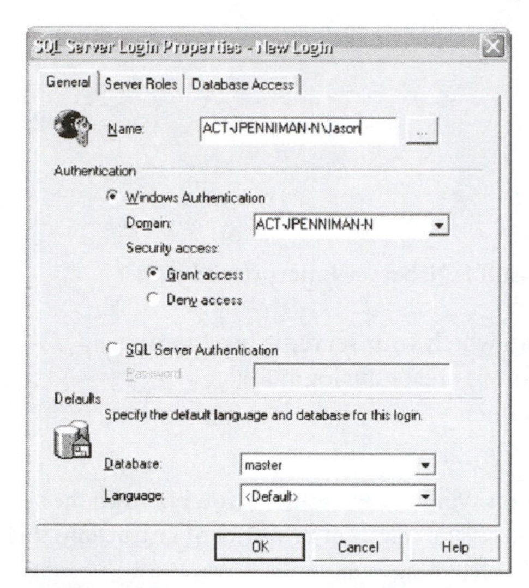

FIGURE 3-17 SQL Server login properties

SQL Server Login

The second type of login is a SQL Server login, sometimes called a SQL Server active login. This login is not associated with a Windows account; instead, it is a security account created within SQL Server itself.

Creating SQL Server Logins

You create SQL Server logins from either the command line or from Enterprise Manager.

From the Command Line

To create a SQL Server login from the Query Analyzer, you use the SP_ADDLOGIN system stored procedure. The syntax is as follows:

```
sp_addlogin [ @loginame = ] 'login'
    [ , [ @passwd = ] 'password' ]
    [ , [ @defdb = ] 'database' ]
    [ , [ @deflanguage = ] 'language' ]
    [ , [ @sid = ] sid ]
    [ , [ @encryptopt = ] 'encryption_option' ]
```

where

- `@loginame` is the name you choose for the login and `@passwd` is the password for the login. The default is NULL.
- `@defdb` is the name of the default database for the user. A user can access his default database after login without having to issue a USE command. The default is NULL.
- `@deflanguage` is the default language for the user. The default is the current default language of the SQL Server instance. Changing the current language of the SQL Server instance does not change the language of existing logins.
- `@sid` is the security identification number (SID) for the user. It is stored as a **varbinary (16)**. The default is NULL. If @sid is left NULL, the SQL Server automatically generates a SID for the login. If a value is supplied for @sid, it must be 16 bytes in length, and it cannot already exist in the instance of SQL Server with which you are working.
- `@encryptopt` specifies whether or not to encrypt the password in the database. The values represented in the preceeding code block work this way.

In a typical situation, you do not use the encryption option because the default is to encrypt, and this is a best practice. You'll learn more about why in later chapters.

To create a SQL Server login named `sam` with password `Str0ng!P@ssw0rd` you issue the following command. (You'll learn why this password was chosen in the next chapter.)

```
exec sp_addlogin 'sam', 'Str0ng!P@ssw0rd'
```

Typically, you specify a default database, especially when dealing with application users. To specify a default database of Northwind for sam, enter the following:

```
exec sp_addlogin 'sam', 'Str0ng!P@ssw0rd', 'Northwind'
```

From Enterprise Manager

To create a new SQL Server login in Enterprise Manager, follow these steps:

1. Open Enterprise Manager.
2. Expand the server group your server is in.
3. Expand the server you want to create the login for.
4. Expand the **Security** container.
5. Click **Logins.**
6. On the menu bar, click **Action**, then click **New Login**.
7. Type the name of the user, in this case, **Sam.**
8. Click the **SQL Server Authentication** option button.
9. Provide a password for the user in the Password text box. The password is masked as you type.
10. Click **OK** (see Figure 3-18).

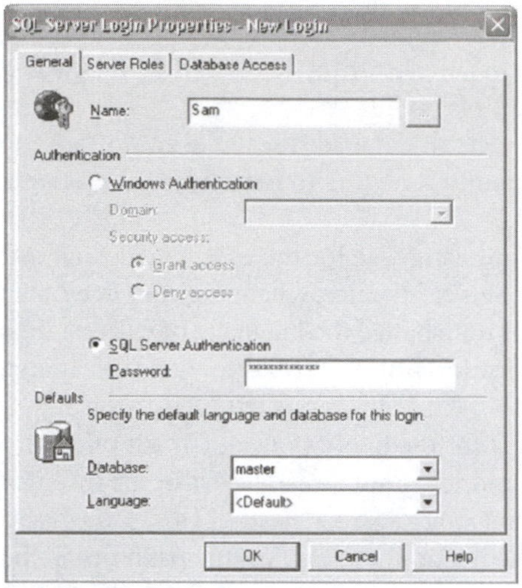

FIGURE 3-18 Server login properties—new login screen

As with Oracle10*g*, SQL Server stores all logins in the SYSLOGINS table found in the Master database, whereas users are stored in SYSUSERS table found in the current database. See Figure 3-19 for an illustration of the SYSUSERS table.

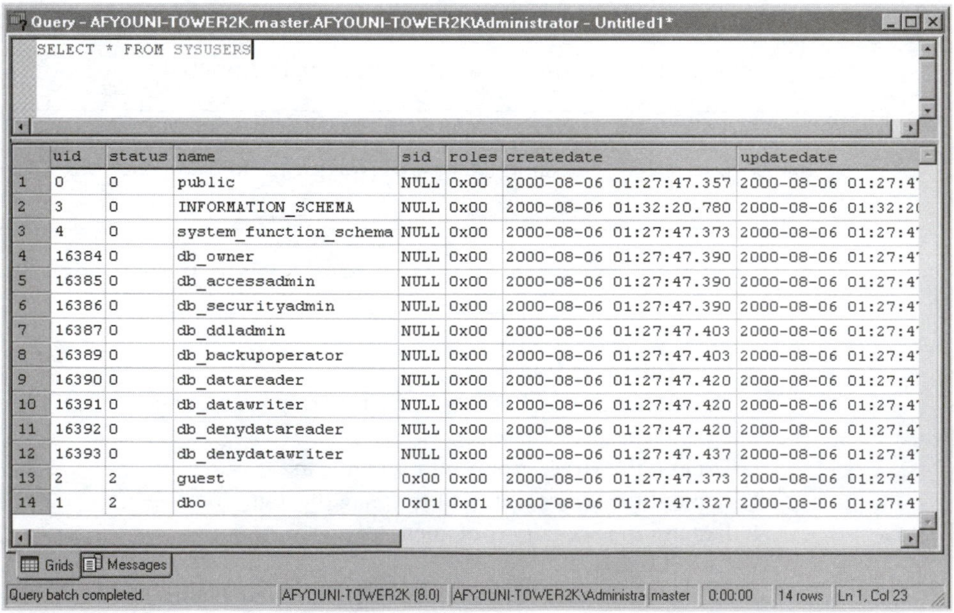

Query - AFYOUNI-TOWER2K.master.AFYOUNI-TOWER2K\Administrator - Untitled1*

```
SELECT * FROM SYSUSERS
```

	uid	status	name	sid	roles	createdate	updatedate
1	0	0	public	NULL	0x00	2000-08-06 01:27:47.357	2000-08-06 01:27:4
2	3	0	INFORMATION_SCHEMA	NULL	0x00	2000-08-06 01:32:20.780	2000-08-06 01:32:2
3	4	0	system_function_schema	NULL	0x00	2000-08-06 01:27:47.373	2000-08-06 01:27:4
4	16384	0	db_owner	NULL	0x00	2000-08-06 01:27:47.390	2000-08-06 01:27:4
5	16385	0	db_accessadmin	NULL	0x00	2000-08-06 01:27:47.390	2000-08-06 01:27:4
6	16386	0	db_securityadmin	NULL	0x00	2000-08-06 01:27:47.390	2000-08-06 01:27:4
7	16387	0	db_ddladmin	NULL	0x00	2000-08-06 01:27:47.403	2000-08-06 01:27:4
8	16389	0	db_backupoperator	NULL	0x00	2000-08-06 01:27:47.403	2000-08-06 01:27:4
9	16390	0	db_datareader	NULL	0x00	2000-08-06 01:27:47.420	2000-08-06 01:27:4
10	16391	0	db_datawriter	NULL	0x00	2000-08-06 01:27:47.420	2000-08-06 01:27:4
11	16392	0	db_denydatareader	NULL	0x00	2000-08-06 01:27:47.420	2000-08-06 01:27:4
12	16393	0	db_denydatawriter	NULL	0x00	2000-08-06 01:27:47.437	2000-08-06 01:27:4
13	2	2	guest	0x00	0x00	2000-08-06 01:27:47.373	2000-08-06 01:27:4
14	1	2	dbo	0x01	0x01	2000-08-06 01:27:47.327	2000-08-06 01:27:4

Query batch completed. | AFYOUNI-TOWER2K (8.0) | AFYOUNI-TOWER2K\Administra | master | 0:00:00 | 14 rows | Ln 1, Col 23

FIGURE 3-19 All users in SYSUSERS table

Removing Users

Dropping a user is simple, but whenever you are about to drop a user, remember the following scenario.

A new database administrator was hired by a manufacturing company to be a part of a team of four other administrators. On his first week he received a request to remove an account belonging to one of the consultants who was working on a current project. The request stated that the consultant contract had ended. The next day, the DBA received this urgent question from the lead developer: "When an account is removed, does the database remove all objects that the account owns?" When the DBA answered that this was true, he heard a groan on the other line. The lead developer said: "The consultant wrote code that had not been checked into version control, and we need it." The DBA laughed and told the developer that he had a backup of the account. He had made it a practice never to remove an account without making a backup of it, and you should too.

You also need to remember that before removing an account, you should obtain a written request approved by the immediate manager of the account holder for auditing purposes. Next you examine how to remove an account first using Oracle10*g* and then SQL Server.

Removing an Oracle User

SQL provides a command called DROP that removes a user account from the database. In Oracle10*g* you can issue this command from SQL*Plus by simply typing the following:

```
SQL> DROP USER SCOTT;

User dropped.
```

If the user does not own any objects, the command is successfully executed. If the user owns one or more objects, the following error is displayed:

```
SQL> DROP USER SCOTT;
DROP USER SCOTT
*
ERROR at line 1:
ORA-01922: CASCADE must be specified to drop 'SCOTT'
```

This message informs you that if you want to drop a user who owns objects you must use the CASCADE option as shown below.

```
SQL> DROP USER SCOTT CASCADE;

User dropped.
```

As demonstrated in the scenario, it is important to back up the account and save it for one to three months in case it is needed. Another good practice for removing accounts is to get a listing of all objects that the account owns and send it to the immediate manager of the account holder for verification. Yet another good practice is to lock the account or revoke the CREATE SESSION privileges, especially when an employee leaves the company or the employee does not require access to the database. Now that you understand how to remove an Oracle user, you are ready to learn how to remove an account and login ID from a SQL server.

SQL Server: Removing Windows Integrated Logins

You can remove an account and login ID from a SQL server by using either the command line or Enterprise Manager.

From the Command Line

To drop an existing Windows Integrated login, use the SP_DENYLOGIN system-stored procedure.

```
sp_denylogin [ @loginame = ] 'login'
```

In the code above, `@loginname` is the name of the Windows Integrated login to be removed. The following statement drops the login account `jason`:

```
exec sp_denylogin 'myserver\jason'
```

From Enterprise Manager

To drop a login in Enterprise Manager, simply highlight the desired login and choose Delete from the Action menu.

Modifying Users

You will often be asked to change the attributes of an existing database user. Modifications can involve changing a password, locking an account, or increasing a storage quota. You need to know how to follow best practices to make these changes. A written approval must be provided and must state reasons for the change. You can use the ALTER USER DDL statement of SQL to modify any attribute of a user. To see how this works in practice, the following sections present an Oracle10*g* demonstration on how to modify an existing user followed by a SQL Server demonstration.

Modifying an Oracle User

Using SQL*Plus you can issue an ALTER USER statement as shown in the following code listing. In this code, the issued statement changed the password from the existing value to LION.

```
SQL> ALTER USER SCOTT IDENTIFIED BY LION
  2  /

User altered.
```

Figure 3-20 illustrates another example, this time using Oracle Enterprise Manager to change the default tablespace for the SCOTT account from USERS to EXAMPLE.

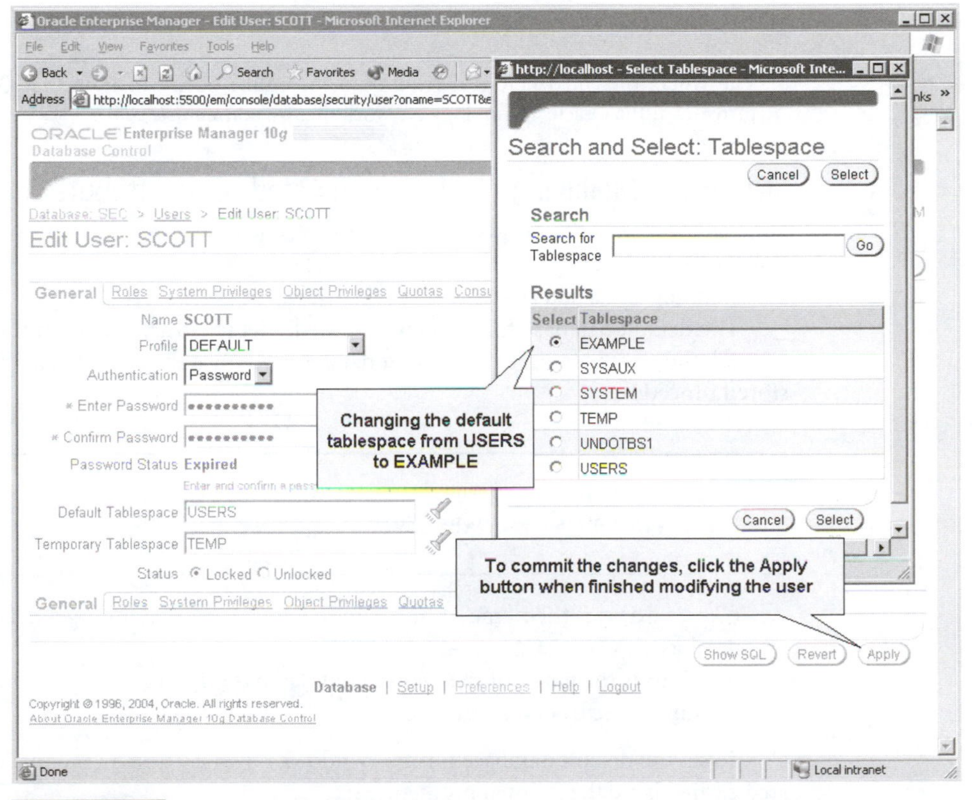

FIGURE 3-20 Illustration of modifying an existing Oracle user account

As a reminder, the following code block represents the full syntax needed to modify an Oracle user account using the ALTER statement.

```
ALTER USER username
   [ IDENTIFIED { BY password | EXTERNALLY | GLOBALLY AS
'external_name' } ]
   [ DEFAULT TABLESPACE tablespace]
   [ TEMPORARY TABLESPACE  { tablespace |
tablespace_group_name } ]
   [ QUOTA { integer { K | M } ON tablespace | UNLIMITED ]
   [ PROFILE profile ]
   [ PASSWORD EXPIRE ]
   [ ACCOUNT { LOCK | UNLOCK } ]
```

The syntax presented here is derived from the online documentation that Oracle provides at the Oracle Technology Network site: *www.otn.oracle.com.*

SQL Server: Modifying Windows Integrated Login Attributes

You can modify the Windows Integrated login attributes from the command line or from Enterprise Manager.

From the Command Line

The default database for the login is initially set to `master`. To set the default database for the trusted login to a different database, you use the SP_DEFAULTDB system stored procedure.

```
sp_defaultdb      [ @loginame = ] 'login' ,
    [ @defdb = ] 'database'
```

- `@loginame` is the name of the login to which you're assigning a default database.
- `@defdb` is the name of the new default database. The database must already exist in the SQL Server instance.

To set the default database for the Windows Login `mydomain\sam` that was created earlier, issue the following statement:

```
exec sp_defaultdb 'mydomain\sam', 'Northwind'
```

The default language for the login is initially set to the current default language for the SQL Server instance. To change the login's default language, you use the SP_DEFAULTLANGUAGE system stored procedure.

```
sp_defaultlanguage [ @loginame = ] 'login'
                   [ , [ @language = ] 'language' ]
```

- `@loginame` is the login name for which you are setting the default language. The value can be an existing Microsoft SQL Server login or a Microsoft Windows NT user or group.
- `@language` is the default language of the login. The optional default is NULL. If left NULL, the default language for the login is set to the current default language of the SQL Server instance.

```
exec sp_defaultlanguage 'mydomain\sam', 'spanish'
```

From Enterprise Manager

To modify an existing SQL Server login using Enterprise Manager, follow these steps:

1. Open Enterprise Manager.
2. Expand the Security Container and click the **Logins** node.
3. Select the desired login.
4. Click **Properties** on the Action Menu to open the SQL Server Login Properties dialog box.

Default Users

This section outlines the accounts that are created in a typical installation of an Oracle or SQL Server. The main reason to list these users is to differentiate the essential users from the optional users.

Oracle Server Default Users

No matter what type of installation you choose, Oracle10*g* creates two main users. The first one is SYS, which is the owner of the data dictionary and has privileges to perform any task on the database. The other user is SYSTEM, which has privileges to perform almost all tasks on the database. Other users created are examples and sample schemas for practice purposes. Rule number one is to change the default password or select a password that conforms to the company password policy. (See Chapter 4 for more details on this topic.) Figure 3-21 shows a list of all users created when a General Database option was selected at the time of an Oracle10*g* installation and database creation.

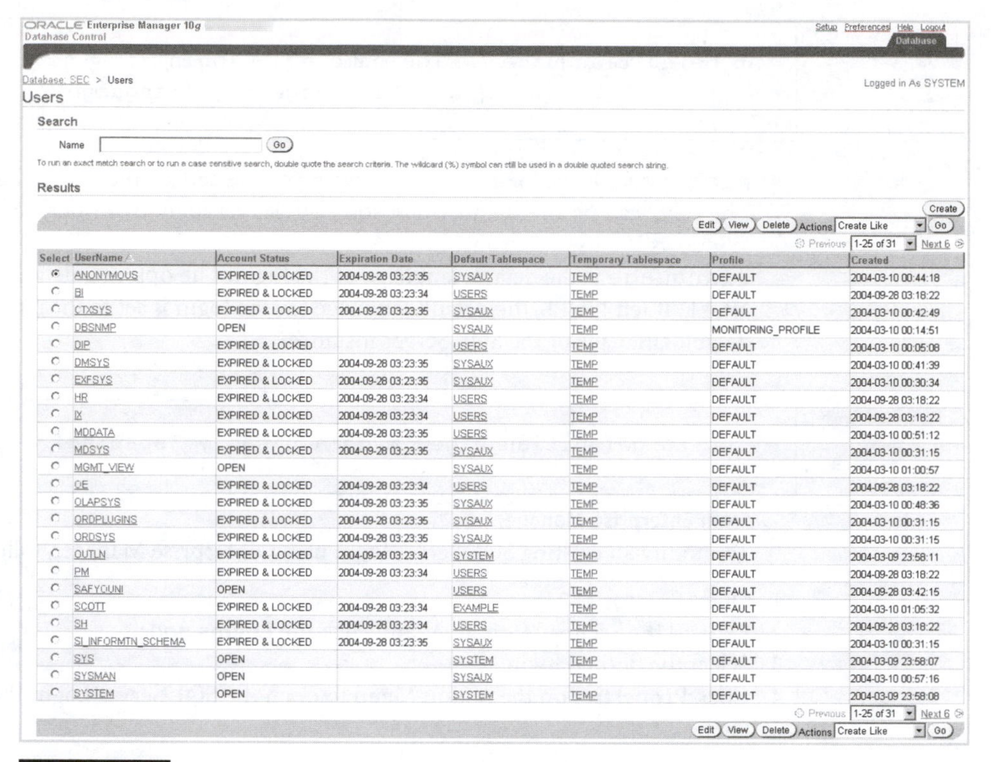

Select	UserName	Account Status	Expiration Date	Default Tablespace	Temporary Tablespace	Profile	Created
◉	ANONYMOUS	EXPIRED & LOCKED	2004-09-28 03:23:35	SYSAUX	TEMP	DEFAULT	2004-03-10 00:44:18
○	BI	EXPIRED & LOCKED	2004-09-28 03:23:34	USERS	TEMP	DEFAULT	2004-09-28 03:18:22
○	CTXSYS	EXPIRED & LOCKED	2004-09-28 03:23:35	SYSAUX	TEMP	DEFAULT	2004-03-10 00:42:49
○	DBSNMP	OPEN		SYSAUX	TEMP	MONITORING_PROFILE	2004-03-10 00:14:51
○	DIP	EXPIRED & LOCKED		USERS	TEMP	DEFAULT	2004-03-10 00:05:08
○	DMSYS	EXPIRED & LOCKED	2004-09-28 03:23:35	SYSAUX	TEMP	DEFAULT	2004-03-10 00:41:39
○	EXFSYS	EXPIRED & LOCKED	2004-09-28 03:23:35	SYSAUX	TEMP	DEFAULT	2004-03-10 00:30:34
○	HR	EXPIRED & LOCKED	2004-09-28 03:23:34	USERS	TEMP	DEFAULT	2004-09-28 03:18:22
○	IX	EXPIRED & LOCKED	2004-09-28 03:23:34	USERS	TEMP	DEFAULT	2004-09-28 03:18:22
○	MDDATA	EXPIRED & LOCKED	2004-09-28 03:23:35	USERS	TEMP	DEFAULT	2004-03-10 00:51:12
○	MDSYS	EXPIRED & LOCKED	2004-09-28 03:23:35	SYSAUX	TEMP	DEFAULT	2004-03-10 00:31:15
○	MGMT_VIEW	OPEN		SYSAUX	TEMP	DEFAULT	2004-03-10 01:00:57
○	OE	EXPIRED & LOCKED	2004-09-28 03:23:34	USERS	TEMP	DEFAULT	2004-09-28 03:18:22
○	OLAPSYS	EXPIRED & LOCKED	2004-09-28 03:23:35	SYSAUX	TEMP	DEFAULT	2004-03-10 00:48:36
○	ORDPLUGINS	EXPIRED & LOCKED	2004-09-28 03:23:35	SYSAUX	TEMP	DEFAULT	2004-03-10 00:31:15
○	ORDSYS	EXPIRED & LOCKED	2004-09-28 03:23:35	SYSAUX	TEMP	DEFAULT	2004-03-10 00:31:15
○	OUTLN	EXPIRED & LOCKED	2004-09-28 03:23:34	SYSTEM	TEMP	DEFAULT	2004-03-09 23:58:11
○	PM	EXPIRED & LOCKED	2004-09-28 03:23:34	USERS	TEMP	DEFAULT	2004-09-28 03:18:22
○	SAFYOUNI	OPEN		USERS	TEMP	DEFAULT	2004-09-28 03:42:15
○	SCOTT	EXPIRED & LOCKED	2004-09-28 03:23:34	EXAMPLE	TEMP	DEFAULT	2004-03-10 01:05:32
○	SH	EXPIRED & LOCKED	2004-09-28 03:23:34	USERS	TEMP	DEFAULT	2004-09-28 03:18:22
○	SI_INFORMTN_SCHEMA	EXPIRED & LOCKED	2004-09-28 03:23:35	SYSAUX	TEMP	DEFAULT	2004-03-10 00:31:15
○	SYS	OPEN		SYSTEM	TEMP	DEFAULT	2004-03-09 23:58:07
○	SYSMAN	OPEN		SYSAUX	TEMP	DEFAULT	2004-03-10 00:57:16
○	SYSTEM	OPEN		SYSTEM	TEMP	DEFAULT	2004-03-09 23:58:08

FIGURE 3-21 List of all default users created during an Oracle software installation

Note that the account status of most of the accounts is password expired and locked. You must unlock these accounts before they can be used.

SYS and ORAPWD Utility

Oracle10*g* provides the password utility ORAPWD to create a password file, which stores the password for the user SYS and also serves as a placeholder for any future users that will have the SYSDBA or SYSOPER privilege. The password file is used to authenticate the SYS user when the SYS user is logging in with SYSDBA or SYSOPER privileges.The ORAPWD utility is used mainly when creating a new database or when the SYS password is forgotten. The following statement shows the syntax for this utility. For more details on this utility, consult Oracle documentation.

```
C:\>orapwd
Usage: orapwd file=<fname> password=<password> entries=<users> force=<y/n>

  where
    file - name of password file (mand),
    password - password for SYS (mand),
    entries - maximum number of distinct DBA and     force - whether to
overwrite existing file (opt),
OPERs (opt),
  There are no spaces around the equal-to (=) character.
```

SQL Server Default Users

There are two default logins in SQL Server: SA and BUILT-IN\Administrators.

SA is the system administrator. It's equivalent to the SYS user in Oracle. Permissions for SA cannot be modified, nor can the login be deleted. SA has access to all databases and objects in the instance of SQL Server. Its database user name is DBO.

BUILT-IN\Administrators is associated with the local administrators' group on the Windows server. Windows users placed in this group have security access equivalent to SA by default, but you can change its permissions or remove the login completely.

Remote Users

All database user accounts are created and stored in the database regardless of whether they are connected locally or remotely. When a user logs on to the database through the machine where the database is located, the database is called a local database. However, a user could log on to the database using a different host machine than the machine where the database is located. In this case, this user is referred to as a remote user, as illustrated in Figure 3-22. In Oracle10*g*, remote users can be authenticated by the operating system of the computer on which the user is located. Oracle10*g* allows this authentication provided the REMOTE_OS_AUTHENT initialization parameter is set to TRUE. When this parameter is set to TRUE, Oracle10*g* checks the operating system user name and tries to find a matching user name in the database. When this parameter is set to FALSE, Oracle10*g* disallows operating system authentication for remote users. Unlike Oracle10g, SQL Server does not support this type of remote user authentication.

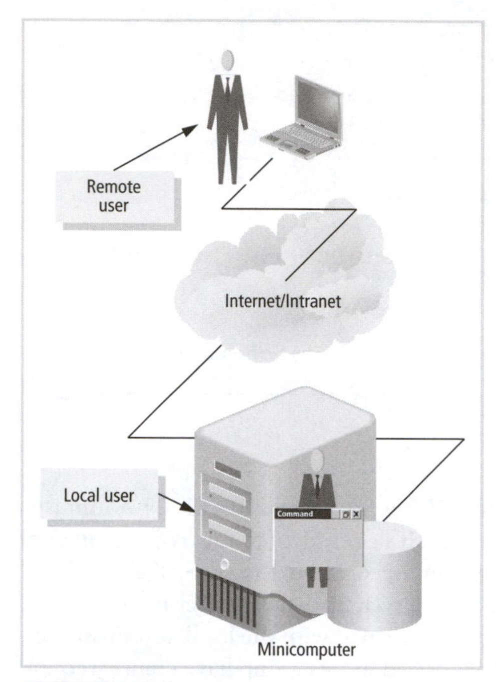

FIGURE 3-22 Local and remote users

CAUTION

You should never use remote user authentication because it exposes the database to security violation risks such as unintended or illegal access to the hosted server or database itself.

Database Links

A prominent retail company just launched a new retail system using a distributed database architecture. The platform of this architecture was based on an Oracle10*g* database hosted by a Windows 2000 server. While tending to a customer request, a customer service representative received an application error. Immediately, he reported the error to the application support team and the team instantly started investigating the problem. The application support person working on this issue realized that all the rows for a specific table in the central database had been deleted—they had vanished into thin air.

After recovering the rows from an existing backup, a full investigation was initiated to find out exactly what happened. The investigation went on for two weeks but to no avail. The investigation concluded in a report that there was an internal security breach but they could not determine who caused it.

Months went by until one day, while an application DBA was tuning a query using the QA (Quality Assurance) environment, he noticed that the query results and behavior were exactly like those of production. This was strange, because QA had less data than production, and the data in QA had been scrubbed (changing private and sensitive data to unrecognized values). After further investigation, he found that a database link to production existed on the QA server. When the DBA examined the property of the database link, he realized that it was a public link using the FIXED USER method (described later). This link allowed any user authenticated to the database to perform any query on production without authentication.

This incident actually occurred, and don't be surprised how often it may reoccur, especially because QA environments are usually refreshed from production. This internal breach of security was caused by lack of stringent procedures for creating database links.

What exactly is a database link? It's is connection from one database to another database. The linked databases could both be Oracle10*g*, or both SQL Server, or a mix of Oracle10*g* and SQL Server. A database link enables a user to perform Data Manipulation Language (DML) statements or any other valid SQL statement on a database while logged onto a different database, as shown in Figure 3-23. Both Oracle10*g* and SQL Server provide this functionality. (In Figure 3-23, SEC is a database name commonly used in this book, and PRD is another arbitrary database name.)

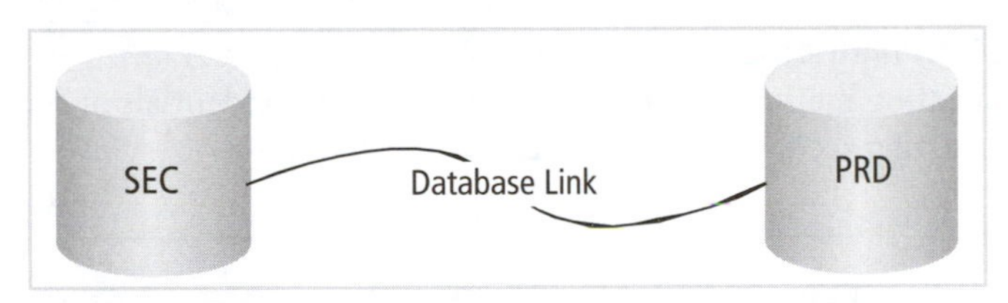

FIGURE 3-23 Database link architecture

In Oracle10*g*, database links can be created in two ways: as PUBLIC, which makes the database links accessible to every user in the database, or as PRIVATE, which gives ownership of the database to a user. With PRIVATE, the database is not accessible by any other user unless the user has been given access by the owner.

There are three types of authentication methods when creating a database link. The following sections provide details on how to create a database link for each type.

Authentication Methods

This section describes authentication methods for connecting to Oracle10*g* databases using database link mechanisms.

Authentication Method 1: CURRENT USER

This authentication method orders Oracle10*g* to use the current user credentials for authentication to the database to which the user is trying to link.

```
SQL> CONNECT SYSTEM@BOOK
Enter password: ******
Connected.

SQL> CREATE PUBLIC DATABASE LINK PRD
  2     CONNECT TO CURRENT_USER
  3     USING 'PRD'
  4  /

Database link created.
```

Authentication Method 2: FIXED USER

This authentication method orders Oracle10*g* to use the user and password provided in this clause for authentication to the database to which the user is trying to link.

```
SQL> CREATE PUBLIC DATABASE LINK PRD
  2     CONNECT TO SCOTT IDENTIFIED BY TIGER
  3      USING 'PRD'
  4  /

Database link created.
```

Authentication Method 3: CONNECT USER

This authentication method orders Oracle10*g* to use the credentials of the connected user who has an existing account in the database to which the user is trying to link.

```
SQL> CREATE PUBLIC DATABASE LINK PRD
  2     USING 'PRD'
  3  /

Database link created.
```

Which one of these methods should you use? The answer is not easy and depends on the situation. However, it is safe to say that you should stay away from PUBLIC database links and always use PRIVATE links.

You can view all existing database links by querying the data dictionary view DBA_DB_LINKS. As usual, Oracle10*g* provides the initialization parameter DBLINK_ENCRYPT_LOGIN that you can configure to notify Oracle10*g* of whether or not to use encryption when connecting to other databases.

In Microsoft SQL Server 2000, if a login has access to more than one database, you can access one database from another by referencing the object in the form of a DATABASE.OWNER.OBJECT. For example, if the login jason has a user in both the Northwind and Pubs databases, jason can execute and retrieve all rows from the author's table in the Pubs database as follows:

```
use Northwind
select * from pubs.dbo.authors
```

A final note of clarification: Oracle10*g* database links and SQL Server linked servers are the same concept but use different architecture and implementation. Both allow users to connect from one database to another database to perform different SQL statements.

Linked Servers

Microsoft SQL Server 2000 also uses the concept of linked servers. Linked servers allow you to connect to almost any Object Linking and Embedding Database (OLEDB) or Open Database Connectivity (ODBC). OLEDB is a Microsoft component that allows Windows applications to connect and access different database systems. ODBC is a Microsoft protocol used for connecting Windows applications to different database systems, including other SQL servers and Oracle10*g* servers, as shown in Figure 3-24. For example, say you have an Oracle10*g* server with the database ACME_ORA, and you want to access it directly from a SQL server. You can set up a linked server to the database, and then access tables and views in the form of LINKED_SERVER.OWNER.OBJECT as shown in the following command:

```
select * from acme_db.sam.employees
```

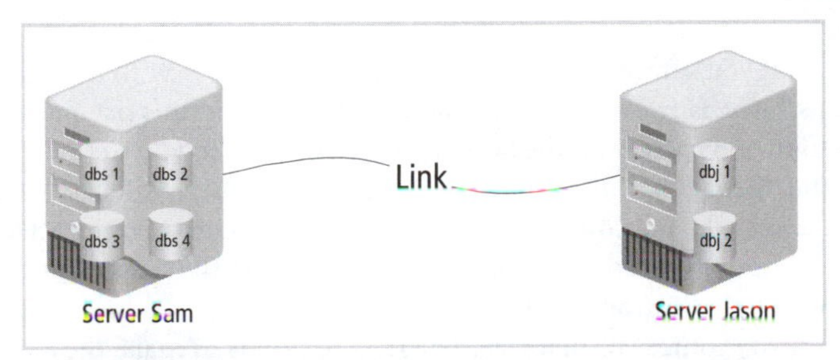

FIGURE 3-24 Linked servers architecture using SQL Server

For connections that do not support the syntax shown in the previous code block, you can use the OPENQUERY function:

```
select * from openquery(acme_db 'select * from sam.employees')
```

Of course, for any of this to work, you need to map logins in your SQL Server instance to users in the linked database. The linked servers concept is beyond the scope of this text, but there are a few security points of which you need to be aware.

Look at the Security tab in the Linked Servers Properties dialog box shown in Figure 3-25.

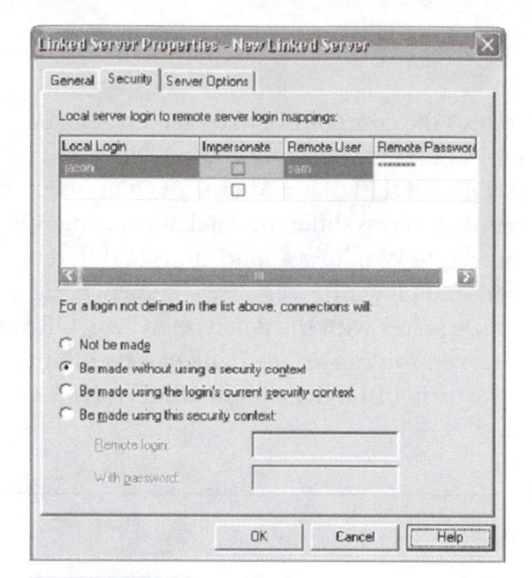

FIGURE 3-25 Creating a new linked server with SQL Server

Notice, that you can map a login on the local SQL server with a user on the remote data source. However, if a user logs on who is not in this list and he or she tries to connect to the linked server, the SQL server uses the rules you define in the lower portion of the dialog box shown in Figure 3-25. This is where you can get into trouble. The default is to make the connection without using a security context. Depending on your setup, the user might not have access to anything on the server; however, the user could activate an application role (assuming the remote is a SQL server) and operate under the security context of that role. The next option is to make the connection using the login's current context. In other words, if you are using a trusted Windows login, and that login has permissions in the remote server or database, the login connects and operates under that security context. The last option is to connect using a specified context. Basically, this option means that if the login isn't in the list shown in Figure 3-25, the user can connect to the remote database with these credentials. You should avoid this option because it is the least secure and can get you into trouble. The ideal configuration is to map your logins appropriately, and choose "Not be made."

Remote Servers

Along the same line as linked servers, you can communicate with another SQL server by creating a remote server. Instead of using OLEDB, communication occurs across a remote procedure call (RPC), eliminating a layer of code. As with a linked server, you map logins on the local server to logins on the remote. Again, stay way from the Map all remote logins to option and map your individual logins appropriately, as shown in Figure 3-26.

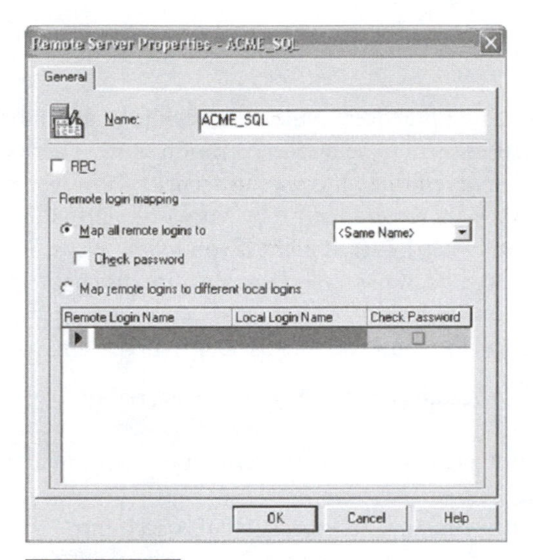

FIGURE 3-26 SQL Server remote server properties

Practices for Administrators and Managers

The database administration job is never ending and very challenging. Database administrators are responsible for managing the database, which includes managing user accounts, data files, and memory, to name a few of many tasks. In addition to administrating users, the database administrator is constantly performing other administrative tasks, such as backup, recovery, and performance tuning. The database mission does not end as long as a database is in continuous use. With that said, the database administrator should be supported with all available means to perform his or her tasks in a consistent and efficient manner. As indicated earlier, one of these supports is the documentation of procedures and processes that the administrator must employ to ensure consistency. This requires a unified effort in which almost everyone in the company must participate. However, the main champion of this effort is the database manager who should mandate that all tasks to be performed by administrators be documented and logged.

To make wise decisions, database managers have the sizable responsibility of keeping up with database best practices, database technology, and database security issues. Unfortunately, in many cases the database manager is responsible for many groups, including system administrators, network administrators, operations, help desk personnel, application support, and others. The issue is this: when database managers are responsible for more than the database group, their attention is so divided that they cannot provide the support that the database administrators need.

How is this related to user administration? You must always remember that the user is your client, and the database your client is using must meet integrity requirements. This means that when the database administrator does not get the required support and his work is thereby compromised, the database is at risk and system performance for the user (client) is affected.

Best Practices

You might question the term "best practices," which has been used a number of times in this chapter. Often best practices seem to be merely common sense. There's always more than one solution to a problem, even when it comes to security. However, best practices are worth learning because they are recommended by Microsoft, DBAs, and security professionals, and because you will sleep better at night if you follow them. Some of the best practices apply to the Oracle10*g* and SQL servers described in this chapters and to other servers as well.

These are the best practices for administering users, privileges, and roles:

- Follow your company's procedures and policies to create, remove, or modify database users.
- Always change the default password and never write it or save it in a file that is neither encrypted nor safe.
- Never share user accounts with anyone, especially DBA accounts.
- Always document and create logs for changes to and removals of database user accounts.
- Never remove an account even if it is outdated. Instead disable or revoke connection privileges of the account.
- Give access permissions to users only as required and use different logins and passwords for different applications.
- Educate users, developers, and administrators on user administration best practices, as well as company policies and procedures.
- Keep abreast of database and security technology. You should be aware of all new security vulnerabilities that may increase database security risks.
- Constantly review and modify procedures as necessary to be in line with your company's policies and procedures. Keep procedures up to date with the dynamic nature of database and security technology.
- As mentioned earlier in this chapter, by default, the SQL server installs in the Windows Only security model. The best practice is to change this to the Mixed Mode model (the only other option). This can be done at the time of installation or later through the server configuration utilities.

For a SQL server, it is preferable to mimic Oracle's recommended installation for the UNIX environment. Basically, you create a local Windows group on the server called DBA, grant login access for this group to the SQL Server instance, and make the trusted login a SYSADMIN. Granted, if your network administrator wants to, he can still put himself in the group because he can manage the Windows server. However, it does prevent the network administrator from logging into the console and accidentally doing harm, and it still allows you to use a Windows Integrated security model.

While on the subject of default installs, you should note that by default, the SQL Server services run as the local SYSTEM Windows account. This is not good. You should always change this to a local Windows or Windows domain account. Microsoft provides articles on how to do this. If the system is compromised, the attacker is running the security context of the SQL Server service for the instance. If the attacker runs it as SYSTEM, your company's entire network belongs to the attacker. On the other hand, if the culprit is a user who has permission only to run the SQL Server services, he has to work harder to get any further than your server.

You learned earlier that SQL Server automatically creates a login identity associated with the local Windows group BUILT-IN\Administrators. In a typical Windows installation, the local administrator account on the Windows server is a member of this group, and if the server is a Windows domain member, the DOMAIN\Domain Admins domain group is also a member of this group. What does this mean to you, the DBA? It means that your non-DBA network administrator has full system administrator privileges to your SQL Server instance. Don't panic. Microsoft allows you to change that. There are two approaches you can take depending on your business model. You could take the BUILT-IN\Administrators out of SYSADMIN and put them in a more practical role such as SECURITYADMIN, so they can be responsible for creating the initial logins and users when a new employee starts. Or, you can drop this login altogether.

When possible, you should also block direct access to the database tables. All access to the data should be through stored procedures and views. This provides a level of encapsulation and helps maintain data integrity. You'll learn more about this technique in Chapter 5.

While you are selecting and limiting users who have access to the server, you should also restrict access to the server and databases. Typically, users do not log into a database and start running queries; instead, users use a business application that in turn accesses the data. In this case, the application logs into the server, and users are authenticated and authorized within the application, not the database. You'll learn more about this process in Chapter 5.

Last, but not least, no matter what your security model, use strong passwords and best practice policies that have been demonstrated to be effective. It doesn't matter how meticulous and clever you get with the permissions. If the password to SA, or any user for that matter, is simple, or worse, blank, your work is worthless. (It may surprise you to know that a blank SA password was the default installation before SQL Server 2000.) You'll learn more about password and security policies in the next chapter.

There is one more important point: *patches, patches, patches*. Keep up to date with all the latest service packs and security patches for SQL Server and the Windows server it's running on.

Chapter Summary

- You should document tasks and procedures to provide a paper trail for retracing exactly what happened when a breach of security occurs and to ensure administration consistency.
- The CREATE USER statement enables database administrators to create a database user account.
- The CREATE USER statement is a SQL statement that is supported by all relational databases.
- The IDENTIFIED clause instructs Oracle10*g* on how to authenticate a user account.
- There are three methods of authentication available in Oracle10*g*: the BY PASSWORD option, the EXTERNALLY option, and the GLOBALLY as `external_name` option.

- In Oracle10*g*, the DEFAULT TABLESPACE clause of the CREATE USER statement specifies the default storage for the user.
- In Oracle10*g*, the TEMPORARY TABLESPACE clause of the CREATE USER statement takes two options: the name of temporary tablespace or the name of a temporary tablespace group.
- In Oracle10*g*, the QUOTA clause of the CREATE USER statement informs Oracle of how much storage space a user is allowed for a specified tablespace.
- In Oracle10*g*, the PROFILE clause of the CREATE USER statement indicates the profile used for limiting database resources and enforcing password policies.
- In Oracle10*g*, the PASSWORD EXPIRE clause of the CREATE USER statement tells Oracle to expire the user password and prompts the user to enter a new password.
- In Oracle10*g*, the ACCOUNT clause of the CREATE USER statement tells Oracle whether or not to disable or enable an account.
- Oracle10*g* external authentication depends on an external party to authenticate the user.
- With SQL Server, a login ID controls access to the SQL Server system.
- You cannot log in to a SQL Server database without first supplying a valid login ID and password.
- When creating a new user account, you should obtain a written request approved by the immediate manager of the account holder for auditing purposes.
- SQL provides a command called DROP that removes a user account from the database.
- To drop an existing Windows Integrated login, you use the `SP_DENYLOGIN` system stored procedure.
- Use the ALTER USER DDL statement of SQL to modify any attribute of a user.
- When a user logs on to the database through the computer on which the database is located, that database is called a local database.
- A remote user is a user who logs onto the database using a different host computer from the computer on which the database is located.
- A database link is a connection from one database to another database.
- Linked servers allow you to connect to most any OLEDB or ODBC data source, including other SQL servers and Oracle servers, for access by users in the current instance of SQL Server.

Review Questions

1. Using Enterprise Manager, create a Windows Integrated login and set the default database to Pubs.
2. Print an Oracle report that displays a list of all existing users in the database sorted by the name of the users who were created last month.
3. Develop a script that prompts the administrator for an authentication type and then lists all users that are using that entered authentication.
4. List all the different account statuses in your Oracle database.
5. List all Oracle initialization parameters that are related to authentication.
6. A lead developer called the other day to lock an account. Outline the steps that you would take to comply with the request.
7. Outline the steps you would take to close a SQL Server user account.
8. List the system tables you would use to view all users in an Oracle database and SQL Server database.

9. Create a user using Oracle and another user using SQL Server and try to connect to the database. Explain what happened for each user.
10. Log on to an Oracle database as SYS and lock the SYS and SYSTEM accounts. Explain what happened.

Hands-on Projects

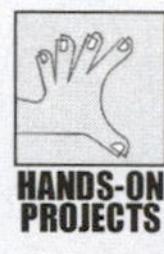

Hands-on Project 3-1

You are the new DBA for Acme Corporation. Management is concerned about the security of the corporation's data and has asked you to work with the network security architect in securing the company's SQL servers. You look at the current configuration, and find:

The MSSQLSERVER service is running SYSTEM.

There is no password on SA.

BUILT-IN\Administrators is set up with its default security settings.

With the assistance of the security architect, what steps can you take to make Acme Corporation's SQL Server instance more secure?

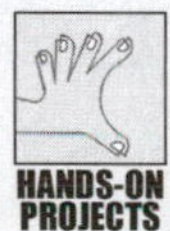

Hands-on Project 3-2

Write a script to create two SQL Server logins and users for them in the Pubs and Northwind databases and make their default database Northwind.

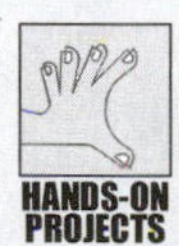

Hands-on Project 3-3

Your company is finally upgrading from SQL Server 6.5 to SQL Server 2000. To avoid having to create all the logins by hand, you execute a query against SYSLOGINS on the 6.5 system and retrieve all the login names and encrypted passwords for a script you will write to recreate them in the SQL Server 2000 system. Describe your SP_ADDLOGIN.

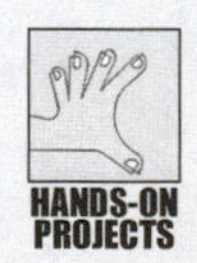

Hands-on Project 3-4

You were just hired as a new junior Oracle database administrator to assist a team of five senior database administrators. In your first week, you were handed the task of writing a script as a template for creating database users. The script is to prompt the administrator for a user name and password, and the script creates the user according to the following guidelines:

The default tablespace is TOOLS.

The temporary tablespace is TEMP_TOOL_GROUP.

The account is initially locked.

The user account will be assigned to the DEFAULT profile.

The TOOLS tablespace has a 10 MB quota.

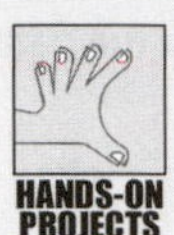

Hands-on Project 3-5

Write a series of steps to demonstrate the security risk when using an Oracle database link.

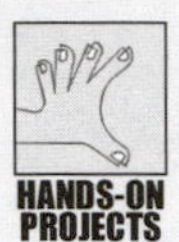

Hands-on Project 3-6

Develop an Oracle report that displays to all users when they are nearing 80% of their quota limits.

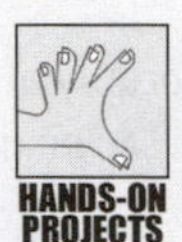

Hands-on Project 3-7

Use Oracle to create a public database link and a private database link, and discuss two of the security implications. Provide an example for each implication.

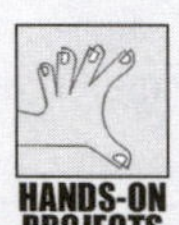

Hands-on Project 3-8

Using a SQL server, show how server links can impact database security.

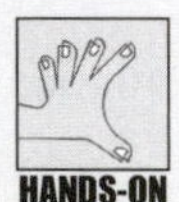

Hands-on Project 3-9

Outline the steps to create a database user account.

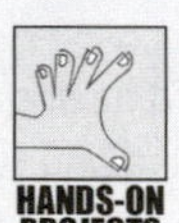

Hands-on Project 3-10

What procedures would you follow if you were told that an employee was terminated?

Case Projects

Case 3-1 Training Case

You were hired as a database security expert to train the database administration team. At the first training session, you ask the trainees to perform the following tasks using Oracle Enterprise Manager. Describe what you expect your trainees to produce.

- Create a user named CH0306 with the password CH0306. Use all default values for all options.
- Modify the user options using SQL*Plus to change the default tablespace and temporary tablespace to TOOLS and TEMP respectively.
- Lock the account.
- Using SQL*Plus change the user password to DBSEC.
- View the account using the data dictionary.

- View user quotas using the data dictionary.
- Using Oracle Enterprise Manager, provide unlimited storage quotas to TOOLS and USERS.

At the second training session, you ask the students to outline a list of the best three practices of user administration. List what you expect your students to produce.

Case 3-2 Access Is Needed

Your development team wants to get access to the production SQL Server to pull some data subsets for an analysis project. Your partner sets up a linked server on the development server that links to the production server.

a. Describe your recommendation for setting up the remote login.

b. Using Enterprise Manager, give the user created in the preceding step access to the pubs database.

c. Using Query Analyzer, and the login from the first step (a), change the access from pubs to Northwind.

4

Profiles, Password Policies, Privileges, and Roles

LEARNING OBJECTIVES

Upon completion of this material, you should be able to:

- Define and use a profile
- Design and implement password policies
- Implement password policies in Oracle and SQL Server
- Grant and revoke user privileges
- Create, assign, and revoke user roles
- List best practices for securing a network environment

Introduction

After a long search for a dream home, you and your significant other purchase the perfect house. On closing day, the previous owner hands you the key to the house, and you can't wait to move in. When moving day finally arrives, you work tirelessly to unpack and settle in. One of the first things you do is change the door locks so that you feel secure in your new home.

Now put this scenario into the context of computer passwords. The house is your user account. The key to the house is the password. For home security, in addition to changing the key, you might install an alarm and motion detector, get a watch dog; and even install fences. A company's user accounts should have equal protection. The company needs to protect its assets and enforce stringent guidelines to protect the keys to computer accounts. This key is the password.

This chapter is a continuation of Chapter 3 in which you learned how to create and administer users. It discusses the four aspects of user administration and user security. These aspects are profiles, passwords, privileges, and roles. Learning the four aspects of user administration and security is essential background for the coming chapters in this book.

Defining and Using Profiles

A **profile** is a security concept that describes the limitation of database resources that are granted database users. A profile is a way of defining database user behavior to prevent users from wasting resources such as memory or CPU consumption. For example, consider the following scenario:

> A financial database application went into production after thorough analysis, design, and development. Initially, the database ran without any problems, but before long clients started complaining that the database was slow to respond to their requests. A database administrator was assigned to investigate this sudden surge of complaints. He conducted a detailed inspection of all queries submitted to the database but could not locate any query that might have caused performance issues. He was able to confirm that all queries submitted by the application were fully optimized and were not the source of the problem.
>
> Being a stellar troubleshooter, he created a profile that would limit the use of CPU and memory. He also assigned this profile to all users other than application users and then reported the problem solved. A few days later one of the developers called the application support team complaining about an error he got while running a query. He said, "This is the first time I have seen this type of error. It is not query related, it is database related." The application support person asked the developer for the error number and reported it to the database administrator. The database administrator immediately said, "I knew it! It is a long-running query that consumes memory and CPU."

This is type of incident occurs often in both small and large companies. For this reason, some database management systems have implemented the profile concept. Not every database system offers the profile concept—Oracle does and Microsoft SQL Server 2000 doesn't. In this section you discover how profiles are implemented in Oracle.

Creating Profiles in Oracle

A profile in Oracle helps define two elements of security: restrictions on resources and the implementation of password policy. This section outlines restrictions on resources. The password aspect of a profile is discussed later in this chapter. Figure 4-1 shows these two aspects of a profile in Oracle.

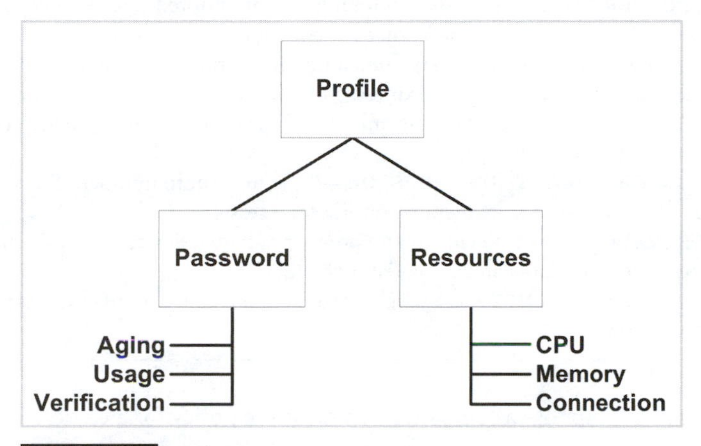

FIGURE 4-1 Profile feature in Oracle

Oracle allows you to create profiles using the CREATE PROFILE statement. The full syntax of this statement follows:

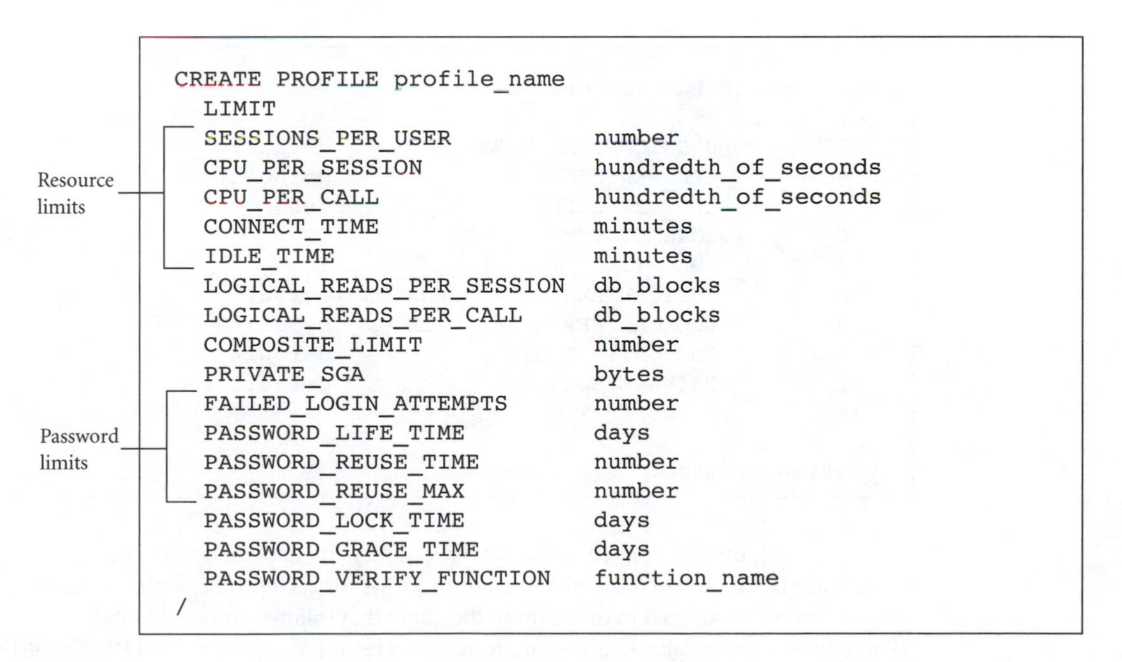


```
CREATE PROFILE profile_name
  LIMIT
  SESSIONS_PER_USER            number
  CPU_PER_SESSION              hundredth_of_seconds
  CPU_PER_CALL                 hundredth_of_seconds
  CONNECT_TIME                 minutes
  IDLE_TIME                    minutes
  LOGICAL_READS_PER_SESSION    db_blocks
  LOGICAL_READS_PER_CALL       db_blocks
  COMPOSITE_LIMIT              number
  PRIVATE_SGA                  bytes
  FAILED_LOGIN_ATTEMPTS        number
  PASSWORD_LIFE_TIME           days
  PASSWORD_REUSE_TIME          number
  PASSWORD_REUSE_MAX           number
  PASSWORD_LOCK_TIME           days
  PASSWORD_GRACE_TIME          days
  PASSWORD_VERIFY_FUNCTION     function_name
/
```

Where:

- SESSIONS_PER_USER—Is the maximum number of allowed concurrent open sessions per user
- CPU_PER_SESSION—Is the maximum number in hundredths of seconds of CPU time allowed per session (for the duration of the session)
- CPU_PER_CALL—Is the maximum number in hundredth of seconds of CPU time allowed per call (for the duaration of statement call)
- CONNECT_TIME—Is the maximum amount of time a user connection is allowed; the value of this parameter is expressed in minutes.
- IDLE_TIME—Is the maximum amount of idle time in minutes before a user connection is disconnected
- LOGICAL_READS_PER_SESSION—Is the maximum number of database blocks allowed to be read from memory or disk for a session.
- PRIVATE_SGA—Is the maximum number of bytes allowed to be allocated for the user; this is available only in shared server mode.
- COMPOSITE_LIMIT—Is a weighted sum of all resource limits expressed in service units

For example, suppose a company has a **resource** policy for a clerk user as follows:

- Connection time should not exceed two hours.
- Session should be disconnected if idle time exceeds 15 minutes.
- CPU time allowed per call is 10 seconds.

Here is the implementation for this scenario:

```
SQL> CREATE PROFILE CH04_PROF
  2     LIMIT
  3        SESSIONS_PER_USER          default
  4        CPU_PER_SESSION            default
  5        CPU_PER_CALL               1000
  6        CONNECT_TIME               120
  7        IDLE_TIME                  15
  8        LOGICAL_READS_PER_SESSION  default
  9        LOGICAL_READS_PER_CALL     default
 10        COMPOSITE_LIMIT            default
 11        PRIVATE_SGA                default
 12  /

Profile created.
```

To view all profiles created in the database, query the data dictionary view, DBA_PROFILES. Note that this view contains 16 rows for each profile, with one row for each limitation as outlined in the result of the query that follows. Notice that the RESOURCE column value KERNEL indicates it is a resource restriction and PASSWORD indicates it is password policy implementation. Note that in Oracle there is a tool called Resource Manager that enables you to create different CPU usage policies to set limits on CPU utilization.

```
SQL> SELECT *
  2    FROM DBA_PROFILES
  3   WHERE PROFILE = 'CH04_PROF'
  4  /

PROFILE      RESOURCE_NAME                RESOURCE  LIMIT
---------    -------------------------    --------  -----
CH04_PROF    COMPOSITE_LIMIT              KERNEL    DEFAULT
CH04_PROF    SESSIONS_PER_USER            KERNEL    DEFAULT
CH04_PROF    CPU_PER_SESSION              KERNEL    DEFAULT
CH04_PROF    CPU_PER_CALL                 KERNEL    1000
CH04_PROF    LOGICAL_READS_PER_SESSION    KERNEL    DEFAULT
CH04_PROF    LOGICAL_READS_PER_CALL       KERNEL    DEFAULT
CH04_PROF    IDLE_TIME                    KERNEL    15
CH04_PROF    CONNECT_TIME                 KERNEL    120
CH04_PROF    PRIVATE_SGA                  KERNEL    DEFAULT
CH04_PROF    FAILED_LOGIN_ATTEMPTS        PASSWORD  DEFAULT
CH04_PROF    PASSWORD_LIFE_TIME           PASSWORD  DEFAULT
CH04_PROF    PASSWORD_REUSE_TIME          PASSWORD  DEFAULT
CH04_PROF    PASSWORD_REUSE_MAX           PASSWORD  DEFAULT
CH04_PROF    PASSWORD_VERIFY_FUNCTION     PASSWORD  DEFAULT
CH04_PROF    PASSWORD_LOCK_TIME           PASSWORD  DEFAULT
CH04_PROF    PASSWORD_GRACE_TIME          PASSWORD  DEFAULT

16 rows selected.
```

To modify a limit for a profile, you use ALTER PROFILE as follows:

```
SQL> ALTER PROFILE CH04_PROF
  2      LIMIT IDLE_TIME  30
  3  /

Profile altered.
```

You assign a profile to a user using ALTER USER as follows:

```
SQL> ALTER USER SAFYOUNI PROFILE CH04_PROF
  2  /

User altered.
```

Of course, to view the profile to which each user is assigned, you issue the following query:

```
SQL> SELECT USERNAME, PROFILE
  2     FROM DBA_USERS
  3   WHERE USERNAME LIKE 'S%'
  4  /

USERNAME                  PROFILE
------------------        ---------
SH                        DEFAULT
SI_INFORMTN_SCHEMA        DEFAULT
SCOTT                     DEFAULT
SYSMAN                    DEFAULT
SYS                       DEFAULT
SYSTEM                    DEFAULT
SAFYOUNI                  CH04_PROF
```

Note that user SAFYOUNI is assigned to profile CH04_PROF, whereas other users are assigned to a default profile called DEFAULT, which has no limits as shown in the following query:

```
SQL> SELECT *
  2     FROM DBA_PROFILES
  3   WHERE PROFILE = 'DEFAULT'
  4  /

PROFILE     RESOURCE_NAME                RESOURCE     LIMIT
-------     -------------------------    ---------    ---------
DEFAULT     COMPOSITE_LIMIT              KERNEL       UNLIMITED
DEFAULT     SESSIONS_PER_USER            KERNEL       UNLIMITED
DEFAULT     CPU_PER_SESSION              KERNEL       UNLIMITED
DEFAULT     CPU_PER_CALL                 KERNEL       UNLIMITED
DEFAULT     LOGICAL_READS_PER_SESSION    KERNEL       UNLIMITED
DEFAULT     LOGICAL_READS_PER_CALL       KERNEL       UNLIMITED
DEFAULT     IDLE_TIME                    KERNEL       UNLIMITED
DEFAULT     CONNECT_TIME                 KERNEL       UNLIMITED
DEFAULT     PRIVATE_SGA                  KERNEL       UNLIMITED
DEFAULT     FAILED_LOGIN_ATTEMPTS        PASSWORD     UNLIMITED
DEFAULT     PASSWORD_LIFE_TIME           PASSWORD     UNLIMITED
DEFAULT     PASSWORD_REUSE_TIME          PASSWORD     UNLIMITED
DEFAULT     PASSWORD_REUSE_MAX           PASSWORD     UNLIMITED
DEFAULT     PASSWORD_VERIFY_FUNCTION     PASSWORD     NULL
DEFAULT     PASSWORD_LOCK_TIME           PASSWORD     UNLIMITED
DEFAULT     PASSWORD_GRACE_TIME          PASSWORD     UNLIMITED

16 rows selected.
```

Of course, you can view all details about users and profiles using the Oracle Enterprise Manager Security tool as illustrated in Figure 4-2.

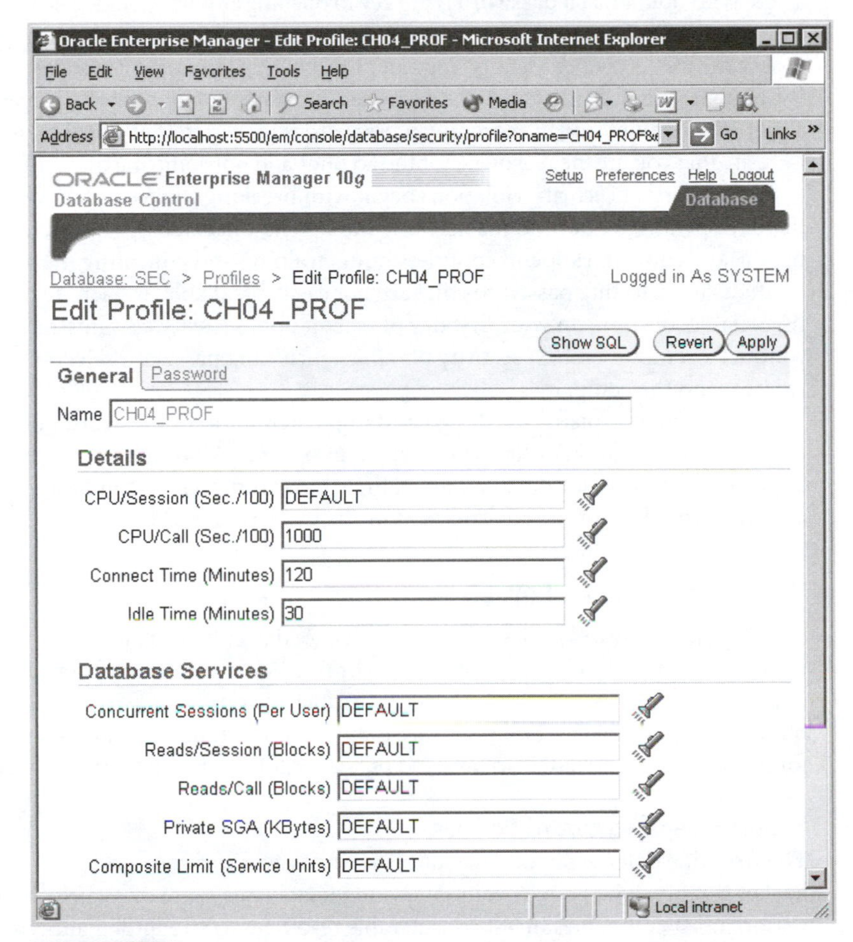

FIGURE 4-2 Oracle Enterprise Manager showing profile CH04_PROF through the Security Manager tool

Creating Profiles in SQL Server 2000

At the time of this writing, profiles or similar objects are not available in Microsoft SQL Server 2000 or 2005.

You may recall that Oracle profiles limit the resources available to the user within the database. This allows you to tune connections and resource utilization to prevent a user from crashing a server with a bad query. Query and connection time-outs in a SQL server-based application are handled at the application level within OLEDB. These settings keep connection time-outs and bad queries from crashing your SQL server.

Designing and Implementing Password Policies

There is no doubt that a **password** is the key to opening a user account. The stronger the password, the longer it takes a hacker to break in. User authentication depends on a password to ensure the user account's identity. You can probably recall phoning a company to discuss an issue about your account. Often, the customer care representative asks you several questions about your account, and in most cases a password is requested. The purpose of this request is to verify that you are the real owner of the account and not someone pretending to be you.

Many hacker security violations begin with breaking the password to an account and thereby opening the door to the network, the system, the database, and password-protected files. Many companies spend countless hours training and educating their employees on methods for selecting passwords that are not easily breakable. In fact, if you join any large financial institution, on your first day of orientation, security administration is likely to conduct a seminar on best security practices, including password selection, password storage, and the company's policy on passwords.

This section explains the design and implementation of password policies that decrease the likelihood of hackers breaking passwords. Although the discussion is aimed at password policies for database users, these password policies can be applied to any account, regardless of the environment or the platform.

What Is a Password Policy?

Simply put, a **password policy** is a set of guidelines that enhances the robustness of a password and reduces the likelihood of its being broken. Most guidelines deal with various aspects of passwords, such as password complexity, frequency of password changes, and password reuse. These guidelines not only enhance password protection, they also establish standards for institutions to increase employee and public confidence in their security measures.

Importance of Password Policies

The frontline defense of your account is your password; if your password is weak, a hacker can break in, destroy your data, and violate your sense of security. For this specific reason, most companies invest considerable resources to strengthen authentication by adopting technological measures that protect their assets. Of course, you already know that a password is a main component of the authentication process.

A company that does not take password policies seriously is most likely vulnerable to security risks and violations. When a company adopts a password policy, it forces employees to abide by the guidelines set by the company and raises employee awareness of password protection. Another benefit is that having solid password policies helps ensure that a company does not fail audits. Audits are conducted to ensure that the company is carrying out business operations in an acceptable manner and to identify operational faults and security gaps.

Designing Password Policies

Most companies use a standard set of guidelines for their password policies. These guidelines can comprise one or more of the following:

- **Password complexity**—A set of guidelines used when selecting a password. For example, a company could require that a password be eight characters in length and contain at least one digit and a symbol. The purpose of password complexity is to decrease the chances of a hacker guessing or breaking a password. For example, if you select your son's name as a password, it can be easily guessed and broken. In addition, if you select a password that is short and contains only alphabetic characters, a hacker can use a dictionary of words as a tool to break your password.

- **Password aging**—Indication of how long a password can be used before it expires
- **Password usage**—Indication of how many times the same password can be used
- **Password storage**—A method of storing a password in an encrypted manner

Of course, you must understand that in some companies a password might not be a good method of authentication, especially if the company deals with highly sensitive operations. In these cases, the company often resorts to biometric safeguards as discussed in Chapter 2.

Implementing Password Policies

How you implement a password policy depends on whether or not a database management system provides functions that support password security. For instance, Oracle has invested heavily in providing mechanisms to enforce security, including implementation of password policies, whereas a Microsoft SQL Server depends on the operating system to implement password policies. This section of the book shows you how to develop password policies in Oracle. For SQL servers, you learn how to apply password policies in the Windows 2000 operating system later in this chapter.

Password Policies in Oracle

As discussed earlier in the chapter, Oracle provides a profile concept to impose resource usage restrictions and build password policies. In the previous section of this chapter, you were shown how to create a profile specifically to restrict resource usage. In this section you focus on the password aspect of the profile. To show you how to create such a profile policy, the end of this section supplies an example of putting the two aspects of a profile together.

The CREATE PROFILE statement consists of two parts as shown earlier in this chapter. The following is the password policy section of the statement:

```
CREATE PROFILE PASSWORD_POLICY
LIMIT
{ { FAILED_LOGIN_ATTEMPTS
  | PASSWORD_LIFE_TIME
  | PASSWORD_REUSE_TIME
  | PASSWORD_REUSE_MAX
  | PASSWORD_LOCK_TIME
  | PASSWORD_GRACE_TIME
  }
  { expr | UNLIMITED | DEFAULT }
| PASSWORD_VERIFY_FUNCTION
    { function | NULL | DEFAULT }
}
```

Where:

- FAILED_LOGIN_ATTEMPTS—Is the number of failed login tries allowed before the account is locked
- PASSWORD_LIFE_TIME—Is the number of days the password is valid before it is aged out
- PASSWORD_REUSE_TIME—Is the number of days before a password can be reused; this parameter works with PASSWORD_REUSE_MAX parameter.
- PASSWORD_REUSE_MAX—Is the number of times a password can be reused
- PASSWORD_LOCK_TIME—Is the number of days an account is locked due to failed login attempts
- PASSWORD_GRACE_TIME—Is the number of days ahead of expiration the user is warned that the password expires
- PASSWORD_VERIFY_FUNCTION—Is an indication to Oracle to use a custom-made function to validate password complexity

Now you put these concepts into practice. Suppose you want to create a profile to enforce your company's policy which states that:

- A user password cannot be reused.
- A password must expire every 15 days.
- Only one login attempt is allowed.

To create such a profile, you issue the following statement:

```
SQL> CREATE PROFILE ACME_PASSWORD_PROFILE
  2   LIMIT
  3   FAILED_LOGIN_ATTEMPTS 1
  4   PASSWORD_LIFE_TIME    15
  5   PASSWORD_REUSE_TIME   DEFAULT
  6   PASSWORD_REUSE_MAX    1
  7  /

Profile created.
```

Figure 4-3 shows the previous statement in Oracle's Create Profile dialog box. You can use Oracle Enterprise Manager to create profiles very easily as shown in Figure 4-3.

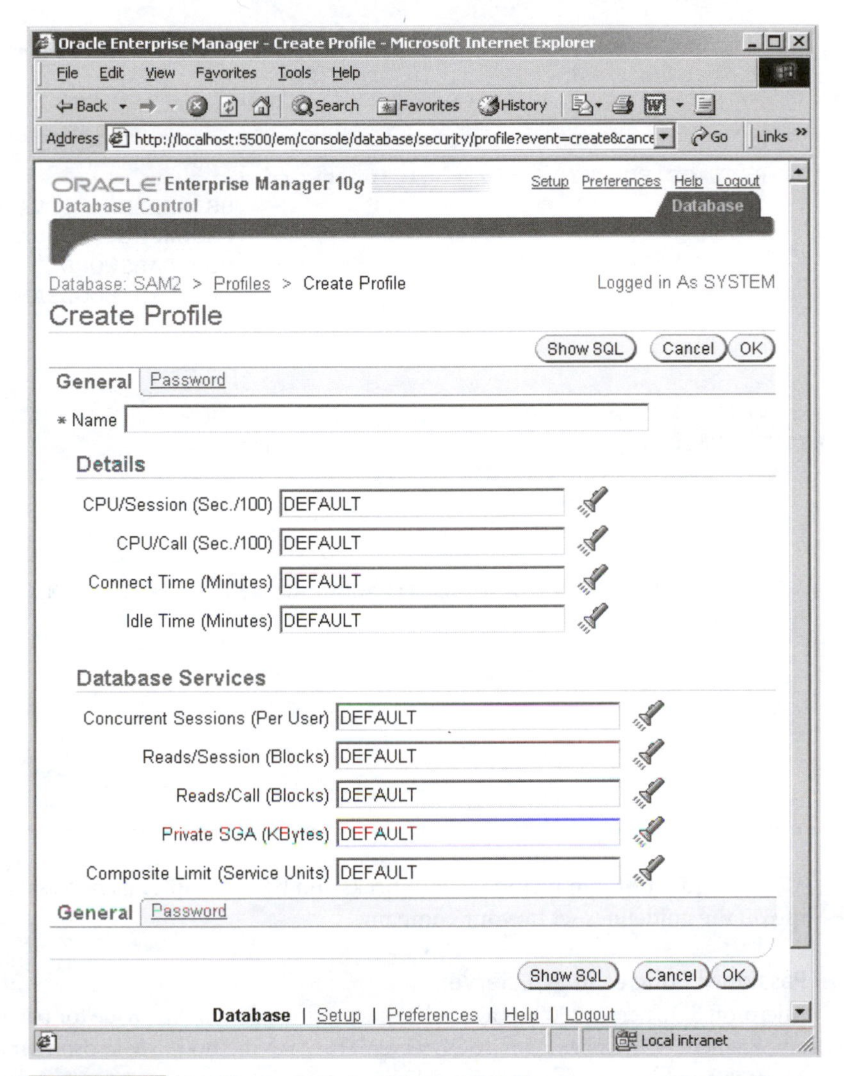

FIGURE 4-3 Profile creation using the Oracle Enterprise Manager Security tool

Here is another example to demonstrate the use and implementation of the PASSWORD_VERIFY_FUNCTION parameter. Suppose your company decides that password complexity requires the password to be 10 characters and only digits are allowed. To enforce this, you must write your own PL/SQL function to validate the password when it is selected. The prototype of this function must be as follows:

```
CREATE OR REPLACE FUNCTION verify_function (
                                  username     varchar2,
                                   password     varchar2,
                                    old_password varchar2
                                     ) RETURN Boolean;
```

The following function is used to enforce password complexity such as that stated in the preceding example:

```
CREATE OR REPLACE FUNCTION digitonly_password (
                                               USERNAME      VARCHAR2,
                                               PASSWORD      VARCHAR2,
                                               OLD_PASSWORD      VARCHAR2
                                              ) RETURN BOOLEAN IS

BEGIN

   IF LENGTH(PASSWORD) < 10 THEN
      RETURN FALSE;
   END IF;

   FOR I IN 1..LENGTH(PASSWORD) LOOP

      IF INSTR('1234567890', SUBSTR(PASSWORD, I,1)) = 0 THEN
         RETURN FALSE;
      END IF;

   END LOOP;

END;
/
```

Your function can include many checks and be as complex as necessary, depending on the guidelines set by your company.

Password Policies in SQL Server

Microsoft SQL Server 2000, as a stand-alone product, does not provide for password policy enforcement when logging on a SQL server. There is a method to Microsoft's madness. Microsoft architecture follows a model known as an integrated server system. In this model, all server applications and the resources they provide are tightly integrated with the Windows server system and its security architecture. At first glance, in comparison to other models, this approach may seem insecure. When implemented correctly, however, it's very secure and provides other benefits such as ease of management, and most importantly, the ability to authenticate to the database without sending passwords across the network.

That being said, password policy enforcement in a SQL Server environment is handled by implementing SQL Server in Windows authentication mode and applying policies within the Windows server system. Before examining implemented techniques and best practices in hardening this type of implementation, first look at the two authentication protocols supported by Windows—NTLM and Kerberos5. To secure a SQL server in this method, you need an understanding of how Windows authenticates a user.

As of Windows 2000, NTLM is no longer used for domain authentication. It is, however, still the method used to authenticate a local user to the Windows server.

NTLM

NTLM authenticates using a challenge/response methodology. Basically, this means that when you attempt to access a resource, the server hosting the resource "challenges" you to prove your identity. You then issue a "response" to that challenge, and if you respond correctly, you are authenticated to the server. Once authenticated, the server goes through an authorization process for the resource you've requested. Taking a closer look, you see the authentication process consists of three messages. Message 1 is sent from the client to the server and is the initial request for authentication. Message 2 is then sent from the server to the client and contains the challenge. The challenge is eight bytes of random data. Message 3 is sent from the client to the server, and it contains the response to the challenge. The response is a 24-byte DES-encrypted hash of the eight-byte challenge that can be decrypted only by a set of DES keys created using the user's password. The algorithm that does all this is not detailed here, but one piece of it is important when it comes to creating password policies. The password is padded to 14 bytes, then split into two seven-byte segments as shown in Figure 4-4. You'll see why this is important later.

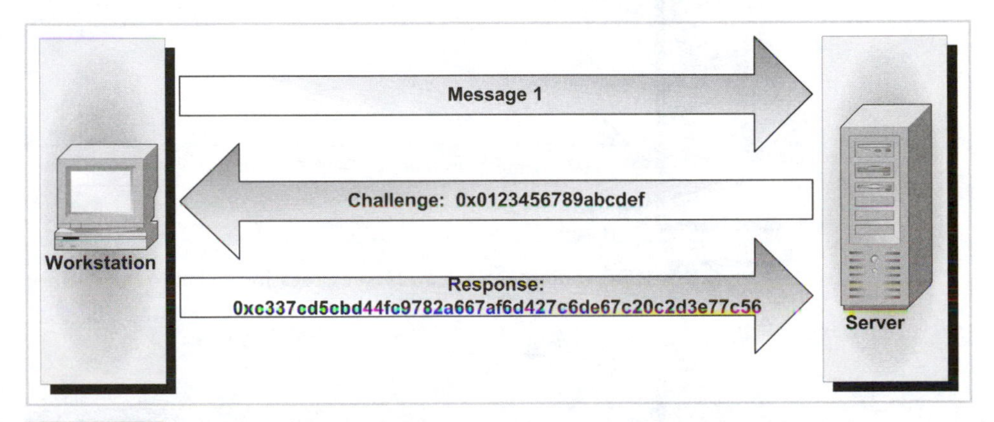

FIGURE 4-4 NTLM authentication

The benefit to NTLM is that passwords are verified without ever actually sending the password across the Web.

Kerberos

Kerberos authentication differs from NTLM in many ways. Instead of using the password to encrypt and decrypt challenge/response messages, a secret key, known only to the server and client and unique to the session, is used to encrypt the handshake data. This allows not only for the server to validate the authenticity of the client, but for the client to validate the authenticity of the server. This is an important difference and is one of the reasons Kerberos is more secure than NTLM. Just as an attacker can impersonate a client, an attacker can also impersonate a server; Kerberos helps mitigate this danger by requiring the client to also verify that the server is authentic.

Kerberos authentication requires a trusted third resource (in this transaction, the two parties communicating are considered the first and second resources) known as a Key Distribution Center (KDC). The KDC generates the secret keys for each session established. The new session ticket, containing the new key, has a time-out value associated with it; this is another security measure. Should a secret key ever be cracked, the time-out value ensures that it doesn't work the next time it's used, that is, when the attacker tries to use the key.

In a nutshell, what happens during authentication is that once the secret key is obtained from the KDC, the client encrypts its request for a resource with the secret key. The server then decrypts the message using the same key, encrypts just the time stamp on the message, and sends that time stamp back to the client. This tells the server that the client has the same key for that session that the server has, and it tells the client that the server has the correct key. See Figures 4-5 and 4-6.

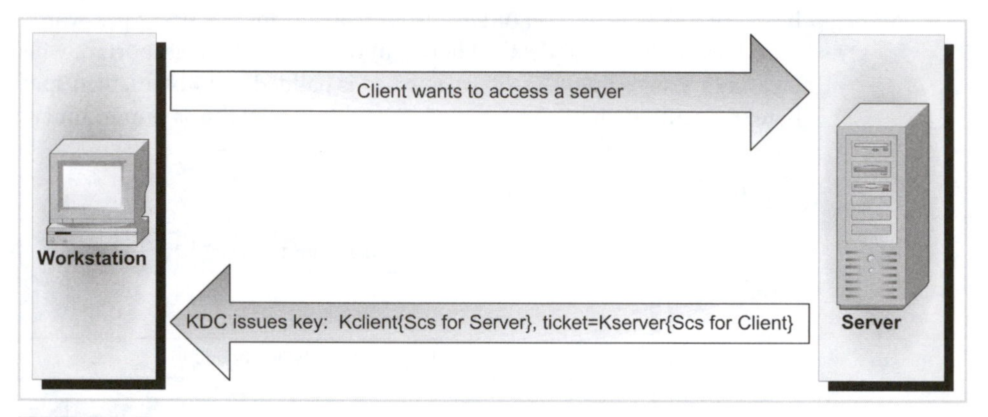

FIGURE 4-5 KDC generates a key and issues a session ticket to the client

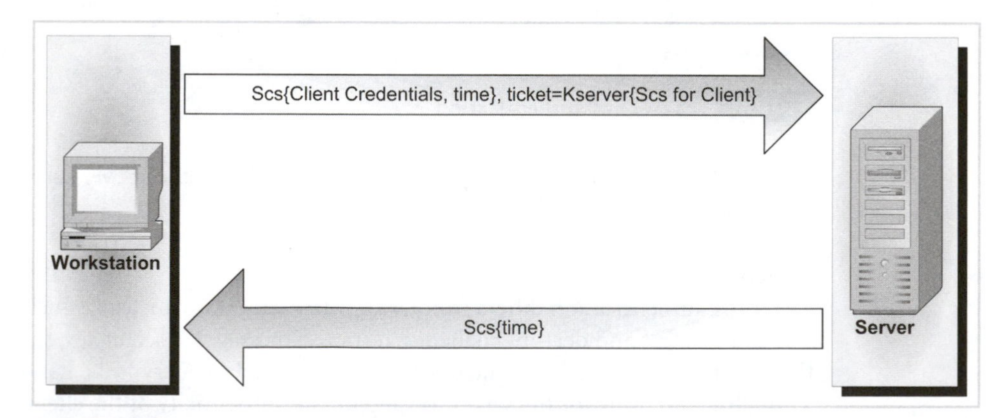

FIGURE 4-6 Client sends authentication proof to the server

The client sends proof of authenticity to the server; the server then sends the encrypted time on the ticket to the client, proving its authenticity.

An important note on Kerberos: because time is used in the authentication process, it is imperative that the time be synchronized networkwide.

For more information on the Kerberos implementation within the Microsoft Windows server system, you can download and read the white paper from *http://technet.microsoft.com*.

Another Look at Windows Integrated (Trusted) Logons

What does all this mean to you, the DBA? Remember, there is a method to Microsoft's madness. You've just learned that with both NTLM and Kerberos, the user's password is never actually sent across the network. This makes it impossible for an attacker to analyze the data being sent across the network to easily obtain a user's password. With native SQL Server logons, however, this is not the case. User names and passwords are usually stored in connection strings, or prompted for at the time of connection to the SQL server, and are sent across the network in clear text. In this scenario, an attacker can easily analyze traffic on the network and obtain a user's credentials for SQL Server or for Oracle. With Windows authentication, the initial handshake is encrypted. In fact, the user has already been authenticated at the OS level, and the SQL server has only to authorize the user within the SQL Server instance and its databases. Also, in the case of Kerberos, the client is sure that the server is authentic. This makes it very difficult for an attacker to impersonate your SQL server in an attempt to gain knowledge of your SQL Server implementation.

That is the reasoning behind Microsoft recommending you use Windows integrated authentication, instead of SQL Server logons when securing a Microsoft SQL Server 2000 implementation.

Setting Password Policies

Now that you know why Microsoft chose to implement security policies for a SQL Server implementation using Windows logons, take a look at how it is done.

Password policies for your Windows environment are set in one of two places depending on your setup: in the Local Security Policy management console, for policies enforced on local users on the server itself; or in the Domain Security Policy management console, for policies enforced on domain users. Because you don't have a domain setup for this course, this text outlines local security policies. However, local and domain policies are identical, and the only difference in the tools is where the policies are stored on the back end.

Open up the Local Security Policy management console on your machine, and take a look at what types of policies you can enforce.

1. Click **Start**, **All Programs**, and **Administrative Tools**. Your view should resemble Figure 4-7.

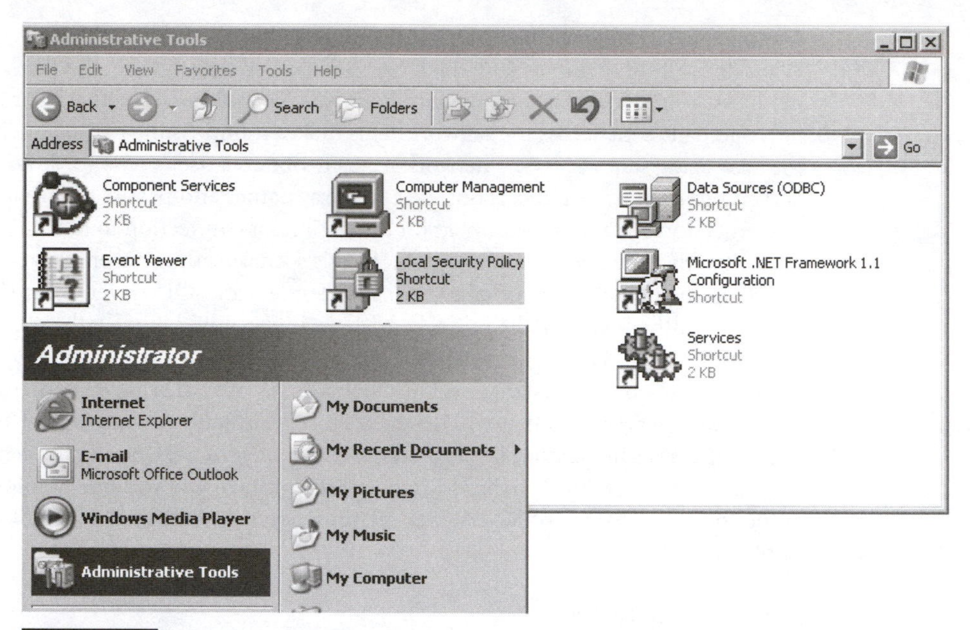

FIGURE 4-7 Windows Administrative Tools

2. Double-click **Local Security Policy** to see the window shown in Figure 4-8.

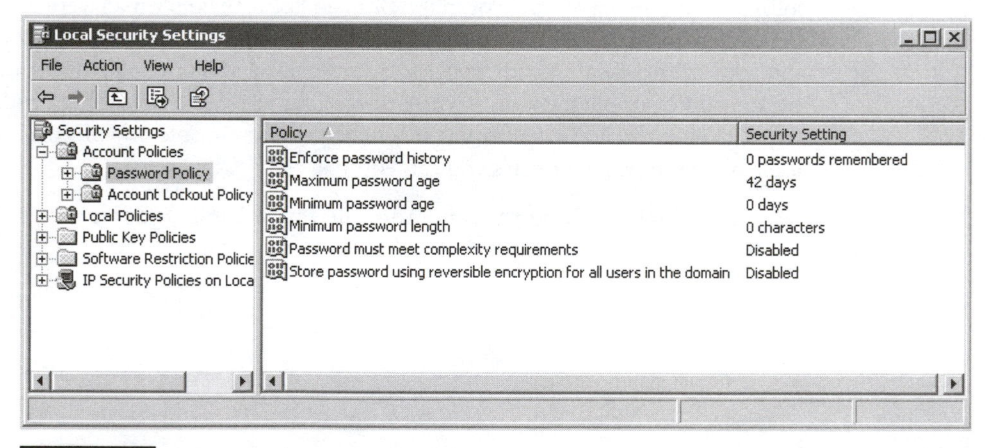

FIGURE 4-8 Windows Local Security Settings

3. Expand **Account Policies.**
4. Click **Password Policy.** There are six policy parameters that you can enforce, as described in Table 4-1.

Table 4-1 Password policy selections

Policy	Description
Enforce password history	Indicates that when users change passwords, the new password must be different from the last *n* passwords
Maximum password age	Indicates how many days must pass before a new password expires and must be changed
Minimum password age	Indicates how many days must pass before a password can be changed
Minimum password length	Indicates that a user's password must be at least *n* characters in length
Password must meet complexity requirements	Indicates whether or not a password must meet a predetermined level of complexity, e.g., it must use mixed case (capital and noncapital letters) and must contain one or more letters, numbers, and symbols
Store passwords using reversible encryption	Indicates whether or not to store the password as a hash that can be decrypted*

* Passwords are always stored encrypted. Typically, when a user enters a password to be validated, the user's entry is encrypted and the hashes are compared. By default, passwords are encrypted using an asymmetric encryption (different key for encryption versus decryption) with no decryption key. Some applications, such as Microsoft Exchange Server 2003, support authentication methods that require the encrypted password have the ability to be decrypted.

5. Click the **Account Lockout Policy** node at the left, and take a look at what policies can be set, as illustrated if Figure 4-9.

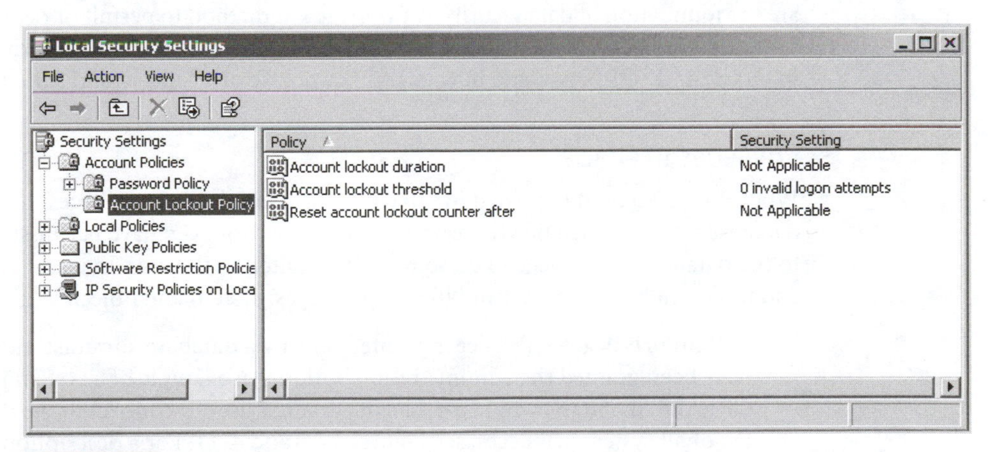

FIGURE 4-9 Account lockout selections

Figure 4-9 shows three policies that can be configured as described in Table 4-2.

Table 4-2 Account lockout policies

Policy	Description
Account lockout duration	Specifies how long an account is locked out (cannot be used when it is locked)
Account lockout threshold	Specifies how many invalid logon attempts can be made before Windows locks the account
Reset account lockout counter after	Specifies when the counter that tracks the number of invalid logon attempts should be reset to 0

Granting and Revoking User Privileges

Suppose you were issued a user account to a production database supporting a payroll application. You were told that you had been given all permissions necessary to access data. Therefore, you started navigating through the database to see what tables you could access, and you were surprised to find that you could access all of them. Then you attempted to view the data within these tables and surprise—you were able to view the data! Then, to push it the extra mile, you tried to modify some of the data with the intention to rollback your changes. Again, you were able to. You tried to insert a new row and you did so easily. Then you decided to commit the ultimate database sin—delete a row. Nothing stopped you, so you wondered why as an ordinary employee you had been given all these privileges and such wide access?

Indeed, there is something wrong when privileges are assigned without proper planning and without following a robust, well-considered application security model. Privileges are the foundation of data security. A **privilege** is a method to permit or deny access to data or to perform a database operation. In this section you learn the types of privileges available in Oracle and SQL Server and how to grant or revoke privileges to and from users.

Privileges in Oracle

When a user account is created in Oracle, the account has no privileges to perform any database task. In fact, a newly created account does not even have permission to connect to the database. A privilege to do so must be granted to the user. In Oracle, as in many database systems, there are two types of privileges: system and object.

- **System privileges**—Privileges granted only by a database administrator or users who have been granted the administration option. An example of a system privilege is CREATE SESSION, which permits the granted user to connect to the database. For a list of all system privileges, see Figures 4-10 and 4-11. For a description of each system privilege you should consult with Oracle documentation found on *otn.oracle.com.*
- **Object privileges**—Privileges granted to an Oracle user account by the schema owner of a database object or a user who has been granted the GRANT option (see the list in Figure 4-12). An example of an object privilege is DELETE, which permits the granted user to delete rows from a table. Suppose, Jason owns a table called EMPLOYEE and he grants DELETE privilege to Sam on the EMPLOYEE table by issuing the following statement:

```
SQL> GRANT DELETE ON EMPLOYEE TO SAM;
```

After this statement is issued, Sam can delete rows from the EMPLOYEE table.

ADMINISTER ANY SQL TUNING SET
ADMINISTER DATABASE TRIGGER
ADMINISTER RESOURCE MANAGER
ADMINISTER SQL TUNING SET

ADVISOR

ALTER ANY CLUSTER
ALTER ANY DIMENSION
ALTER ANY EVALUATION CONTEXT
ALTER ANY INDEX
ALTER ANY INDEXTYPE
ALTER ANY LIBRARY
ALTER ANY MATERIALIZED VIEW
ALTER ANY OUTLINE
ALTER ANY PROCEDURE
ALTER ANY ROLE
ALTER ANY RULE
ALTER ANY RULE SET
ALTER ANY SEQUENCE
ALTER ANY SQL PROFILE
ALTER ANY TABLE
ALTER ANY TRIGGER
ALTER ANY TYPE

ALTER DATABASE
ALTER PROFILE
ALTER RESOURCE COST
ALTER ROLLBACK SEGMENT
ALTER SESSION
ALTER SYSTEM
ALTER TABLESPACE
ALTER USER

ANALYZE ANY
ANALYZE ANY DICTIONARY

AUDIT ANY
AUDIT SYSTEM

BACKUP ANY TABLE

BECOME USER

COMMENT ANY TABLE

CREATE ANY CLUSTER
CREATE ANY CONTEXT
CREATE ANY DIMENSION
CREATE ANY DIRECTORY
CREATE ANY EVALUATION CONTEXT
CREATE ANY INDEX
CREATE ANY INDEXTYPE
CREATE ANY JOB
CREATE ANY LIBRARY
CREATE ANY MATERIALIZED VIEW
CREATE ANY OPERATOR
CREATE ANY OUTLINE
CREATE ANY PROCEDURE
CREATE ANY RULE
CREATE ANY RULE SET
CREATE ANY SEQUENCE
CREATE ANY SQL PROFILE
CREATE ANY SYNONYM
CREATE ANY TABLE
CREATE ANY TRIGGER
CREATE ANY TYPE
CREATE ANY VIEW

CREATE CLUSTER
CREATE DATABASE LINK
CREATE DIMENSION
CREATE EVALUATION CONTEXT
CREATE INDEXTYPE
CREATE JOB
CREATE LIBRARY
CREATE MATERIALIZED VIEW
CREATE OPERATOR
CREATE PROCEDURE
CREATE PROFILE
CREATE PUBLIC DATABASE LINK
CREATE PUBLIC SYNONYM
CREATE ROLE
CREATE ROLLBACK SEGMENT
CREATE RULE
CREATE RULE SET
CREATE SEQUENCE
CREATE SESSION
CREATE SYNONYM
CREATE TABLE
CREATE TABLESPACE
CREATE TRIGGER
CREATE TYPE
CREATE USER
CREATE VIEW

FIGURE 4-10 List of Oracle system privileges, part 1

DEBUG ANY PROCEDURE
DEBUG CONNECT SESSION

DELETE ANY TABLE

DEQUEUE ANY QUEUE

DROP ANY CLUSTER
DROP ANY CONTEXT
DROP ANY DIMENSION
DROP ANY DIRECTORY
DROP ANY EVALUATION CONTEXT
DROP ANY INDEX
DROP ANY INDEXTYPE
DROP ANY LIBRARY
DROP ANY MATERIALIZED VIEW
DROP ANY OPERATOR
DROP ANY OUTLINE
DROP ANY PROCEDURE
DROP ANY ROLE
DROP ANY RULE
DROP ANY RULE SET
DROP ANY SEQUENCE
DROP ANY SQL PROFILE
DROP ANY SYNONYM
DROP ANY TABLE
DROP ANY TRIGGER
DROP ANY TYPE
DROP ANY VIEW

DROP PROFILE
DROP PUBLIC DATABASE LINK
DROP PUBLIC SYNONYM
DROP ROLLBACK SEGMENT
DROP TABLESPACE
DROP USER

ENQUEUE ANY QUEUE

EXECUTE ANY CLASS
EXECUTE ANY EVALUATION CONTEXT
EXECUTE ANY INDEXTYPE
EXECUTE ANY LIBRARY
EXECUTE ANY OPERATOR
EXECUTE ANY PROCEDURE
EXECUTE ANY PROGRAM
EXECUTE ANY RULE
EXECUTE ANY RULE SET
EXECUTE ANY TYPE

FLASHBACK ANY TABLE
FLASHBACK ANY

TRANSACTION

FORCE ANY TRANSACTION
FORCE TRANSACTION

GLOBAL QUERY REWRITE

GRANT ANY OBJECT PRIVILEGE
GRANT ANY PRIVILEGE
GRANT ANY ROLE

IMPORT FULL DATABASE

INSERT ANY TABLE

LOCK ANY TABLE

MANAGE ANY QUEUE
MANAGE SCHEDULER
MANAGE TABLESPACE

ON COMMIT REFRESH

QUERY REWRITE

RESTRICTED SESSION
RESUMABLE

SELECT ANY DICTIONARY
SELECT ANY SEQUENCE
SELECT ANY TABLE

UNDER ANY TABLE
UNDER ANY TYPE
UNDER ANY VIEW

UNLIMITED TABLESPACE

UPDATE ANY TABLE

FIGURE 4-11 List of Oracle system privileges, part 2

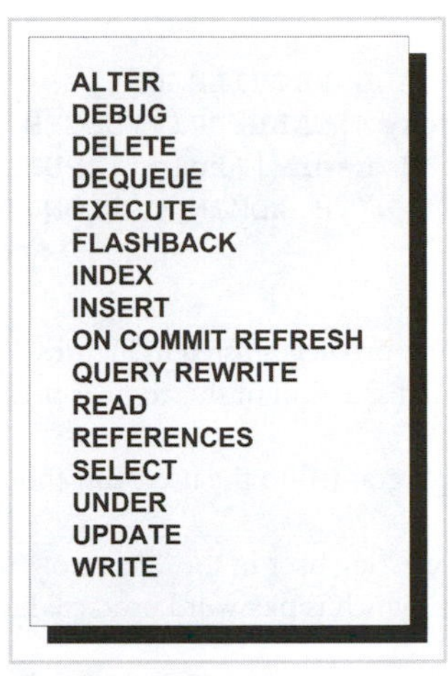

FIGURE 4-12 List of Oracle object privileges

Granting and Revoking System Privileges

The easiest task for a database administrator is granting a privilege, and yet it is a very critical job and should be considered seriously. You should not grant users any privileges without following a fully developed process for granting privileges. The privileges process should also include revoking privileges, which is a sensitive task that is too easy to perform without knowing the consequences. You should not revoke a privilege for a user without knowing the consequences, especially if it is an application user. In Oracle you can grant a privilege by using the data control language (DCL) GRANT statement. The syntax for the GRANT statement is listed in the code that follows. Remember that the GRANT statement is used to grant system privileges, object privileges, and roles. As you examine the syntax, notice that you can grant several privileges in the same statement to one or many users.

A **role** is a concept used to organize and administer privileges in an easy manner. A role is like a user, except it cannot own objects. A role can be assigned privileges and then assigned to users. Roles are treated in more detail later in the chapter.

```
GRANT { system_privilege | role | ALL PRIVILEGES }
        [, { system_privilege | role | ALL PRIVILEGES } ]...
   TO { user | role | PUBLIC } [, { user | role | PUBLIC } ]...
      [ IDENTIFIED BY password ] [ WITH ADMIN OPTION ]
```

Where:

- `System_privilege`—Is one of the system privileges listed in Figures 4-10 and 4-11.
- `Role`—Is the name of an existing role; a full discussion of this topic is presented later in this chapter.
- `ALL PRIVILEGES`—Is all system privileges granted to the user; you should never use this clause.
- `PUBLIC`—Is the privilege granted to every existing user in the database.
- `IDENTIFIED BY`—Is the granted privilege, which is password protected; you must provide a password to revoke it.
- `WITH ADMIN OPTION`—Is an indication to Oracle that the grantee has a privilege to administer granted privilege(s); in other words, the grantee can grant given privileges to other users.

An example of granting a user a privilege is presented in the next query. In this statement you are granting the privilege SELECT ANY TABLE to user SAFYOUNI, which enables this user to query any table in the database.

```
SQL> GRANT SELECT ANY TABLE TO SAFYOUNI
  2  /

Grant succeeded.
```

If you decide to revoke a privilege from a user, you use the REVOKE statement which is a DCL statement. The following syntax shows how to revoke a system privilege, followed by an example:

```
REVOKE { { system_privilege | role | ALL PRIVILEGES }
           [, { system_privilege | role | ALL PRIVILEGES } ]...
         FROM { user | role | PUBLIC } [, { user | role | PUBLIC } ]...
       } ;
```

```
SQL> REVOKE SELECT ANY TABLE FROM SAFYOUNI
  2  /

Revoke succeeded.
```

You can also use the Oracle Enterprise Manager Security tool to grant and revoke system privileges. Just follow these steps:

1. Open **Enterprise Manager** and select the database to which you want to connect.
2. Click the **Users** link and locate the user to whom you want to grant a privilege.
3. Click the **System Privileges** link, then click **Modify**, and apply the privilege you want to grant the user, as shown in Figure 4-13. You may grant as many privileges as you want.
4. When ready, click the **OK** button to apply these privileges.

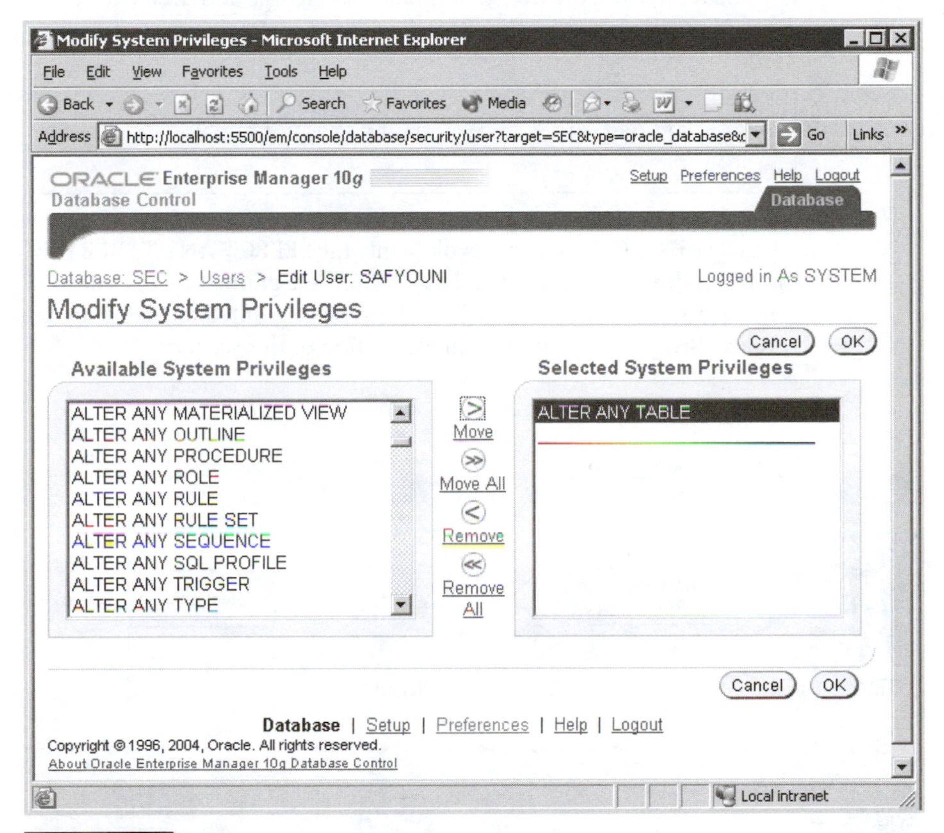

FIGURE 4-13 Granting a system privilege to a user

Before moving on to the next section, you must learn an important concept: how Oracle treats the revoking of the ADMIN option. Consider the following scenario: Suppose you are a database administrator for a small company. You want to impress a new developer who was just hired, so you grant her the SELECT ANY TABLE privilege with ADMIN option.

```
SQL> GRANT SELECT ANY TABLE TO LINDA WITH ADMIN OPTION
  2  /
```

Linda, the developer, really likes another developer and wants to impress him, so she grants him the SELECT ANY TABLE privilege as shown in the following statement:

```
SQL> GRANT SELECT ANY TABLE TO GEORGE
  2  /
```

A day later you discover that Linda likes George and decide to revoke her SELECT ANY TABLE privilege, so you issue the following statement:

```
SQL> REVOKE SELECT ANY TABLE FROM LINDA
  2  /
```

The REVOKE statement revokes only the SELECT ANY TABLE privilege from Linda—it does not cascade to all users who were granted this privilege by Linda. In other words, the SELECT ANY TABLE privilege granted to George by Linda is not revoked. See Figure 4-14 for a full illustration of this scenario.

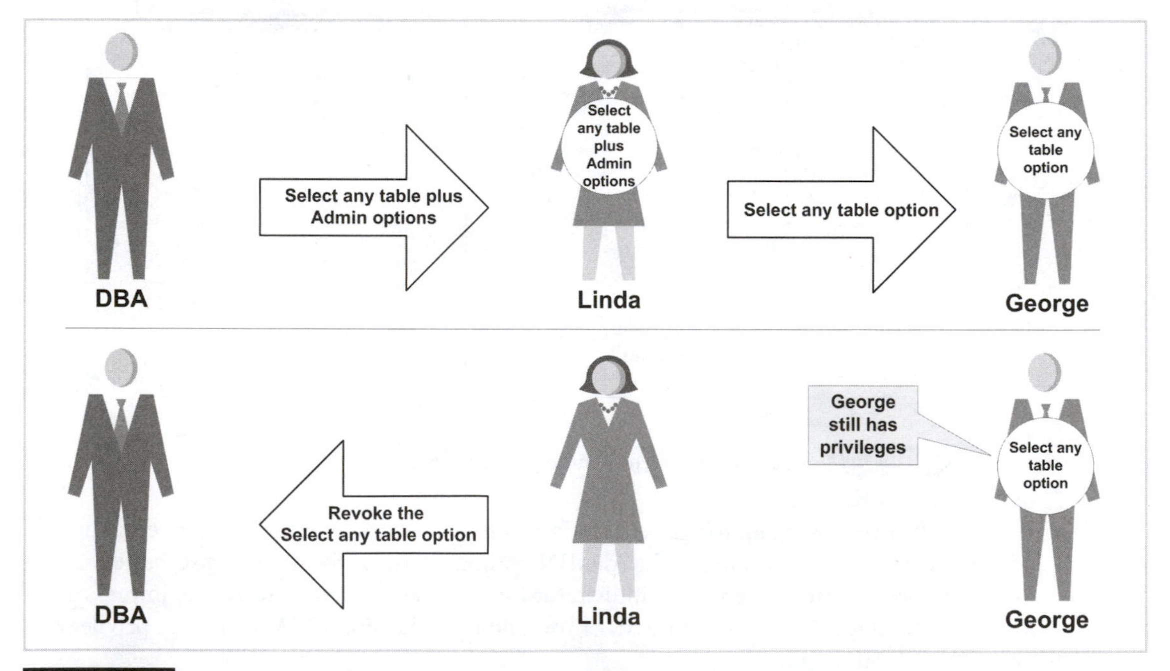

FIGURE 4-14 Revoking a system privilege with the ADMIN option

You are not finished with this section yet. You still need to know how to view privileges granted to a user. Oracle provides several useful data dictionary views.

- **DBA_SYS_PRIVS**
- **ALL_SYS_PRIVS**
- **USER_SYS_PRIVS**

Normally, as a DBA, you use the DBA_SYS_PRIVS data dictionary view, which contains the following columns:

```
SQL> DESC DBA_SYS_PRIVS
 Name                 Null?          Type
 ------------         --------       ------------
 GRANTEE              NOT NULL       VARCHAR2(30)
 PRIVILEGE            NOT NULL       VARCHAR2(40)
 ADMIN_OPTION                        VARCHAR2(3)
```

If you want to know all system privileges granted to user Scott, you issue the following query:

```
SQL> SELECT *
  2    FROM DBA_SYS_PRIVS
  3   WHERE GRANTEE = 'SCOTT'
  4  /

GRANTEE          PRIVILEGE                    ADMIN
-------          --------------------         -----
SCOTT            UNLIMITED TABLESPACE         NO
```

Granting and Revoking Object Privileges

In a manner similar to system privileges, Oracle uses the same GRANT statement to provide database object privileges to another user. However, the grantor is either the owner of the object or a user who has been granted the GRANT option. Before delving into the GRANT action, consider Table 4-3 which presents all privileges for the most common object types.

Table 4-3 Privileges for common object types

Privilege	Table	View	PL/SQL Stored Procedure*	Sequence
ALTER	✓	✗	✗	✓
DEBUG	✓	✓	✓	✗
DELETE	✓	✓	✗	✗
EXECUTE	✗	✗	✓	✗
FLASHBACK	✓	✓	✗	✗
INDEX	✓	✗	✗	✗
INSERT	✓	✓	✗	✗
ON COMMIT				

Table 4-3 Privileges for common object types (continued)

Privilege	Table	View	PL/SQL Stored Procedure*	Sequence
REFRESH	✓	✗	✗	✗
QUERY	✓	✗	✗	✗
REWRITE	✓	✗	✗	✗
READ	✗	✗	✗	✗
REFERENCES	✓	✓	✗	✗
SELECT	✓	✓	✗	✓
UNDER	✗	✗	✗	✗
UPDATE	✓	✗	✗	✗
WRITE	✗	✗	✗	✗

* The information in this table is derived from the online documentation that Oracle provides at the Oracle Technology Network site: *otn.oracle.com*.

Use the following syntax to grant an object privilege to a user:

```
GRANT { object_privilege | ALL [ PRIVILEGES ] }
        [ (column [, column ]...) ]
        [, { object_privilege | ALL [ PRIVILEGES ] }
        [ (column [, column ]...) ] ]...
   ON { schema.object | { DIRECTORY directory_name |
        JAVA { SOURCE | RESOURCE } [ schema. ]object
      }
   TO { user | role | PUBLIC }
        [, { user | role | PUBLIC } ]...
      [ WITH HIERARCHY OPTION ]
      [ WITH GRANT OPTION ]
```

Where:

- `WITH HIERARCHY OPTION`—Informs Oracle that the grantee is granted privilege to all subobjects
- `WITH GRANT OPTION`—Informs Oracle that the grantee has permission to grant privileges to other users

As an example of a user, Scott is granting a SELECT privilege on the EMP table to a user SAFYOUNI. In this case, you must log on as SCOTT and issue the following statement:

```
SQL> CONN SCOTT@SEC
Enter password: *****
Connected.
SQL> GRANT SELECT ON EMP TO SAFYOUNI
  2  /

Grant succeeded.
```

If Scott decided to revoke this privilege from SAFYOUNI, Scott must use the REVOKE statement. The following syntax shows how to revoke a system privilege and is followed by an example:

```
REVOKE { { object_privilege | role | ALL PRIVILEGES }
         [, { object_privilege | role | ALL PRIVILEGES } ]...
       FROM { user | role | PUBLIC } [, { user | role | PUBLIC } ]...
     } ;
```

```
SQL> REVOKE SELECT ON EMP FROM SAFYOUNI
  2  /

Revoke succeeded.
```

As always, you can use the Oracle Enterprise Manager Security tool to grant and revoke object privileges. Just follow these steps:

1. Open **Enterprise Manager** and select the database to which you want to connect.
2. Click the **Users** link and select the user to whom you want to grant the privilege.
3. Click the **Object Privileges** link and apply the privilege you want to grant to the user, as shown in Figure 4-15. You may grant as many privileges as you want.
4. When ready, click the **OK** button to apply these privileges.

When using the Oracle Enterprise Manager, the steps you take to grant privileges, revoke privileges, and many other tasks are similar and basically obvious. To avoid repetition of steps in this chapter, therefore, the steps are detailed only when you need guidance to successfully perform a task.

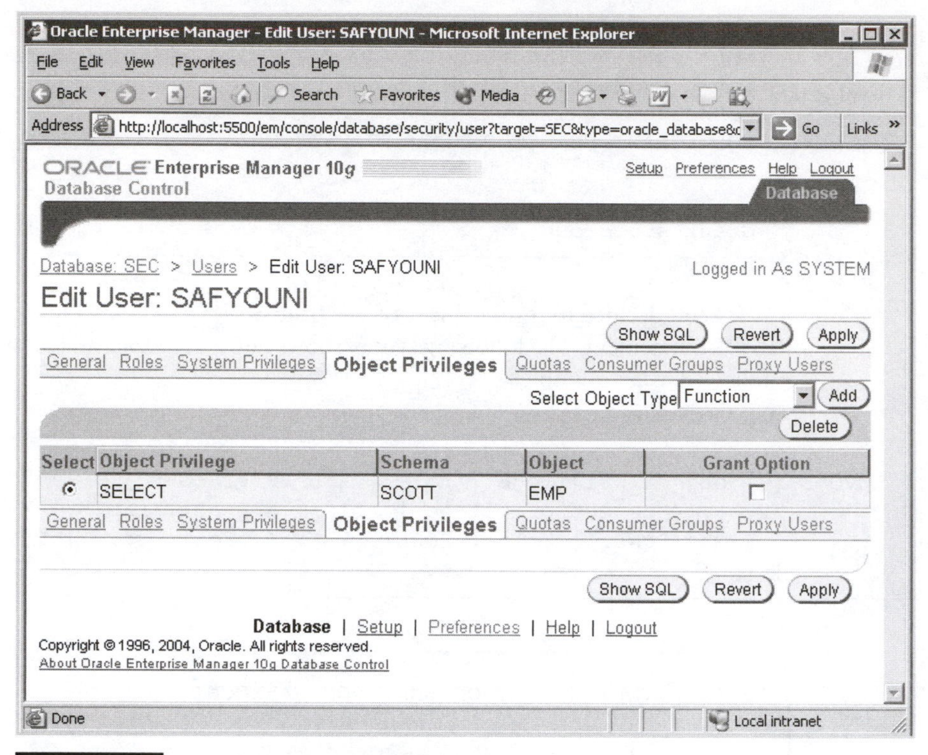

FIGURE 4-15 Granting an object privilege to a user

The procedure for revoking object privileges with the GRANT option is different than granting system privileges. Remember the earlier scenario of the DBA who likes Linda? Consider the following situation.

Suppose Scott is a developer for a small company and a new developer, Linda, has just been hired. Scott wants to impress Linda, so he grants her the SELECT privilege on the EMP table with the GRANT option.

```
SQL> GRANT SELECT ON EMP TO LINDA WITH GRANT OPTION
  2  /
```

Linda likes another developer and wants to impress him, so she grants him the SELECT on EMP table privilege, as shown in the following statement:

```
SQL> GRANT SELECT ON EMP TO GEORGE
  2  /
```

A day later Scott finds out that Linda does not like him, so he decides to revoke her SELECT on EMP table privilege, and he issues the following statement:

```
SQL> REVOKE SELECT ON EMP FROM LINDA
  2  /
```

The REVOKE statement revokes the SELECT privilege on the EMP table from Linda and cascades to all users that were granted this privilege by Linda. In other words, the SELECT on EMP table privilege granted to George by Linda is also revoked. See Figure 4-16 for a full illustration of this scenario.

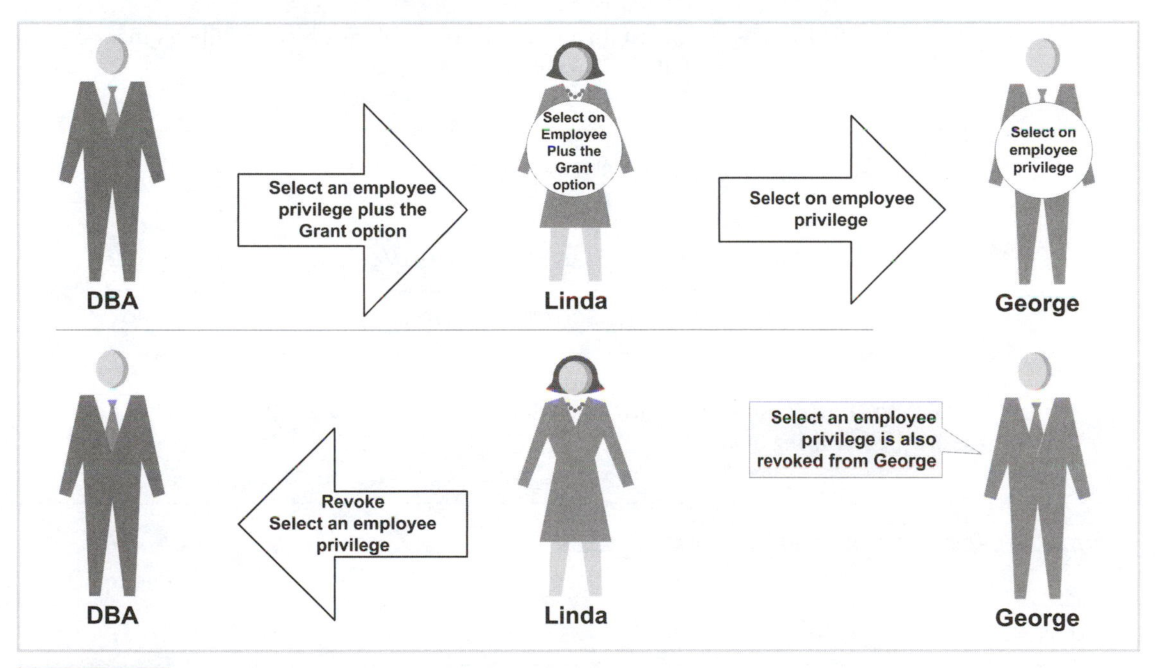

FIGURE 4-16 Revoking an object privilege with the GRANT option

As with system privileges, you can view object privileges granted to a user. Oracle provides several useful data dictionary views:

- USER_TAB_PRIVS
- USER_TAB_PRIVS_MADE
- USER_TAB_PRIVS_RECD
- ALL_TAB_PRIVS
- ALL_TAB_PRIVS_MADE
- ALL_TAB_PRIVS_RECD
- DBA_TAB_PRIVS

For example, if you want to see what object privileges user SAFYOUNI has, you issue the following query:

```
SQL> SELECT *
  2     FROM DBA_TAB_PRIVS
  3    WHERE GRANTEE = 'SAFYOUNI'
  4   /

GRANTEE   OWNER  TABLE_NAME  GRANTOR  PRIVILEGE  GRANTABLE HIERARCHY
--------  -----  ----------  -------  ---------  -------------------
SAFYOUNI  SCOTT  EMP         SCOTT    SELECT     NO
```

If user SAFYOUNI is logged on and wants to see all the object privileges that he has received, he issues the following query:

```
SQL> SELECT * FROM USER_TAB_PRIVS_RECD
  2  /

OWNER    TABLE_NAME    GRANTOR    PRIVILEGE    GRA    HIE
-----    ----------    -------    ---------    ---    ---
SCOTT    EMP           SCOTT      SELECT       NO     NO
```

Also, if Scott wants to see what privileges he has granted to other users, he can issue the following query:

```
SQL> SELECT * FROM USER_TAB_PRIVS_MADE
  2  /

GRANTEE    TABLE_NAME    GRANTOR    PRIVILEGE    GRA    HIE
-------    ----------    -------    ---------    ---    ---
SAFYOUNI   EMP           SCOTT      SELECT       NO     NO
```

Privileges in SQL Server

SQL Server has four types or levels of permissions: system or server level, database level, table (object) level, and column level. It's important to note that having server- or database-level permissions doesn't mean you have access to subordinate objects. For example, as you learned earlier, there is a fixed server role that gives its members privileges to manage database security. This, however, does not give the member access to manipulate data tables.

Server Privileges

You give logons access to perform certain tasks at the server level by placing them in fixed server roles. You learned how to do this and were introduced to the function of each role earlier in the chapter. This section takes a closer look at these roles.

Sysadmin

Members of the sysadmin fixed server role can perform any function within the system. There is zero limitation to what sysadmin can do.

Serveradmin

Members of the serveradmin role can perform certain server-level functions. Members of the serveradmin role can:

- Add members to the serveradmin role (members are automatically serveradmin role owners).
- Execute the DBCC FREEPROCCACHE command
- Execute the SP_CONFIGURE system-stored procedure (displays or modifies server setting configuration).
- Execute the SP_FULLTEXT_SERVICE system-stored procedure (sets search service to full text).
- Execute the SP_TABLEOPTION system-stored procedure (sets options for tables).
- Execute the RECONFIGURE command.
- Shut down the server.

The definitions of the SP commands contained in this chapter are adapted from Microsoft SQL Server online documentation found at *http://msdn.microsoft.com/library*.

Setupadmin

Members of the setupadmin role can manage linked servers and start-up procedures. Specifically, they can:

- Add members to the setupadmin role.
- Add, drop, and configure linked servers.
- Mark a stored procedure as a start-up procedure.

Securityadmin

Members of the securityadmin role can manage logons, use the CREATE DATABASE PERMISSIONS statement, read error logs, and change passwords. Specifically, members of the securityadmin role can:

- Add members to the securityadmin role.
- Grant, deny, and revoke the CREATE DATABASE command.
- Read the error log.
- Execute SP_ADDLINKEDSVRLOGIN (adds a link between the local server and the remote server).
- Execute SP_ADDLOGIN (adds new login name that allows a user to connect to a SQL Server instance).

- Execute SP_DROPLOGIN (removes a login name).
- Execute SP_GRANTLOGIN (allows a Windows user or group account to connect to SQL Server).
- Execute SP_DENYLOGIN (prevents a Windows user or group account from connecting to SQL Server).
- Execute SP_REVOKELOGIN (removes a login entry in SQL Server for Windows user or group account).
- Execute SP_DEFAULTDB (sets default database for current login name).
- Execute SP_DEFAULTLANGUAGE (sets default language for current login name).
- Execute SP_DROPLINKEDSVRLOGIN (removes a link from the local server to the linked server for current login name).
- Execute SP_DROPREMOTELOGIN (removes link from the remote server to the linked local server for current login name).
- Execute SP_HELPLOGINS (provides login information).
- Execute SP_PASSWORD (adds or changes password for a login name).
- Execute SP_REMOTEOPTION (sets options for tables).

Processadmin

Members of the processadmin role can manage processes running in the SQL server. Specifically, they can:

- Add members to the processadmin role.
- Execute the KILL command to terminate a process.

Dbcreator

Members of the dbcreator role can create, alter, and drop databases. Specifically, they can:

- Add members to the dbcreator role.
- Execute the CREATE DATABASE DDL command.
- Execute the ALTER DATABASE DDL command.
- Execute the DROP DATABASE DDL command.
- Execute the RESTORE DATABASE DDL command.
- Execute the RESTORE LOG DDL command.
- Extend a database.
- Execute the SP_RENAMEDB system-stored procedure (renames a database).

Diskadmin

Members of the diskadmin role can manage the disk files for the server and databases. Specifically, members of the diskadmin role can:

- Add members to the diskadmin role.
- Execute the DISK INIT command.
- Execute the SP_ADDUMPDEVICE system-stored procedure (adds a backup device to Microsoft SQL).
- Execute the SP_DROPDEVICE system-stored procedure (drops a database device or backup device from Microsoft SQL Server).

Bulkadmin

Members of the bulkadmin role can perform bulk insert operations. Specifically, they can:

- Add members to the bulkadmin role.
- Execute the BULK INSERT DML statement.

Database Privileges

There are two ways to grant users the ability to perform certain tasks at the database level: add them to fixed database roles or grant them specific permissions. You were introduced to the fixed database roles earlier in the chapter. Now, take a closer look at the following list that provides all fixed database roles accompanied with a brief description of each role.

Fixed Database Roles

- db_owner—Members of the db_owner role have complete access to the database and can perform any and all functions within.
- db_accessadmin—Members of the db_accessadmin role can add and remove users from the database. As a db_accessadmin, you do not have permission to execute any statements, only the system-stored procedures used to manage users.
- db_securityadmin—Members of the db_securityadmin role can change all permissions, object ownership, roles, and role membership. Members can execute GRANT, DENY, and REVOKE statements, as well as the system-stored procedures used to manage ownership and role membership.
- db_ddladmin—Members of the db_ddladmin role can execute all DDL statements except GRANT, DENY, and REVOKE. Members cannot execute data manipulation language (DML) statements.
- db_backupoperator—Members of the db_backupoperator role can execute DBCC statements pertaining to backup, CHECKPOINT, and BACKUP statements. (DBCC is a SQL Server tool used for database performance and other database diagnosis tasks.)
- db_datareader—Members of the db_datareader role can issue SELECT and READTEXT statements on any table in the database.
- db_datawriter—Members of the db_datawriter role can issue INSERT, UPDATE, DELETE, and UPDATETEXT statements on any table in the database.
- db_denydatareader—Members of the db_denydatareader role are explicitly denied SELECT and READTEXT permissions on all tables in the database.
- db_denydatawriter—Members of the db_denydatawriter role are explicitly denied INSERT, UPDATE, UPDATETEXT, and DELETE permissions on all tables in the database. You'll see where this role and the db_denydatareader role can come in handy in Chapter 5.

Statement Permissions

To grant more detailed privileges than those offered by the fixed database roles, you can also grant specific statement permissions to the user. These permissions are listed in Table 4-4.

Table 4-4 SQL Server statement permissions

Statement	Permits the User to
CREATE TABLE	Create tables in the database
CREATE VIEW	Create views in the database
CREATE PROCEDURE	Create stored procedures in the database
CREATE FUNCTION	Create functions in the database
CREATE DEFAULT	Create defaults in the database
CREATE RULE	Create rules in the database
BACKUP DATABASE	Back up the database
BACKUP LOG	Back up the database transaction log(s)

Granting Statement Permissions

To grant statement permissions using Query Analyzer, you use the GRANT statement, which has the following syntax structure:

```
GRANT { ALL | statement [ ,...n ] } TO security_account [
,...n ]
```

Where:

- `ALL`—Grants all statement permissions; it can be used only by members of the *sysadmin* fixed server role.
- `Statement`—Is the statement or list of statements on which you're granting permission; the statement values are listed in Table 4-4
- `TO`—Specifies the database user(s) or role(s) to whom you're granting the statement permissions.

To grant all permissions to the role admins, use this code:

```
GRANT ALL TO admins
```

To grant CREATE TABLE and CREATE VIEW to users sam and jason, use this code:

```
GRANT CREATE VIEW, CREATE TABLE TO sam, jason
```

Using Enterprise Manager, you can grant statement permissions by following these steps:

1. Open **Enterprise Manager**.
2. Expand the server that contains the database in which you want to grant the statement permissions.
3. Open the **Properties** dialog box for the database.
4. Click the **Permissions** tab.
5. Place a check mark in the desired permission column for the desired user(s) or role(s). See Figure 4-17 for results.
6. Click **OK**.

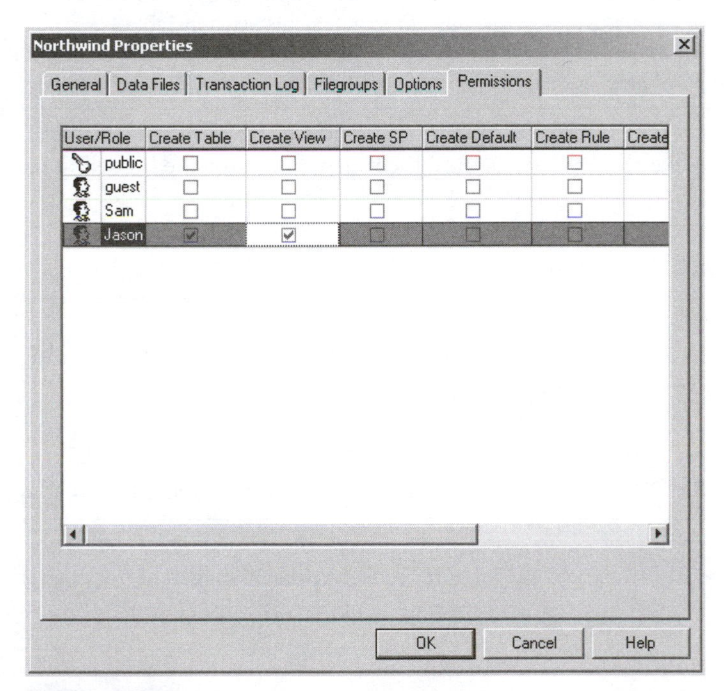

FIGURE 4-17 Granting user permissions

A few rules apply to these permissions and implicit permissions that are inherited. When you give a user permission to CREATE *some object* (for example, CREATE TABLE), that user can create tables in the database. When that user creates a table, the user becomes the object's owner. With that comes the implied ownership rights; the user can alter, drop, and grant permissions to the object which he or she created. If you grant permissions to create an object to a role, the member that creates the object is the object owner. Also, any permission applied to the role "public" affects all users in the database. Statements that can be executed by members of the public role by default are:

- BEGIN TRANSACTION
- COMMIT TRANSACTION
- ROLLBACK TRANSACTION
- SAVE TRANSACTION
- PRINT, RAISERROR
- SET

Revoking Statement Permissions

To revoke these permissions using Query Analyzer, you use the REVOKE statement.

```
REVOKE { ALL | statement [ ,...n ] }
FROM security_account [ ,...n ]
```

Where:

- `ALL`—Revokes all statement permissions; it can be used only by members of the sysadmin fixed server role
- `Statement`—Is the statement, or list of statements on which you're granting permission; the value(s) are any of the statements listed in Table 4-3
- `FROM`—Specifies the database user(s) or role(s) from whom you're revoking the statement permissions

To revoke all permissions from the role public, issue the following command:

```
REVOKE ALL FROM public
```

To revoke CREATE TABLE and CREATE VIEW from users Sam and Jason, issue the following command:

```
REVOKE CREATE VIEW, CREATE TABLE FROM sam, jason
```

To revoke statement permissions in Enterprise Manager, perform a series of steps similar to those you used to grant statement permissions. Figure 4-18 shows the results.

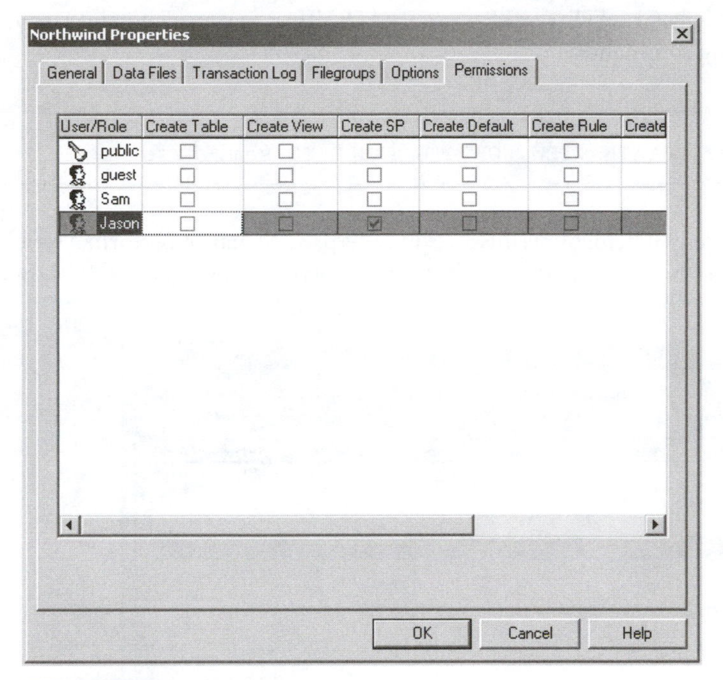

FIGURE 4-18 Granting database permissions with SQL Server

Denying Statement Permissions

To deny these permissions using Query Analyzer, you use the DENY statement.

```
DENY { ALL | statement [ ,...n ] }
TO security_account [ ,...n ]
```

Where:

- `ALL`—Denies all statement permissions; it can be used only by members of the sysadmin fixed server role
- `Statement`—Is the statement, or list of statements, on which you're denying permission; the value(s) are any of the statements listed in Table 4-3
- `TO`—Specifies the database user(s) or role(s) to whom you're denying the statement permissions

To deny all permissions to the role public, issue the following command:

```
DENY ALL TO public
```

To deny CREATE TABLE and CREATE VIEW to users Sam and Jason, issue the following command:

```
DENY CREATE VIEW, CREATE TABLE TO sam, jason
```

To deny statement permissions in Enterprise Manager, perform a series of steps similar to those you used to grant statement permissions. Figure 4-19 shows the results.

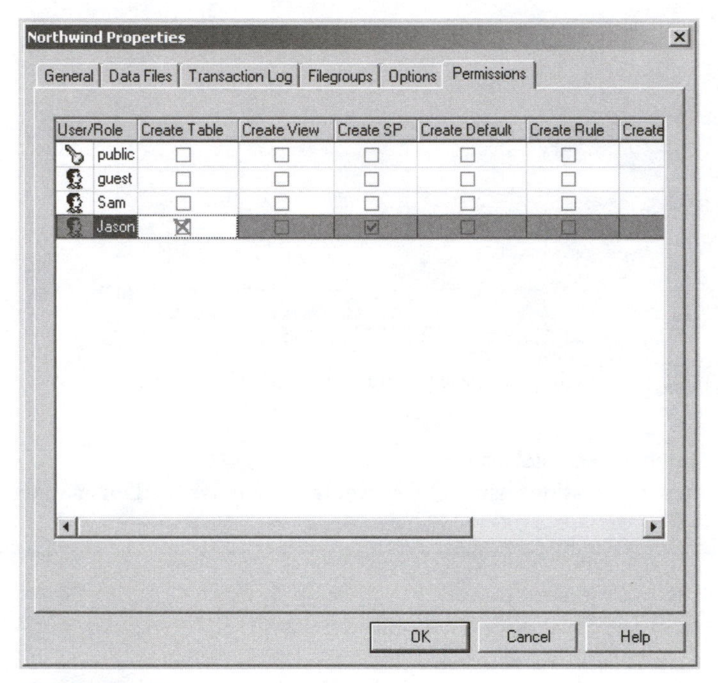

FIGURE 4-19 Revoking database permissions with SQL Server

Table and Database Objects Privileges

You can also set privileges for accessing database objects themselves. The three statements used to manage object-level permissions are: GRANT, REVOKE, and DENY.

Granting Object Permissions

Use the following GRANT syntax to grant object permissions:

```
GRANT
{ ALL [ PRIVILEGES ] | permission [ ,...n ] } ON object
TO security_account [ ,...n ]
[ WITH GRANT OPTION ]
[ AS { group | role } ]
```

Where:

- ALL—Grants all statement permissions; it can be used only by members of the sysadmin fixed server role
- PRIVILEGES—(Optional) Is included for SQL-92 compliance
- Permission—Is the object permission (s) you are granting; permissions you can grant on a table, table-valued function, or a view are listed as follows:
  - SELECT—Reads data from the table (issue the SELECT DML command)
  - INSERT—Inserts data into the table (issue the INSERT DML command)
  - DELETE—Deletes data from the table (issue the DELETE DML command
  - UPDATE—Updates data in the table (issue the UPDATE DML command)
- REFERENCES—Creates a FOREIGN KEY constraint on another table that references this one
- Object—Is the name of the object in the current database on which you are granting permissions
- TO—Specifies the database user(s) or role(s) to which you're granting the object permissions
- WITH GRANT OPTION—Specifies that the user in the TO clause is given the ability to grant the specified object permission to the other users
- AS {group | role}—Specifies the optional name of the role in the current database that has the authority to execute the GRANT statement; you use the AS clause when permissions on an object are granted to a role and the object permissions need to be further granted to users who are not members of the group or role; because only a user can execute a GRANT statement, a specific member of the role grants permissions on the object under the authority of the group or role, in other words, "AS" that role

You can grant the EXECUTE permission only on stored procedures. You can grant EXECUTE and REFERENCES permissions on a scalar-valued function. The following example should clarify this concept. Assume you have three database users: Jason, Sam, and Debbie. Jason built a view V_QUARTERLYSALES and wants to grant SELECT permissions to Sam and give Sam permission to grant SELECT permissions on the view to anyone who needs it. Jason executes the following code:

```
GRANT SELECT ON v_QuarterlySales TO sam WITH GRANT OPTION
```

Sam then wants to grant Debbie access to the view so that she can help him with some analysis but does not want her to give access to anyone else. Sam executes:

```
GRANT SELECT ON v_QuarterlySales TO Debbie
```

Both Sam and Debbie can now select data from the view V_QUARTERLYSALES. Now, look at it a bit differently. Take the same three users and view. This time, however,

Jason wants to give SELECT permissions to the sales team as well as give the team the ability to grant SELECT permissions to others, so he executes:

```
GRANT SELECT ON v_QuarterlySales TO sales WITH GRANT OPTION
```

Sam is a member of the sales role. He wants to give Debbie SELECT permissions to help him with some analysis. So, Sam executes the following and gets an error:

```
GRANT SELECT ON v_QuarterlySales TO Debbie
```

Sam was not given explicit SELECT permissions to the view, nor was he given permission to grant SELECT permission to others. Sam inherits SELECT permissions because he is a member of the sales role, but the role member does not inherit the grant option. So, Sam had to grant SELECT permissions to Debbie "AS" the role:

```
GRANT SELECT ON v_QuarterlySales TO Debbie AS Sales
```

You cannot grant access to users in another database; you can only grant permission to users of a database for objects in that same database.

An important note on stored procedures and functions: If you have EXECUTE permissions on a stored procedure or function, that stored procedure or function executes normally regardless of your permissions to the objects used by that stored procedure.

Now, examine the use of Enterprise Manager to grant permissions. There are two common methods for granting permissions on objects using Enterprise Manager: from the user and from the object.

Method 1: From the User

1. Open **Enterprise Manager**.
2. Expand the database in which the user and the object exist. Select the desired user.
3. Open the desired user's properties dialog box.
4. Click **Permissions**.
5. Place a check mark in the check box corresponding to the desired object and permission.
6. Click **OK**.
7. Click **OK** again.

Method 2: From the Object

1. Open **Enterprise Manager**.
2. Expand the database in which the object and the user exist. Select the desired object.
3. Open the object's **Properties** dialog box.

4. Click **Permissions.**
5. Place a check mark in the check box corresponding to the desired user and permission.
6. Click **OK.**
7. Click **OK** again.

Method 1 is ideal for assigning permissions on multiple objects for a single user. Method 2 works well for assigning permissions on a single object for multiple users.

Revoking Object Permissions

Just as in Oracle, you can revoke object permissions using the REVOKE statement.

```
REVOKE [ GRANT OPTION FOR ]
{ ALL [ PRIVILEGES ] | permission [ ,...n ] } ON object
{ TO | FROM } security_account [ ,...n ]
[ CASCADE ]
[ AS { group | role } ]
```

Where:

- `GRANT OPTION FOR`—Revokes the grant options given to the user(s) or role(s)
- `ALL`—Revokes all permissions granted to the object
- `PRIVILEGES`—(Optional) Is included for SQL-92 compliance
- `Permission`—Specifies the permission(s) to revoke from the user(s) and/or role(s)
- `ON`—Specifies the database object on which you are revoking permissions
- `TO | FROM`—Specifies user(s) and/or role(s) from whom you are revoking permissions
- `CASCADE`—Must be included if permissions are granted to other users by way of the WITH GRANT OPTION clause; the CASCADE clause revokes permissions for all users in the grant chain
- `AS`—Is the same here as for the GRANT statement.

Looking at an example may help you to understand the previous block of code. Jason gave Sam SELECT permission to his view using the WITH GRANT OPTION. Sam then gave Debbie SELECT permissions as well. Because of this, if Jason tries to revoke permissions from Sam, he'll get an error and the permissions will not be revoked; he must specify the CASCADE clause so that the permissions are revoked from Debbie as well.

```
REVOKE SELECT ON v_QuarterlySales FROM Sam CASCADE
```

In the second GRANT example, Sam had to specify `AS sales` to grant Debbie permissions to the view. For Sam to now revoke those permissions, he needs to specify `AS sales` on his `REVOKE` statement:

```
REVOKE SELECT ON v_QuarterlySales FROM Debbie AS sales
```

Using Enterprise Manager, there are two common methods for revoking permissions on objects: from the user and from the object. Follow a series of steps similar to those presented for granting privileges from the user and from the object.

Revoking permissions from the user is ideal for revoking permissions on multiple objects for a single user. Revoking permissions from the object works well for revoking permission on a single object for multiple users.

Denying Object Permissions

You use the DENY statement to prevent users from gaining permissions through a GRANT to their user accounts.

```
DENY
{ ALL [ PRIVILEGES ] | permission [ ,...n ] } ON object
TO security_account [ ,...n ]
[ CASCADE ]
```

Where:

- `ALL`—Denies all permissions to the object
- `PRIVILEGES`—(Optional) Is included for SQL-92 compliance
- `Permission`—Specifies the permission(s) to deny user(s) and/or role(s) on the object
- `CASCADE`—Is needed to revoke and to deny permissions

Denied permissions take precedence over all other permissions. So, if you grant a user SELECT permissions on a table, and that user belongs to a role that has been denied SELECT permissions to the table, that user cannot issue a SELECT statement against the table. One exception to the rule is application roles. If a user is denied permission to an object, and the user activates an application role, the user operates under the application role's security context. For example, examine the following code block:

```
GRANT SELECT ON v_QuarterlySales TO sam
DENY SELECT ON v_QuarterlySales TO agents
EXEC sp_addrolemember 'agents', 'sam'
```

Sam would not be able to do this:

```
SELECT * FROM v_QuarterlySales
```

Another exception to the rule is stored procedures. If you are explicitly denied access to a table, but you have `Execute` permissions on a stored procedure that uses that table, the stored procedure executes successfully. To remove a denied permission, you issue a GRANT or REVOKE command at the same level. For example, if you want to deny Sam the SELECT permission to the view V_QUARTERLYSALES, you issue this command:

```
DENY SELECT ON v_QuarterlySales TO sam
```

The following statement clears the explicit DENY. Sam still does not have access to the object, but if he belongs to a role that does, he inherits the permissions from the role.

```
REVOKE SELECT ON v_QuarterlySales TO sam
```

Again, there are two common methods for denying permissions on objects using Enterprise Manager: from the user and from the object. Follow a series of steps similar to those presented for granting privileges from the user and from the object. Denying permissions from the user is ideal for denying permissions on multiple objects for a single user. Denying permissions from the object works well for denying permission on a single object for multiple users.

Column-level Privileges

Microsoft SQL Server 2000 also gives you the ability to specify object permissions on individual columns. For instance, if you had a table named EMPLOYEES and it contained basic demographic data, as well as confidential data that only human resources should see, you can limit the columns to which certain users have access.

Granting Permissions on Columns

You can use the GRANT statement, but this time you specify to which columns you want to apply the permissions.

```
GRANT
       { ALL [ PRIVILEGES ] | permission [ ,...n ] }
       {
            [ ( column [ ,...n ] ) ] ON { table | view }
            | ON { table | view } [ ( column [ ,...n ] ) ]
       }
    TO security_account [ ,...n ]
    [ WITH GRANT OPTION ]
    [ AS { group | role } ]
```

Where:

- Column—Specifies the column(s) to which you want to apply the permissions; you can either list the columns with the permission or with the table depending on the granularity you're looking for

For example, consider the table named EMPLOYEES with columns SSN, FIRST_NAME, LAST_NAME, ADDRESS, CITY, STATE, and ZIP. You want to grant SELECT access on FIRST_NAME, LAST_NAME, ADDRESS, CITY, STATE, and ZIP to everyone.

```
GRANT SELECT ON Employees (first_name, last_name, address,
   city, state, zip) TO public
```

The previous GRANT statement lists only specific columns on which to grant SELECT permissions; an attempt to select any other column raises an error. Both of the following statements would fail:

```
SELECT ssn FROM Employees
SELECT * FROM Employees
```

Now, say you want only the role hr_managers to update the Social Security number, but other people in the Human Resources Department should be able to update name and address data. You can give SELECT and UPDATE permissions to all columns for the role hr_managers.

```
GRANT SELECT, UPDATE ON Employees TO hr_managers
```

Then you give the human_resources role SELECT permission on all columns, but limit their UPDATE permissions to the specified columns.

```
GRANT SELECT, UPDATE (first_name, last_name, address, city,
  state, zip) ON Employees
   TO human_resources
```

Using Enterprise Manager, there are two common methods for granting column permissions on objects: from the user and from the object. Follow a series of steps similar to those presented for granting privileges from the user and from the object. Granting column permissions on objects from the user is ideal for assigning permissions on multiple objects for a single user. Granting column permissions on objects from the object works well for assigning permission on a single object for multiple users.

Revoking Permissions on Columns

The same holds true for REVOKE and DENY statements. The syntax is the same; you merely specify the columns on which you want to revoke access.

```
REVOKE [ GRANT OPTION FOR ]
       { ALL [ PRIVILEGES ] | permission [ ,...n ] }
       {
           [ ( column [ ,...n ] ) ] ON { table | view }
           | ON { table | view } [ ( column [ ,...n ] ) ]
       }
   { TO | FROM }
       security_account [ ,...n ]
   [ CASCADE ]
   [ AS { group | role } ]
```

For example:

```
REVOKE ALL ON Employees (ssn) FROM public
```

As with previous steps, you use Enterprise Manager to revoke privileges. Again, there are two common methods for revoking column permissions on objects: from the user and from the object. Follow a series of steps similar to those presented for granting privileges from the user and from the object. Revoking column permissions on objects from the user is ideal for revoking permissions on multiple objects for a single user. Revoking column permissions on objects from the object works well for revoking permission on a single object for multiple users.

Denying Permissions on Columns

Again, you simply execute the DENY statement, specifying the columns and permissions you want to explicitly deny.

```
DENY
       { ALL [ PRIVILEGES ] | permission [ ,...n ] }
       {   [ ( column [ ,...n ] ) ] ON { table | view }
           | ON { table | view } [ ( column [ ,...n ] ) ] }
   TO security_account [ ,...n ] [ CASCADE ]
```

The following is an example of denying permissions on a column:

```
DENY ALL ON Employees (ssn) TO general_users
```

Similar to previous steps for revoking permissions, there are two common methods for denying column permissions on objects using Enterprise Manager: from the user and from the object. Follow a series of steps similar to those presented for granting privileges from the user and from the object. Denying column permissions from the user is ideal for denying permissions on multiple objects for a single user. Denying column permissions from the object works well for denying permission on a single object for multiple users.

Creating, Assigning, and Revoking User Roles

A junior database administrator for a small company was asked to develop a security model for a new database application. Because of his lack of experience, especially in security matters, the DBA was very concerned about the assignment and expressed his fears to management. Management insisted that he could do the project and sent him on his way with no direction other than to complete the job. The DBA didn't know what to do and decided that doing nothing might be the only way to get through this. His only action was to grant every user a default role that came with the system. Without raising questions, he gave the developers and engineers every privilege they asked for. The project involved about 35 developers, 20 quality engineers, and 30 users. On the day the database application was to be implemented, chaos reigned. Privileges were inconsistent among the developers; engineers were calling for privilege levels equal to their colleagues, and users were having a field day with their wide-open access. The DBA realized what confusion he had created, but didn't know how to cope other than spending several days trying unsuccessfully to clean up the process.

The confusion could have been avoided, if management had taken the time to direct the DBA and if the DBA had sought advice from experts, especially on the subject of roles.

As stated previously, a role is a concept used to organize and administer privileges in an easy manner. A role is like a user, except it cannot own objects. A role can be assigned privileges and then assigned to users. Take the scenario of the inexperienced DBA. The DBA should have created a role for developers, (DEV_ROLE), a role for quality assurance engineers, (QA_ROLE), and a role for the other users called OTH_ROLE. The DBA could then assign these roles to the appropriate users. When asked to grant a new privilege, the DBA would grant the privilege to one appropriate role and not to other inappropriate roles. On the day the new application went live, the DBA could simply create the roles and apply them to the appropriate users, which would be much easier than granting specific privileges to each user.

Now that you understand the basics of roles, you next examine how to create roles, assign privileges to these roles, and then assign them to users. You learn role creation in both Oracle and SQL Server.

Creating Roles with Oracle

As mentioned earlier, Oracle supports the concept of a role which can be assigned privileges. Once a role is assigned privileges, you can assign the role to users. In addition, you may also assign a role to another role. However granting roles to roles is not highly recommended, because it complicates the structure of the security model. It is customary to create roles for different functions of the application. If you use the previous scenario as an example, you create three roles: DEV_ROLE, QA_ROLE, and OTH_ROLE. To create a role in Oracle, use the following syntax:

```
CREATE ROLE role
   [ NOT IDENTIFIED
   | IDENTIFIED { BY password
                | USING [ schema. ] package
                | EXTERNALLY
                | GLOBALLY
                }
   ] ;
```

Where:

- NOT IDENTIFIED—Indicates that there is no password (no authentication) required when a role is enabled
- IDENTIFIED—Indicates that the role requires authentication using any of the methods listed previously

Now, create the following roles:

```
SQL> CREATE ROLE DEV_ROLE;

Role created.

SQL> CREATE ROLE QA_ROLE;

Role created.

SQL> CREATE ROLE OTH_ROLE;

Role created.
```

To assign a privilege to a role, issue the following statement:

```
SQL> GRANT CREATE SESSION TO DEV_ROLE
/

Grant succeeded.
```

To assign a role to a user, issue the following statement:

```
SQL> GRANT DEV_ROLE TO SAFYOUNI
/

Grant succeeded.
```

You can accomplish the same tasks using the Oracle Enterprise Manager Roles tool. In Oracle, there are numerous dictionary views to help you find out what roles are assigned and to whom they are assigned. The list that follows displays only the most commonly used views:

- DBA_ROLES—Contains all roles in the database.
- DBA_ROLE_PRIVS—Contains roles that are assigned to a user
- ROLE_ROLE_PRIVS—Contains roles that are granted to other roles
- ROLE_SYS_PRIVS—Contains all system privileges assigned to a role
- ROLE_TAB_PRIVS—Contains all object privileges assigned to a role
- SESSION_ROLES—Contains a list of roles that are enabled in the current session

Oracle comes with default roles that have specific privileges to perform different types of tasks. The following list contains the most common roles:

- DBA—Should be assigned to the DBA user account only; with this role the user can do all sorts of DBA tasks and administration, including starting and shutting down the database, performing backups, and other administrative tasks.
- SELECT_CATALOG_ROLE—When this role is assigned to a user, the user can select any data dictionary view.
- RESOURCE—This role is assigned several privileges to create database objects.
- CONNECT—This role is assigned CREATE SESSION privileges and limited privileges to create basic database objects.

As with privileges, you can revoke a role using the REVOKE statement. You can also drop a role by using the DROP statement.

Creating Roles with SQL Server

SQL Server has three types of roles: fixed server, fixed database, and user defined.

User-defined Roles

Like Oracle, Microsoft SQL Server 2000 allows you to create roles in a database. These roles can then be assigned permissions on objects in the database. In SQL Server, there are two types of roles you can create: standard and application.

Standard roles are like groups. They contain members and can be granted or denied permissions and ownership to objects in the database. Standard role members can be database users or other roles.

Application roles do not and cannot contain members. Application roles are required to have a password. Application roles are activated by an application and determine authorization within the database for that application. You'll learn about application roles in more detail, as well as how to create and manage them, in Chapter 5.

To create a user-defined database role, standard or application, you must be a member of the sysadmin, db_owner, or db_securityadmin group.

Creating Roles

To create a new database role using Query Analyzer, you execute the SP_ADDROLE system-stored procedure.

```
sp_addrole [ @rolename = ] 'role' [ , [ @ownername = ] 'owner' ]
```

Where:

- `@rolename`—Is the name of the new role; the role name must be between 1 and 128 characters in length and can include any letter, number, or symbol except the backslash (\ )
- `@ownername`—Is the owner of the new role; default is dbo; if specified, the value must be a user or role in the current database

To add the role of "sales" to the database Northwind:

```
use northwind
exec sp_addrole 'sales'
```

To create a new role in Enterprise Manager:

1. Open **Enterprise Manager**.
2. Expand the database in which you want to create the role.
3. Select the **Roles** node.
4. On the Action menu, choose **New Database Role**.
5. Type the name for your role in the Name box.
6. Click **OK**.

Adding Members to Roles

To add members to roles in the current database using the Query Analyzer, execute the SP_ADDROLEMEMBER system-stored procedure.

```
sp_addrolemember [ @rolename = ] 'role' ,
       [ @membername = ] 'security_account'
```

Where:

- `@rolename`—Is the name of the SQL Server role in the current database
- `@membername`—Is the user or role being added to the role

To add the user Jason to the role sales:

```
exec sp_addrolemember 'sales', 'jason'
```

There are three common methods for adding a member to a role in Enterprise Manager:

Method 1 uses the following steps:

1. Open **Enterprise Manager**.
2. Expand the **Security** container, and open the **Properties** dialog box of the login **Jason**.
3. Click the **Database Access** tab.
4. Highlight the database that contains the role you're adding.
5. Put a check mark next to **sales.**
6. Click **OK.**

Method 2 uses the following steps:

1. Open **Enterprise Manager**.
2. Expand the database that contains the role to which you are adding members.
3. Click the **Roles** node.
4. Open the **Properties** dialog box for **sales.**
5. Click the **Add** button.
6. Select or assign users or roles to a new role.
7. Click **OK** to add.
8. Click **OK** again.

Method 3 uses the following steps:

1. Open **Enterprise Manager**.
2. Expand the database that contains the user to which you want to add a role.
3. Click the **Users** node.
4. Open the **Properties** dialog box for the user.
5. Place a check mark next to the role(s) to which you want to add the user.
6. Click **OK.**

These three methods are best suited to the following uses:

- *Method 1*—For adding database users to roles at the time of logon and user creation
- *Method 2*—For adding multiple users or roles at one time; also used for adding a role to a role
- *Method 3*—For adding users to multiple roles

Dropping Members from Roles

To drop members from roles in the current database using Query Analyzer, you execute the SP_DROPROLEMEMBER system-stored procedure.

```
Sp_droprolemember   [ @rolename = ] 'role' ,
     [ @membername = ] 'security_account'
```

Where:

- `@rolename`—Is the name of the SQL Server role in the current database
- `@membername`—Is the user or role being dropped from the role

To drop the user Jason from the role sales:

```
Use northwind
exec sp_droprolemember 'sales', 'jason'
```

There are three common methods for removing a member from a role in Enterprise Manager:

Method 1 follows these steps:

1. Open **Enterprise Manager.**
2. Open the **Properties** dialog box of the logon **Jason.**
3. Click the **Database Access** tab.
4. Highlight the database that contains the role you're dropping.
5. Remove the check mark next to **sales.**
6. Click **OK.**

Method 2 follows these steps:

1. Open **Enterprise Manager.**
2. Expand the database that contains the role to which you are dropping members.
3. Click the **Roles** node.
4. Open the **Properties** dialog box for **sales.**
5. Select the users or roles you want to remove from the role.
6. Click **Remove.**
7. Click **OK.**

Method 3 follows these steps:

1. Open **Enterprise Manager.**
2. Expand the database that contains the user to which you want to drop from a role.
3. Click the **Users** node.
4. Open the **Properties** dialog box for the user.
5. Remove the check mark next to the role(s) from which you want to remove the user.
6. Click **OK.**

These three methods are best suited to the following uses:

- *Method 1*—For dropping users from roles for several databases to which the login has access
- *Method 2*—For dropping multiple users or roles at the same time
- *Method 3*—For dropping a single user from multiple roles

Dropping User-defined Roles

Roles cannot be dropped if they own objects or contain members. If a role owns objects, the objects must be dropped first. If a role contains members, all members must be removed, either using the SP_DROPROLEMEMBER system-stored procedure or using Enterprise Manager. SP_DROPROLE returns a list of members, if there are any.
You must be a member of sysadmin, db_owner, or db_securityadmin to drop roles from the database. Remember too that only user-defined roles can be dropped. To drop a new

database role from the current database using Query Analyzer, you execute the SP_DROPROLE system-stored procedure.

```
Sp_droprole [ @rolename = ] 'role'
```

Where:

- @rolename—Is the name of the role to be dropped

To drop the role sales from the database Northwind:

```
use northwind
exec sp_droprole 'sales'
```

To drop a role using Enterprise Manager:

1. Open **Enterprise Manager**.
2. Expand the database containing the role to be dropped.
3. Click the **Roles** node.
4. Select the desired role.
5. Choose **Delete** on the Action menu. Click **Yes** to confirm the deletion.

Default Roles

Fixed Server Roles

Fixed server roles cannot be modified, nor can you create your own. These roles, which are described in Table 4-5, are provided easily and give logons access to certain administrative tasks within the server.

Table 4-5 Description of fixed server roles

SQL Server Fixed Role	Role Description
sysadmin	A super role that allows assigned user to perform any task within SQL Server
serveradmin	Allows assigned user to modify SQL Server configuration.
setupadmin	Allows assigned user to perform specific SQL Server setup, such as linking servers and execution of system stored procedures
securityadmin	Allows assigned user to administer SQL Server logins
processadmin	Allows assigned user to administer SQL Server instance processes
dbcreator	Allows assigned user to create and modify SQL Server database
diskadmin	Allows assigned user to administer database data files
bulkadmin	Allows assigned user to perform BULK INSERT statement

* Information presented in this table is adapted from Microsoft SQL Server 2000 documentation

Because fixed server roles are at the server level, you add logons to the roles. To add logons to fixed server roles, you must be a member of the sysadmin fixed server role. The logon sa (system administrator) and members of the built-in\administrators Windows group are automatically members of sysadmin.

Adding Members to Fixed Server Roles

To add a login to a fixed server role in Query Analyzer, you execute the SP_ADDSRVROLEMEMBER system-stored procedure.

```
sp_addsrvrolemember [ @loginame = ] 'login'
     , [ @rolename = ] 'role'
```

Where:

- `@loginame`—Is the name of the login to be added to the fixed server role; it can be a Microsoft SQL Server login or a Microsoft Windows NT user account; if the Windows integrated login has not already been created, it is granted automatically
- `@rolename`—Is the name of the fixed server role to which you're adding the login; the value must be one of the fixed server role names in Table 4-5

To add the domain user Jason to the sysadmin fixed server role:

```
exec sp_addsrvrolemember 'mydomain\jason', 'sysadmin'
```

To add the SQL Server login sam to the securityadmin fixed server role:

```
exec sp_addsrvrolemember 'sam', 'securityadmin'
```

To add a login to a fixed server role using Enterpriser Manager, click the Server Roles tab on the SQL Server Login Properties dialog box. This can be done while you're creating a new login, or later by modifying an existing login.

The first method for adding a login follows these steps:

1. Open **Enterprise Manager**.
2. Open the **Properties** dialog box of the login **sam**.
3. Click the **Server Roles** tab.
4. Put a check mark next to **Security Administrators**.
5. Click **OK**.

The second method for adding a login follows these steps:

1. Open **Enterprise Manager**.
2. Expand the **Security** container.
3. Click the **Server Roles** node.
4. Open the **Properties** dialog box for **Security Administrators**.
5. Click the **Add** button.

6. Select the desired logins (one or many).
7. Click **OK** to add the logins.
8. Click **OK**.

The first method is ideal for adding a single login to multiple roles or to add a login to a fixed server role at the time of creation. The second method works well for adding multiple logins to a single role.

Dropping Members from Fixed Server Roles

To drop a login from a fixed server role in Query Analyzer, you execute the SP_DROP-SRVROLEMEMBER system-stored procedure.

```
sp_dropsrvrolemember [ @loginame = ] 'login'
     , [ @rolename = ] 'role'
```

Where:

- `@loginame`—Is the name of the login to be dropped from the fixed server role; it can be a Microsoft SQL Server login or a Microsoft Windows NT user account
- `@rolename`—Is the name of the fixed server role from which you're dropping the login; the value must be one of the Fixed Server role names in Table 4-5

To drop the domain user Jason from the sysadmin fixed server role:

```
exec sp_dropsrvrolemember 'mydomain\jason', 'sysadmin'
```

To drop the SQL server login sam to the securityadmin fixed server role:

```
exec sp_dropsrvrolemember 'sam', 'securityadmin'
```

To drop a login from a fixed server role using Enterpriser Manager, you select the Server Roles tab on the SQL Server Login Properties dialog box. This can be done while you're creating a new login, or later by modifying an existing login.

The first method for dropping a login follows these steps:

1. Open **Enterprise Manager**.
2. Open the **Properties** dialog box of the login **sam**.
3. Click the **Server Roles** tab.
4. Remove the check mark next to **Security Administrators**.
5. Click **OK**.

The second method for dropping a login follows these steps:

1. Open **Enterprise Manager**.
2. Expand the **Security** container.

3. Click the **Server Roles** node.
4. Open the **Properties** dialog box for **Security Administrators.**
5. Select the desired logins (one or many).
6. Click **Remove.**
7. Click **OK.**

The first method is ideal for removing a single login from multiple roles. The second method works well for dropping multiple logins from a single role.

Fixed Database Roles

Like fixed server roles, fixed database roles cannot be modified. These roles are provided easily and give database users access to certain administrative tasks within the database. You can also grant or deny data access permissions globally to the database.

Role members can be any valid SQL Server database user or role. To add a member to a fixed database role, you must be a member of sysadmin or db_owner. Table 4-6 provides a list of fixed roles and a brief description of each role.

Table 4-6 SQL Server built-in roles

Fixed Database Role	Description
db_owner	Has all permissions in the database
db_accessadmin	Can add or remove user IDs
db_securityadmin	Can manage all permissions, object ownerships, roles, and role memberships
db_ddladmin	Can issue all DDL, but cannot issue GRANT, REVOKE, or DENY statements
db_backupoperator	Can issue DBCC, CHECKPOINT, and BACKUP statements
db_datareader	Can select all data from any user table in the database
db_datawriter	Can modify any data in any user table in the database
db_denydatareader	Cannot select any data from any user table in the database
db_denydatawriter	Cannot modify any data in any user table in the database

* From Microsoft SQL Server Books Online

Adding Members to Fixed Database Roles

Adding members to a fixed database role is done in the same manner as adding a user-defined database role. In Query Analyzer, you use the SP_ADDROLEMEMBER system-stored procedure. To add the SQL database user Jason to the db_securityadmin fixed server role:

```
use northwind
exec sp_addrolemember 'db_securityadmin', 'jason'
```

To add a user to a fixed database role using Enterprise Manager, follow the steps in any of the three following methods:

Method 1 of adding a user to a fixed database role follows these steps:

1. Open **Enterprise Manager**.
2. Open the **Properties** dialog box of the login **Jason.**
3. Click the **Database Access** tab.
4. Highlight the database that contains the role to which you are adding a user.
5. Put a check mark next to **db_securityadmin**.
6. Click **OK.**

Method 2 of adding a user to a fixed database role follows these steps:

1. Open **Enterprise Manager**.
2. Expand the database that contains the role to which you are adding members.
3. Click the **Roles** node.
4. Open the **Properties** dialog box for **db_securityadmin**.
5. Click the **Add** button.
6. Select the users or roles to which you want to add the role.
7. Click **OK** to add.
8. Click **OK** again.

Method 3 of adding a user to a fixed database role follows these steps:

1. Open **Enterprise Manager**.
2. Expand the database that contains the user to whom you want to add a role.
3. Click the **Users** node.
4. Open the **Properties** dialog box for the user.
5. Place a check mark next to the role(s) to which you want to add the user.
6. Click **OK.**

The three methods are best suited for the following uses:

- *Method 1*—For adding database users to roles at the time of login and user creation
- *Method 2*—For adding multiple users or roles at one time; also used to add a role to a role
- *Method 3*—For adding users to multiple roles

Dropping Members from Fixed Database Roles

Members are dropped from a fixed database role in the same manner as a user-defined database role is dropped. Using Query Analyzer, you use the SP_DROPROLEMEMBER system-stored procedure. To drop the SQL database user Jason from the db_securityadmin fixed server role:

```
use northwind
exec sp_droprolemember 'db_securityadmin', 'jason'
```

To drop a user from a fixed database role using Enterprise Manager, follow a series of steps similar to Methods 1, 2, and 3 for adding a user to a fixed database role. The methods are best suited for the following uses:

- *Method 1*—For dropping database users from roles in multiple databases for that login
- *Method 2*—For dropping multiple users or roles at one time
- *Method 3*—For dropping users from multiple roles

Public Database Role

By default, every database has the public database role that cannot be dropped. Every user in the database is automatically a member of "public" and cannot be removed from the role.

Best Practices

The threat of an attacker is mentioned throughout the chapter and will continue to be mentioned throughout the text. After all, you are learning to guard against the attacker. An interesting and frightening fact is that the security threat of an attack from someone within the company is higher than the threat from someone outside. For this reason, it is important to use the best practices listed in this section to develop a secure environment internally, as well as externally. All the firewalls in the world cannot protect your servers against an inside job. Here is a list of best practices:

- Never store passwords for an application in plaintext; make sure all passwords are encrypted.
- Change passwords frequently.
- Make sure that passwords are at least eight characters long.
- Pick a password that you can remember so that you never have to write it down.
- Use roles to control and administer privileges. You should avoid granting privileges directly to a user.
- You should report the compromise or loss of a password to security.
- You should report to security any violation of company guidelines with respect to roles, privileges, profiles, and passwords.
- Never give your password to anyone, not even your manager. If you must reveal your password, change it immediately after the task that requires the password is completed.
- Never share your password with anyone even if the person is close to you.
- Never give your password over the phone.
- Never type your password in an e-mail, especially if the e-mail is not encoded.
- Make sure your password is complex enough to prevent hackers from breaking it easily.
- Use Windows integrated security mode for securing a SQL server.
- If you are in a Windows 2000/2003 domain, use only domain users, so that you can take advantage of Kerberos.

- When configuring policies:
  - Require complex passwords. Also, from an operations standpoint, have your users put special characters in the first seven bytes of the password. This makes that first seven bit hash more difficult to crack, should anyone try.
  - Require a password length of at least eight characters.
  - Set an account lockout threshold. A typical number of trials before the account is locked is three.
  - Do not allow passwords to automatically reset after failing the allowed number of login attempts. Always make unlocking an action that you or the security administrator performs.
  - Expire end-user passwords, but do not expire application-user passwords.
  - Enforce a password history. Five is a good setting for this policy.

Chapter Summary

- A profile is a way of defining database user behavior to prevent users from wasting resources such as memory or CPU.
- A profile in Oracle has two aspects: limitation on resources and implementation of a password policy.
- Use Oracle DBA_PROFILES dictionary view to display all profiles and their attributes.
- Use the ALTER USER statement to assign a profile to a user.
- A user cannot have more than one profile, but a profile can be assigned to many users.
- SQL Server does not have an implementation of the profile concept.
- In Oracle you use the profile concept to implement password policies.
- A password policy is a set of guidelines that enhances the robustness of a password and reduces the likelihood of it being broken.
- In SQL Server, password policies are implemented at the operating system level.
- NTLM authenticates using a challenge/response methodology.
- Kerberos authentication requires a trusted third resource known as a Key Distribution Center (KDC).
- In Oracle, system privileges are granted only by a database administrator or users who have been granted the administration option.
- In Oracle, object privileges are granted by the owner of the database object or a user who has been granted the GRANT privilege.
- SQL Server has four types or levels of permissions: system or server level, database level, table (object) level, and column level.
- Use the GRANT statement to give permission to perform specific system or object tasks.
- Use the REVOKE statement to take away permissions to perform tasks.
- A role is a concept used to organize and administer privileges in an easy manner.
- A role is like a user, except it cannot own objects.
- A role can be assigned privileges and then assigned to users.

- Use GRANT and REVOKE statements to administer roles.
- Public roles cannot be dropped.
- Use the best practices listed in this chapter to develop a secure environment internally, as well as externally.

Review Questions

1. Password policies in SQL Server are implemented through profiles. True or false?
2. In Oracle, once a profile is assigned to a user, the profile cannot be revoked. True or false?
3. In Oracle, you can assign many profiles to a user. True or false?
4. Fixed roles in SQL Server cannot be modified. True or false?
5. You use the DENY statement to prevent a user from gaining permissions to the user account. True or false?
6. You can drop a public role and recreate it with different privileges. True or false?
7. What is the SQL statement for assigning a profile to a user?
8. What are the two SQL statements that assign and remove privileges to and from users?
9. What are the two SQL statements that assign and remove privileges to and from a database?
10. What Oracle data dictionary view do you use to display a password policy for a specific user?
11. Explain when and why you would use roles.
12. List all system privileges found in either Oracle or SQL Server.
13. Using Oracle, produce a report that shows which user if any is assigned any of the following roles: DBA and SELECT_CATALOG_ROLE.
14. Provide steps to list all fixed roles that come with SQL Server.
15. Demonstrate how to restrict access to a column in a table using SQL Server.
16. Demonstrate in detail the use of the DENY statement in SQL Server.
17. List all object privileges granted to Oracle user OE (Order Entry). OE is a user created by default installation of Oracle10*g*.
18. Provide a scenario describing when you would provide the SELECT ANY TABLE privilege.

Hands-on Projects

Hands-on Project 4-1

Suppose your company's policy on password complexity requires that the password be 10 characters long, must contain digits, and does not contain more than three characters of your user name. Create an Oracle verify function to enforce this password complexity.

Hands-on Project 4-2

John, your network administrator, needs to be able to assign new users to roles on the SQL server. Create an account that allows this security function and only this function.

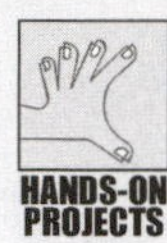

Hands-on Project 4-3

Create the three user-defined roles that are shown in the following table and assign the specified permissions for the CUSTOMERS table:

Role	Select	Insert	Update	Delete
account_managers	√		√	√
customer_service	√		√	
sales_reps	√	√	√	

Hands-on Project 4-4

Create and document a new login and user in the Northwind database. Give this user SELECT permissions to the EMPLOYEES table.

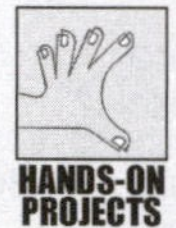

Hands-on Project 4-5

Create and document a password policy that requires at least 12 characters.

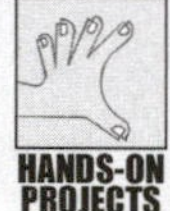

Hands-on Project 4-6

Create and document a password policy that locks the user's account if the password is entered incorrectly four times in a row.

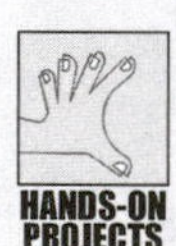

Hands-on Project 4-7

Create and document a password policy that expires the password after 60 days.

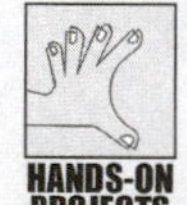

Hands-on Project 4-8

Part of the auditing process your company is conducting requires that you list all users who have been granted system or object privileges ADMIN option or the GRANT option in Oracle.

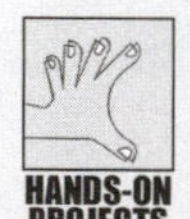

Hands-on Project 4-9

Using Oracle, list all system privileges assigned to user HR.

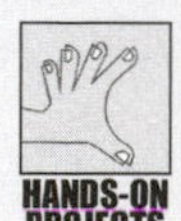

Hands-on Project 4-10

Outline all statements to demonstrate the concept presented in Figure 4-14.

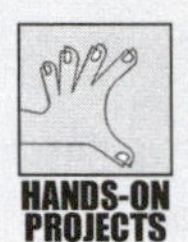

Hands-on Project 4-11

Outline all statements to demonstrate the concept presented in Figure 4-16.

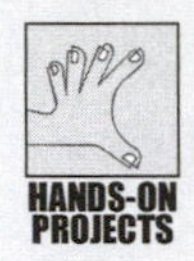

Hands-on Project 4-12

You were hired as an Oracle DBA to work with a team of developers. On your first day, the project lead expressed his concerns about the security aspect of the application. Outline your recommendations for implementing a security policy.

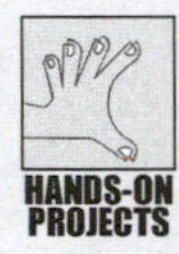

Hands-on Project 4-13

Create and document a SQL Server login that allows you to perform the same functions as a system administrator.

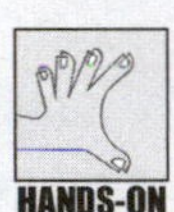

Hands-on Project 4-14

Create and document an Oracle user such as an HR user.

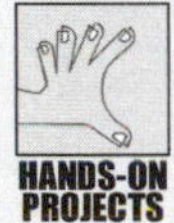

Hands-on Project 4-15

Reverse engineer the resource role in Oracle; that is, document the SQL statements that create the resource role.

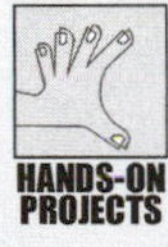

Hands-on Project 4-16

You were hired as an Oracle DBA by a small company for the primary task of creating a security model, which consists of three levels:

- READ level
- UPDATE and INSERT level
- DELETE level

The model will be using tables owned by HR schema. Implement this model.

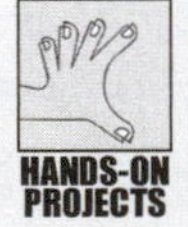

Hands-on Project 4-17

Using Oracle, develop a list of roles or users who have access to the EMPLOYEES tables owned by the HR user.

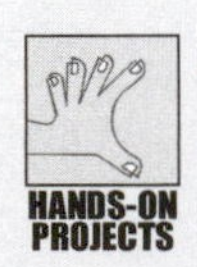

Hands-on Project 4-18

Provide a model in which a role is assigned to another role.

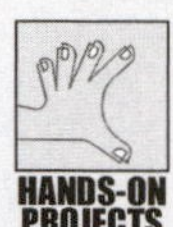

Hands-on Project 4-19

Develop a list that has been requested by your manager. List users and roles that have the DELETE privilege on all tables owned by users in your database except SYS and SYSTEM users.

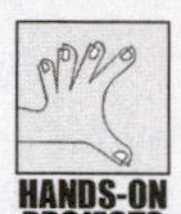

Hands-on Project 4-20

Write an Oracle report listing each user in the database and the roles assigned to the user.

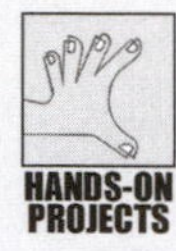

Hands-on Project 4-21

Create an Oracle profile to limit database resources using the following criteria:

- Login time—2 hours
- Idle time—10 minutes
- CPU time required per call—1 second
- 1 block per query

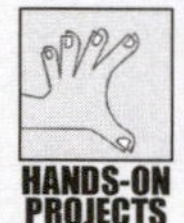

Hands-on Project 4-22

Write the CREATE statement and outline all the steps to test and validate the profile created in Hands-on Project 4-21.

Case Projects

Case 4-1 Oracle Security

You are hired as a database administrator for a small reseller company to implement and administer the database component of their main application. The data model of this application is presented in Figure 4-20. The schema owner is called DBSEC.

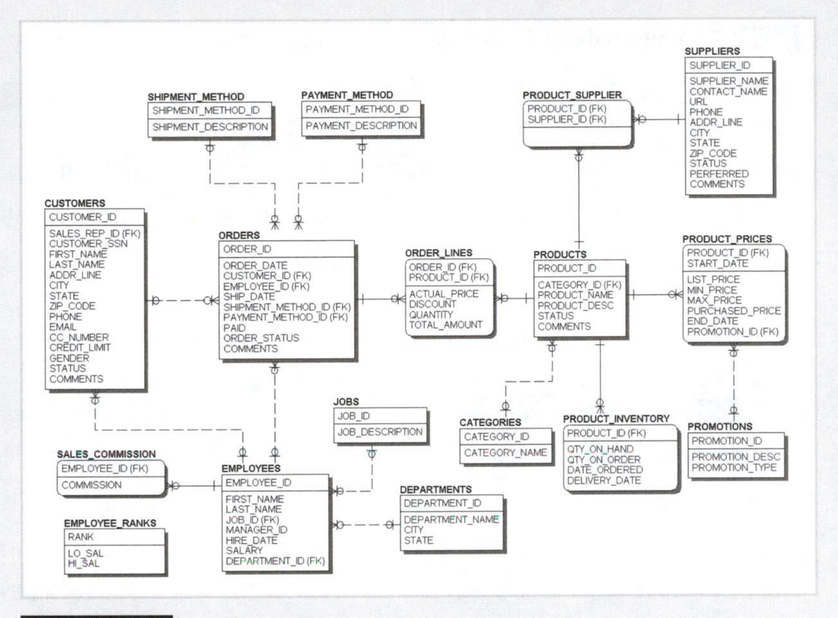

FIGURE 4-20 DBSEC data model

a. Use the scripts provided to create a schema.

b. Create three different profiles based on the criteria listed in Table 4-7.

Table 4-7 Profile configuration requirements

Profile	Resources	Password
DBSEC_ADMIN_PROF	SESSIONS_PER_USER = 5 CONNECT_TIME = 8 hours IDLE_TIME = 1 hour	PASSWORD_LIFE_TIME = 1 month PASSWORD_GRACE_TIME = 7 days
DBSEC_DEVELOPER_ PROF	CONNECT_TIME = 12 hours IDLE_TIME = 2 hours CPU_PER_CALL = 1 minute	PASSWORD_LIFE_TIME = 1 month PASSWORD_GRACE_TIME = 14
DBSEC_CLERK_PROF	SESSIONS_PER_USER =1 CPU_PER_CALL = 5 seconds CONNECT_TIME = 8 hours IDLE_TIME = 30 minutes LOGICAL_READS_PER_CALL = 10 KB	FAILED_LOGIN_ATTEMPTS = 1 PASSWORD_LIFE_TIME = 1 month PASSWORD_LOCK_TIME = 3 days PASSWORD_GRACE_TIME = 14 days

c. Create four different roles that have privileges according to the specification outlined in Table 4-8.

Table 4-8 Roles and privileges configuration requirements

Role Name	Privileges
DBSEC_ADMIN_ROLE	SELECT and ALTER on all DBSEC tables
DBSEC_CLERK_ROLE	SELECT, INSERT, and UPDATE on all DBSEC tables
DBSEC_SUPERVISOR_ROLE	SELECT, INSERT, UPDATE, and DELETE on all DBSEC tables
DBSEC_QUERY_ROLE	SELECT on all DBSEC tables

d. After creating these profiles and roles, create users according to the specifications in Table 4-9.

Table 4-9 User accounts configuration

User Name	Role	Profile
DBSEC_ADMIN	DBSEC_ADMIN_ROLE	DBSEC_ADMIN_PROF
DBSEC_CLERK	DBSEC_CLERK_ROLE	DBSEC_CLERK_PROF
DBSEC_SUPER	DBSEC_SUPERVISOR_ROLE	DBSEC_CLERK_PROF
DBSEC_QUERY1	DBSEC_QUERY_ROLE	DBSEC_CLERK_PROF
DBSEC_QUERY2	DBSEC_QUERY_ROLE	DBSEC_CLERK_PROF
DBSEC_DEVELOPER	DBSEC_ADMIN_ROLE +	DBSEC_DEVELOPER_PROF
	DBSEC_SUPERVISOR_ROLE	

e. Verify your implementation by viewing the data dictionary.

Case 4-2 SQL Server Security

You are the DBA for Acme Corporation. You're asked to implement a new database server using Microsoft SQL Server 2000. In any implementation, security needs to be a theme woven throughout the project, not an afterthought. Design an implementation that allows you to implement policies. The design should also include a role-based security structure.

The business requirements for the database are as follows:

- No user should have delete access to any object.
- Human Resources needs to be able to fully manage employee data.
- Customer Service needs to be able to make changes to existing customers and view order data.
- Customer account managers need to be able to fully manage customer data and orders.
- Sales needs to be able to view customer data and fully manage orders.
- Marketing needs to able to pull quarterly sales numbers to analyze.

Follow these steps to produce a solution:

1. Install SQL Server using the Windows Integrated security mode.
2. Use a Window user for the service accounts.

3. Demote built-in\administrators to security and create a DBA windows group for system administrators.

4. Set up password policies based on best practices.
   a. Use a minimum of eight characters
   b. Use complex passwords
   c. Set account lock-out for after three tries
   d. Keep five passwords in history
   e. Expire passwords after 30 days

5. Set up the roles listed in Tables 4-10 through 4-14.

Table 4-10 Human_Resources

Table	Select	Insert	Update	Delete
Employees	√	√	√	X

Table 4-11 Customer_Service

Table	Select	Insert	Update	Delete
Customers	√		√	X
Orders	√			X
Order Details	√			X

Table 4-12 Account_Managers

Table	Select	Insert	Update	Delete
Customers	√	√	√	X
Orders	√	√	√	X
Order Details	√	√	√	X

Table 4-13 Sales

Table	Select	Insert	Update	Delete
Customers	√			X
Orders	√	√	√	X
Order Details	√	√	√	X

Table 4-14 Marketing

Table	Select	Insert	Update	Delete
Territory	√			X
Orders	√			X
Order Details	√			X

5

Database Application Security Models

Upon completion of this material, you should be able to:

- Describe the different types of users in a database environment and the distinct purpose of each
- Identify and explain the concepts of five security models
- List the most commonly used application types
- Implement the most common application security models
- Understand the use of data encryption within database applications

Introduction

Tony is a database developer hired by a consulting company to implement a database application for a client. The client was extremely concerned about security because of the sensitive and confidential nature of the data. Tony was pleased with this challenging assignment, threw himself into it, and worked long hours, implementing all the modules right up to the last minute before the application went into production. Soon after implementation, management noticed that some of the employees were privy to inappropriate information. Immediately, the director of development launched a full investigation, and a team was put into place to find out what happened. The investigative team quickly drafted a report to management, explaining what had occurred.

The report stated that Tony had implemented a security module that enabled an administrator to create a database user and to grant roles. The report cited that one application administrator created a user for a supervisor that allowed the clerk role to view data only for specific modules within the application. However, because the application user is also a database user, the supervisor was able to log on to the database and see many columns that he was not able to see through the application. That meant that the supervisor was able to access more than he was intended to see.

In previous chapters you learned the necessary background and best practices for database security and its elements. In addition you were shown how to create users and roles and to assign privileges to users. At this point, you can put all the pieces together and mold these concepts and practices into models that can serve as starting points for implementing different application security architectures or designs. This chapter presents an overview of different users and the related security model concepts. It also presents the most frequently used application types. This is followed by a presentation of different application security models. The chapter concludes with a demonstration of password encryption.

Types of Users

The scenario in the introduction of this chapter described a database user being used to log on (be authenticated) to an application. For each application user, a database account must be created and assigned specific privileges. You already know what a database user is and what it is used for. You also understand the concept and use of database roles. What, then, is an **application user**? Before you read the answer to this question, you should know the distinction between an application and a database. For review, here is a quick description of the two.

Application—A program that solves a problem or performs a specific business function; examples of applications include MS Word, Adobe Acrobat Reader, and an inventory program

Database—A collection of related data files used by an application

Database management system—A collection of programs that maintains data files (database)

Now, return to the question posed earlier about an application user. An application user is simply a record created for a user within the application schema to be used for authentication to the application. An application user usually does not have database privileges or roles assigned to the user. All privileges and roles are granted by the application and are specific to the application. The following is a list of the users that are discussed in this chapter.

- **Application administrator**—An application user that has application privileges to administer application users and their roles; application administrators do not require any special database privileges
- **Application owner**—A database user (schema owner) who owns application tables and objects
- **Application user**—The user recorded in the application schema to authenticate the user and to enable the user to perform tasks within the application
- **Database administrator**—A user account that has database administration privileges that enable the user to perform any administration task
- **Database user**—A type of user account that has database roles and/or privileges assigned to it
- **Proxy user**—A database user that has specific roles and privileges assigned to it; the proxy user is employed to work on behalf of an application user; this type is usually useful for adding and isolating application users from the database
- **Schema owner**—A database user that owns database objects
- **Virtual user**—An account that has access to the database through another database user account; a virtual user is referred to in some cases as a proxy user.

In Chapter 4, you were introduced to application roles in Microsoft SQL Server 2000. In this chapter, you learn to use application roles in combination with proxy users to implement client server security models.

Security Models

This section outlines two frequently used security models and should help you understand how application security models are implemented.

Access Matrix Model

This model uses a matrix[1] to represent two main entities that can be used for any security implementation. Figure 5-1 illustrates the access matrix model. As you can see, the columns are represented by objects and the rows are subjects. An object can be a table, view, procedure, or any other database object A subject can be a user, role, privilege, or a module. The intersection of a row and column is an authorization cell, representing the access details on the object granted to the subject. The authorization cell can be access, operation, or commands. The example presented in Figure 5-1 shows that User1 has read/write authorization to Table1, write authorization to Table2, and admin authorization to Table3.

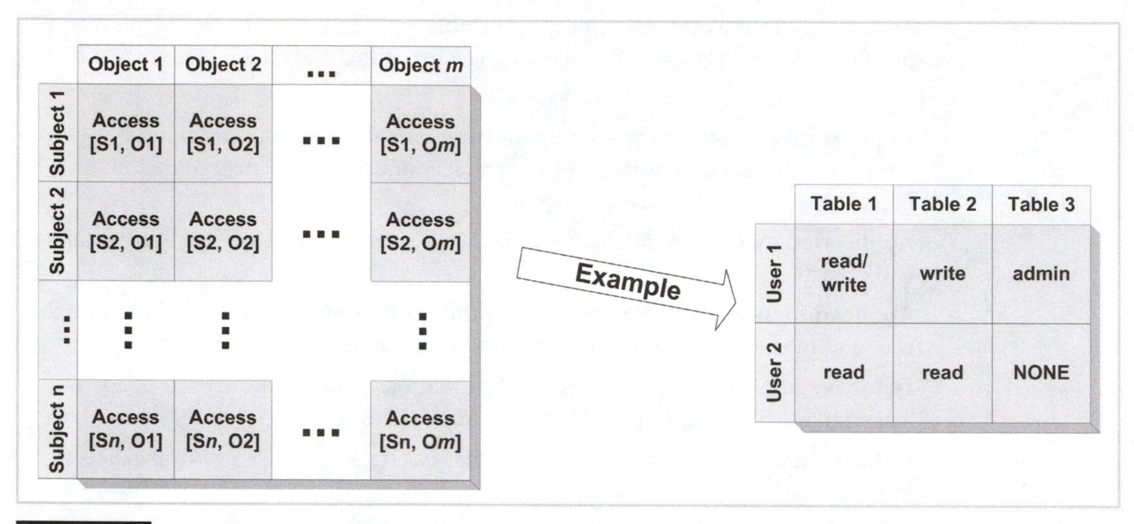

FIGURE 5-1 Access matrix security model

Access Modes Model

This model is based on the Take-Grant model.[2] The model uses both the subject and object entities as the main security entities, and it uses access modes to indicate the tasks that the subject is allowed to perform on the objects. The access modes are divided into static and dynamic modes as listed in Table 5-1. The level shown in the table is a numeric value to indicate the degree of access. A subject with an access level of 4, for example, has access to level 4 and all the levels below it. In this case, access level 4 is the access level of a superuser that has access to all application functionality and privileges. On the other hand, a subject with an access level of 1 has access only to its level and does not have access to any level above it.

Table 5-1 Access modes

Static Modes

Access Mode	Level	Description
use	1	Allows the subject to use the object without modifying the object
read	2	Allows the subject to read the contents of the object
update	3	Allows the subject to modify the contents of the object
create	4	Allows the subject to add instances to the object
delete	4	Allows the subject to remove instances of the object

Table 5-1 Access modes (continued)

Dynamic Modes

Access Mode	Level	Description
grant	1	Allows the subject to grant any static access mode to any other subject
revoke	1	Allows the subject to revoke a granted static access mode from a subject
delegate	2	Allows the subject to grant the grant privilege to other subjects
abrogate	2	Allows the subject to grant the revoke privilege to other subjects

Application Types

This section describes the commonly used application types to help you understand where data security should be enforced. Notice that this theme is repeated over and over: data security enforcement is required where data resides in the database.

Client/Server Applications

Thirty years ago, computing in corporations was centralized in the Management Information System (MIS) Department. The MIS Department was responsible for all information system projects no matter how large or small. With this type of organization, MIS became a bottleneck that held up other departments while it developed and implemented projects. MIS departments developed mainly mainframe (legacy) projects. Figure 5-2 shows the code and the data of a typical mainframe project. Note that the code and data reside in one logical and physical place.

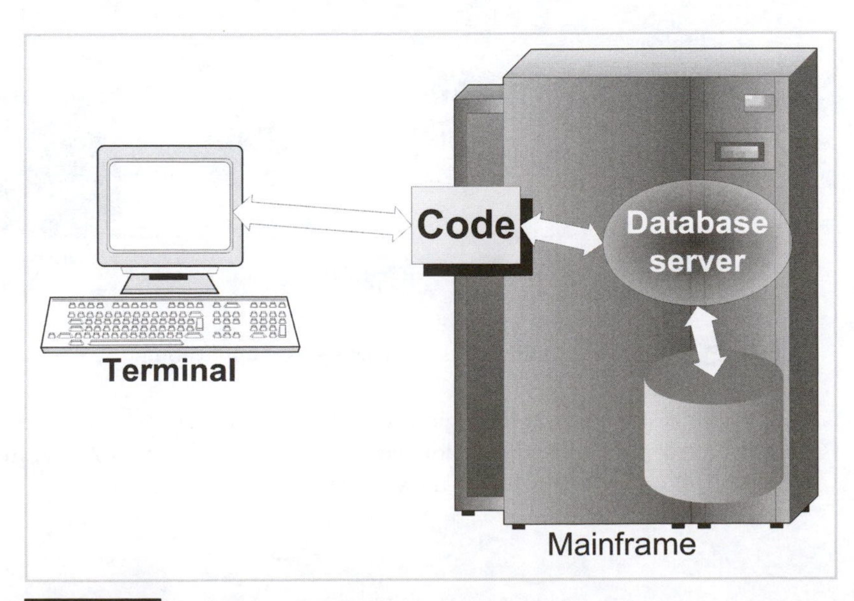

FIGURE 5-2 Mainframe application architecture

When the personal computer was introduced, many departments sought relief from the lengthy and often bureaucratic process of working with the MIS Department. Departments quickly developed their own customized applications, which were designed for a single user and were not scalable. As the number of users and the data grew, so did the performance problems. A better architecture had to be developed that could take advantage of the flexibility of the PC, overcome the bottlenecks of the MIS Department, and overcome the inability of the PC environment to grow with increasing data needs. The answer was client/server architecture, which is based on a business model in which the client submits inquiries and the server responds with answers to these inquiries.

Client/server architecture became a dominating configuration for all applications because it provided a flexible and scalable structure that could take advantage of the processing power of the personal computer, while also utilizing the capacity and power of a dedicated server. Figure 5-3 illustrates the client/server architecture, which is composed of the three main components typically found in a client/server application.[3] These components are usually spread out over several tiers. Each tier has a logical or physical component, which can contain one or more of the following components:

- **User interface component**—Represents all screens, reports, and codes that handle the interaction between the user and the application
- **Business logic component**—Contains all the code that performs data validation and business rules implementation
- **Data access component**—Contains code that retrieves, inserts, deletes, and updates data

A client/server application consists of a minimum of two tiers. Normally four to five tiers is the maximum configuration.

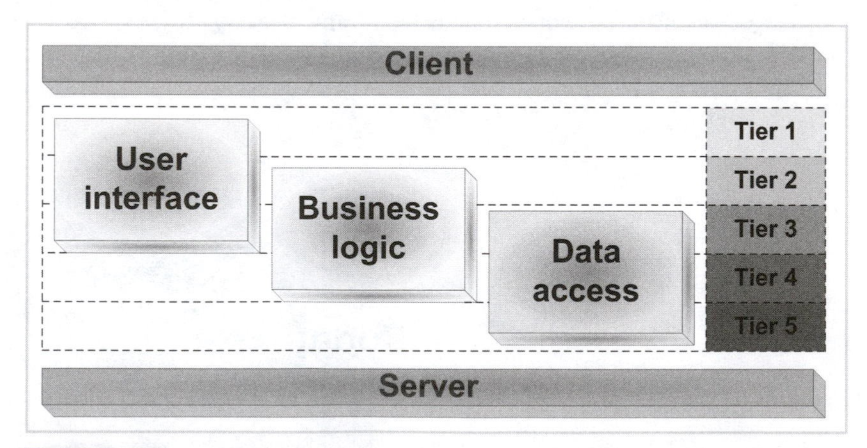

FIGURE 5-3 Logical components of a client/server application

Figure 5-4 shows that the logical components of the client/server architecture can reside on any physical configuration. For example, you can have the user interface and the application server located on the same machine.

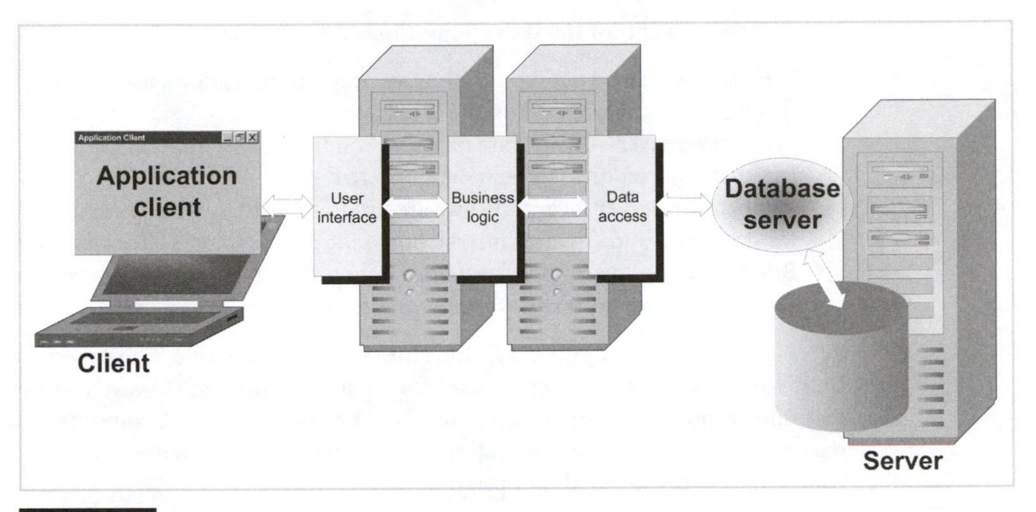

FIGURE 5-4 Physical architecture of a client/server application

The data access component of the client/server architecture is the component responsible for retrieving and manipulating data. The security module should be embedded in this component.

Web Applications

Client/server architecture once dominated business applications, but not for long. Another application architecture evolved with the rise of dot-com and Web-based companies. This new client/server architecture is based on the Web and is therefore referred to as a Web application or a Web-based application. A Web application uses the Web (HTTP protocol) to connect and communicate to the server. As shown in Figure 5-5, a Web browser is the front end of the Web application. A Web application uses HTML pages created using ActiveX, Java applets or beans, or ASP (Active Server Pages). These Web pages are embedded with other Web services.

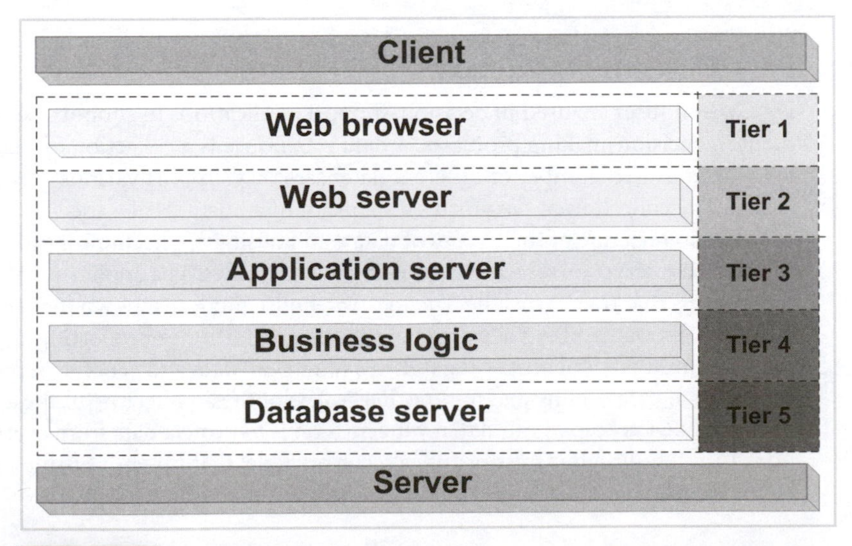

FIGURE 5-5 Logical components of Web application architecture

The components of the Web application shown in Figure 5-5 are:

- **Web browser layer**—A typical browser program that allows users to navigate through Web pages found on the Internet
- **Web server layer**—A software program residing on a computer connected to the Internet that responds to requests submitted by the Web browsers
- **Application server layer**—A software program residing on a computer that is used for data processing and for interfacing to the business logic and database server
- **Business logic layer**—A software program that implements business rules
- **Database server layer**—A software program that stores and manages data

Figure 5-6 shows a physical architecture that is typical for a Web-based application. In this architecture, each layer resides on a separate computer. However, you should note that one or more Web-application layers could be housed on one computer. The main reason for separating Web-application layers to reside on different computers is to distribute the processing load for optimum performance.

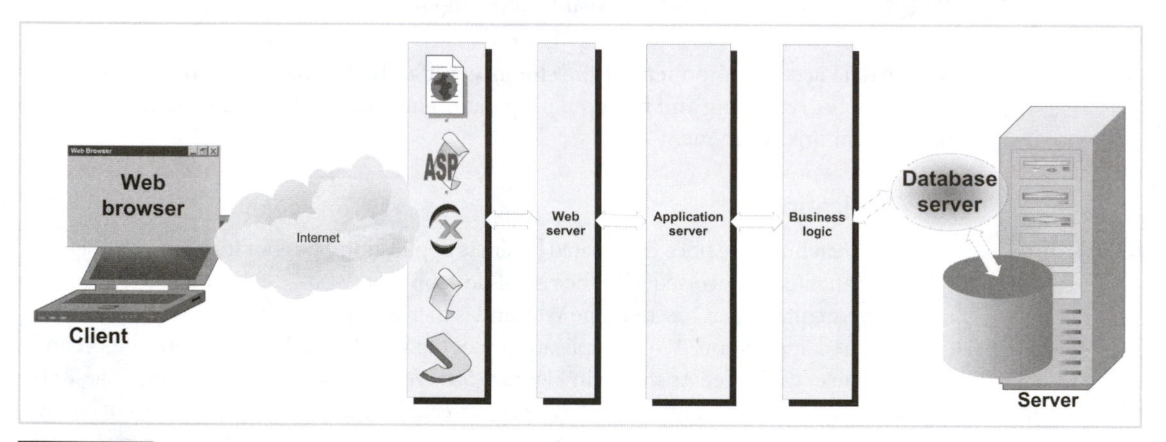

FIGURE 5-6 Physical structure of a Web application

Data Warehouse Applications

Data warehouses are used in decision-support applications to support executive management in decision-making processes. A data warehouse is a collection of many types of data taken from a number of different databases that support various corporate departments. The collection of data forms a snapshot of a business at a specific moment.

The architecture of these types of data warehousing applications is typically composed of a database server on which the application data resides. The application data is extracted by a process that transforms the data to a warehouse model as shown in Figure 5-7. In addition, the data warehouse is accessed by software applications or reporting applications called online analytical processing (OLAP) to retrieve data and generate reports with the capability of data drilling and mining. Regardless of the architecture of the application you are using, always keep in mind that your mission is to protect data from every type of violation. To accomplish this mission, use a security model that fits your business needs. A number of security models are presented in the sections that follow.

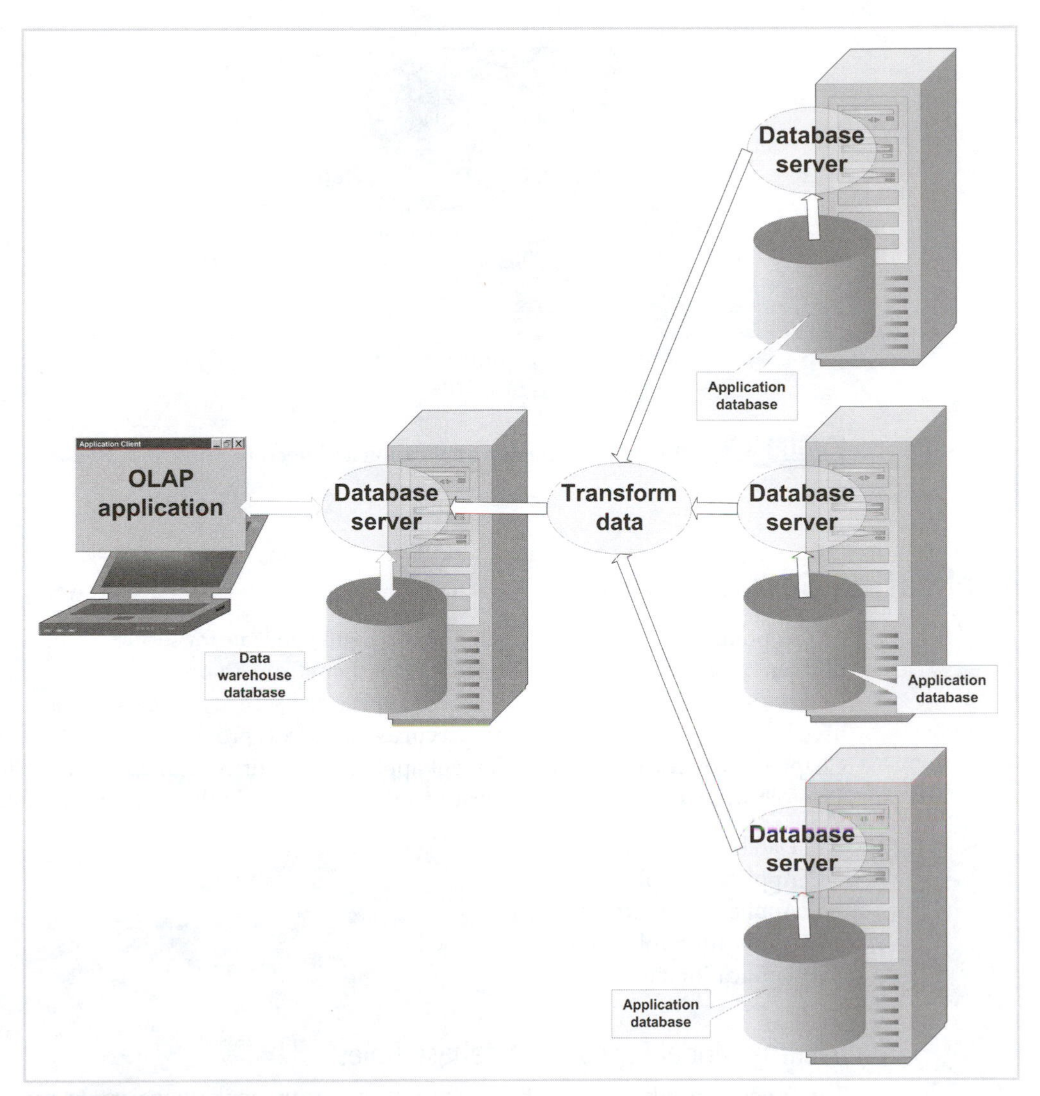

FIGURE 5-7 Physical and logical architecture of a data warehouse

Other Applications

Although this book focuses on database security applications, it is worthwhile to note that there are other types of applications that are not database oriented. These applications may still require a security layer to protect them against inappropriate access and execution of processes. In this case, the security layer should be embedded within the application. In addition, passwords should be protected by using an encryption mechanism similar to those typically used by database systems, regardless of whether the password is stored in a configuration file or operating system registry file. Figure 5-8 illustrates this type of application.

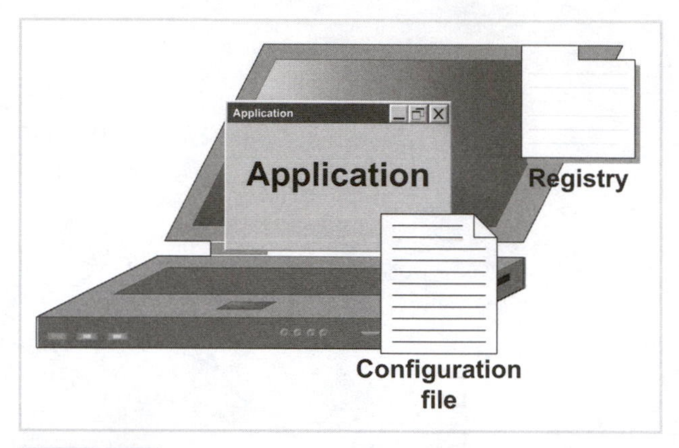

FIGURE 5-8 Typical nondatabase application architecture

Application Security Models

At this point, it is useful to combine the two main concepts for security models and their different types of applications, and to look at them from an implementation perspective. In this section you examine five different application security models that are commonly used by the industry to provide data security and access protection at the table level. In Chapter 6, you learn how to implement application security on the row and column level. The following list outlines the security models presented in the sections that follow:

- Database role based
- Application role based
- Application function based
- Application role and function based
- Application table based

Security Model Based on Database Roles

This model depends on the application to authenticate the application users by maintaining all end users in a table with their encrypted passwords. In this model, each end user is assigned a database role, which has specific database privileges for accessing application tables. The user can access whatever privileges are assigned to the role. In this model, a proxy user is needed to activate assigned roles. All roles are assigned to the proxy user. For example, if the user Scott is assigned to the clerk role, this means that when Scott logs on through the proxy user, the application enables only the clerk role for this session. Figure 5-9 shows the data model for this type of application. Table 5-2 describes the tables for this model. Figure 5-10 shows the architectural view of this model. You might have noticed that the two tables have common control columns prefixed with CTL (short for control). These control columns contain information about the manipulated record. The following list presents a brief description of these columns:

- CTL_INS_DTTM (control insert date time)—Contains the date and time when the record was created
- CTL_UPD_DTTM (control update date time)—Contains the date and time when the record was last updated

- CTL_UPD_USER (control update user)—Contains the user name that created the record or last updated the record
- CTL_REC_STAT (control record status)—Can be used to indicate the status of the record. You can use this column for any purpose; for example, you may set this column to "A" as an indicator that the record is active or "I" to indicate it is inactive

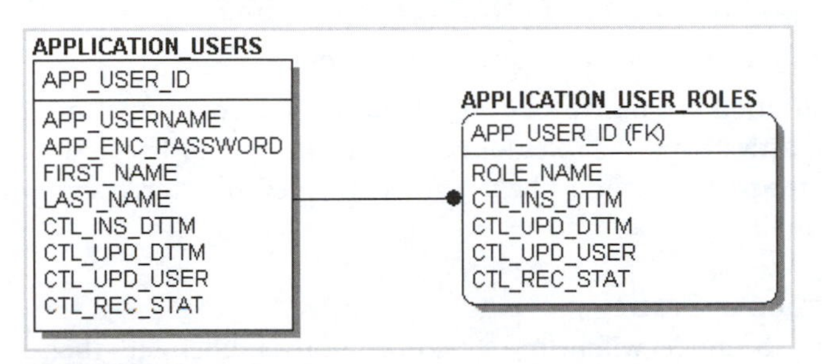

FIGURE 5-9 Security data model based on database roles

Table 5-2 Tables used in a security data model based on database roles

Table Name	Description
APPLICATION_USERS	Stores and maintains all end users of the application with their encrypted passwords
APPLICATION_USER_ROLES	Contains all roles defined by the application and for each role that a privilege is assigned; the privilege can be read, write, or read/write

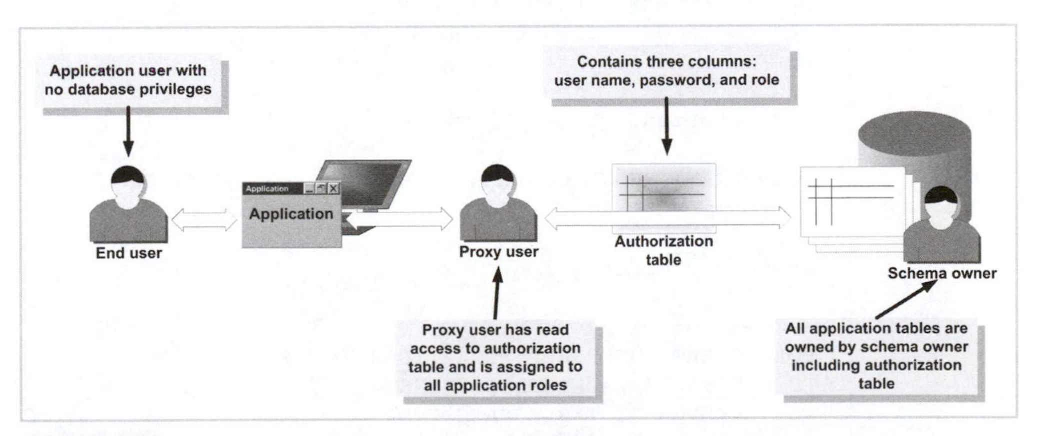

FIGURE 5-10 Architecture of a security data model based on database roles

The following points on this type of security model are worth noting:

- This model uses the database role functionality; therefore it is database dependent. If roles are poorly implemented, the model does not work properly.
- Privileges to tables are also database dependent.

- In this model, you isolate the application security from the database, which makes your implementation database independent.
- Maintenance of the application security does not require specific database privileges. This lowers the risk of database violations.
- Passwords must be securely encrypted, preferably using private and public keys. In this case, you must modify the structure of the APPLICATION_USERS table by adding columns to store public and private keys.
- The application must use proxy users to log on and connect to the application database and activate specific roles for each database session. The user name and password of the proxy user must be securely encrypted and stored somewhere in a configuration file.

Implementation in Oracle

Implementing the model will help you become more familiar with the concept. Be sure you fully understand the concept before you begin implementation. You are trying to create a proxy user called APP_PROXY that will be assigned all application roles and will work on behalf of the application user APP_USER to gain access to all tables owned by the application owner called APP_OWNER. So, you need to create the users, roles, and tables for this application. To implement the model, follow these steps:

1. Create the users by entering the following code:

```
-- Creating Application Owner
SQL> CREATE USER APP_OWNER IDENTIFIED BY APP_OWNER
 2  DEFAULT TABLESPACE USERS
 3  TEMPORARY TABLESPACE TEMP
 4  QUOTA UNLIMITED ON USERS
 5 /

User created.

SQL> GRANT RESOURCE, CREATE SESSION TO APP_OWNER
 2 /

Grant succeeded.

--Creating Proxy user
SQL> CREATE USER APP_PROXY IDENTIFIED BY APP_PROXY
 2   DEFAULT TABLESPACE USERS
 3   TEMPORARY TABLESPACE TEMP
 4 /

User created.

SQL> GRANT CREATE SESSION TO APP_PROXY
 2 /

Grant succeeded.
```

```
--Creating application tables
SQL> CONN APP_OWNER@SEC
Enter password: *********
Connected.
SQL> CREATE TABLE CUSTOMERS
 2 (
 3   CUSTOMER_ID  NUMBER PRIMARY KEY,
 4   CUSTOMER_NAME  VARCHAR2(50)
 5 )
 6 /

Table created.

SQL> CREATE TABLE AUTH_TABLE
 2 (
 3   APP_USER_ID  NUMBER,
 4   APP_USERNAME  VARCHAR2(20),
 5   APP_PASSWORD  VARCHAR2(20),
 6   APP_ROLE      VARCHAR2(30)
 7 )
 8 /

Table created.

--Creating application roles
SQL> CONN SYSTEM@SEC
Enter password: ******
Connected.
SQL> CREATE ROLE APP_MGR;

Role created.

SQL> CREATE ROLE APP_SUP;

Role created.

SQL> CREATE ROLE APP_CLERK;

Role created.

SQL> GRANT APP_MGR, APP_SUP, APP_CLERK TO APP_PROXY
 2 /

Grant succeeded.

SQL> ALTER USER "APP_PROXY" DEFAULT ROLE  NONE
 2 /

User altered.
```

```
-- Assign grants
SQL> CONN APP_OWNER@SEC
Enter password: *********
Connected.
SQL> GRANT SELECT, INSERT, UPDATE, DELETE ON CUSTOMERS TO APP_MGR
/

Grant succeeded.

SQL> GRANT SELECT, INSERT, UPDATE ON CUSTOMERS TO APP_SUP
/

Grant succeeded.

SQL> GRANT SELECT ON CUSTOMERS TO APP_CLERK
 2 /

Grant succeeded.

SQL> GRANT SELECT ON AUTH_TABLE TO APP_PROXY
 2 /

Grant succeeded.
```

2. Add rows to the CUSTOMERS table:

```
SQL> CONN APP_OWNER@SEC
Enter password: *********
Connected.
SQL> INSERT INTO CUSTOMERS VALUES(1, 'Tom Verenda');

1 row created.

SQL> INSERT INTO CUSTOMERS VALUES(2, 'Linda Bella');

1 row created.

SQL> COMMIT
  2  /

Commit complete.
```

3. Add a row for an application user called APP_USER:

```
SQL> INSERT INTO AUTH_TABLE VALUES
  2      (100, 'APP_USER', 'd323deqw4df55fwe', 'APP_CLERK')
  3  /
1 row created.
```

4. Now assume that APP_USER is trying to log in through the PROXY_USER. Your application should look up the role of the user by using the SELECT statement and activating that role:

```
SQL> SELECT APP_ROLE
  2    FROM AUTH_TABLE
  3   WHERE APP_USERNAME = 'APP_USER'
  4  /
APP_ROLE
-------------------
APP_CLERK
```

5. Activate the role for this specific APP_USER session:

```
SQL> CONN APP_PROXY@SEC
Enter password: *********
Connected.
SQL> SET ROLE APP_CLERK
  2  /
Role set.
SQL> SELECT * FROM SESSION_ROLES
  2  /
ROLE
------------------
APP_CLERK
```

You just witnessed implementation of the role-based security model of a database using Oracle10*g*. This section continues with an implementation of the same model in SQL Server.

Implementation in SQL Server

At this point, SQL Server application roles come into play. In Oracle, you added the proxy user to all database roles required by the application and disabled them. At the time of authorization, you enabled the appropriate role based on a role assigned in your authorization table.

In SQL Server 2000, you use application roles. **Application roles** are special roles you create in the database that are then activated at the time of authorization. Application roles require a password and cannot contain members. However, application roles can be members of a standard database, members of fixed database roles, or both.

Application roles are inactive by default. If they are turned off, you can activate them using the SP_SETAPPROLE system-stored procedure. When an application role is activated, any permissions set for the database user that are used to make the connection are ignored and the connection operates under the security context of the activated role. No special permission is required; any user can execute SP_SETAPPROLE to activate an application role.

Creating Application Roles Using the Command Line

To create an application role in the Query Analyzer, use the SP_ADDAPPROLE system-stored procedure. You must be a member of the sysadmin fixed server role or a member of either the db_owner or db_securityadmin fixed database roles to execute SP_ADDAPPROLE.

```
sp_addapprole [ @rolename = ] 'role'
, [ @password = ] 'password'
Where:
@rolename--
The name of the application role; the value must be a valid id
entifier and cannot already exist in the database.
@password--
The password required to activate the role the SQL server stor
es the password as an encrypted hash.
```

To create the application role of clerk for your pharmacy database (see the "Pharmacy Application" section at the end of the chapter), use this command:

```
exec sp_addapprole 'clerk', 'Clerk@ccess'
```

Creating Application Roles Using SQL Server Enterprise Manager

To create an application role named super using Enterprise Manager, follow these steps:

1. Open **Enterprise Manager**.
2. Expand the Roles container for your PHARMACY database. Right-click in the right pane, then select **New Database Role**.
3. Type the name **super** in the name box.
4. Select **Application role** under Database role type.
5. Enter the password **Super@ccess** into the text box shown in Figure 5-11. Click **OK** to create the role.

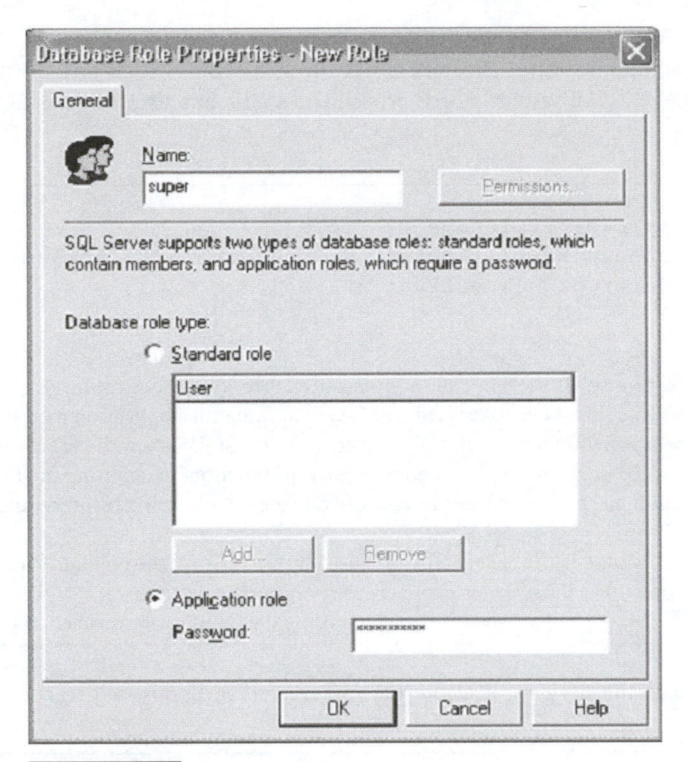

FIGURE 5-11 Database Role Properties using SQL Server Enterprise Manager

Dropping Application Roles Using the Command Line

To drop an application role using the Query Analyzer, you use the SP_DROPAPPROLE system-stored procedure. You must be a member of the sysadmin fixed server role, or the db_securityadmin or db_owner fixed database roles, to execute the stored procedure.

```
sp_dropapprole [@rolename =] 'role'
```

```
@rolename = The application role to drop.
```

Dropping Application Roles Using Enterprise Manager

To drop an application role using Enterprise Manager, you follow the same steps that are used for dropping standard roles.

1. Open **Enterprise Manager.**
2. Expand the **Roles** container of the database from which you are dropping the role.
3. Select and delete the desired role.

Activating Application Roles

To activate an application role, after connecting as a database user you need to execute the SP_SETAPPROLE system-stored procedure. Again, any user can execute this stored procedure:

```
sp_setapprole [@rolename =] 'role'
     ,[@password =] {Encrypt N 'password'} | 'password'
    [,[@encrypt =] 'encrypt_style']
```

Where:

- `@rolename`—Is the name of the application role you are activating
- `@password`—Is the password required to activate the application role; if you are using an ODBC client or the OLEDB provider for SQL Server, the value can be encrypted using the ODBC canonical Encrypt function; when using the Encrypt function, the password must be converted to a Unicode string by preceding the password with N
- `@encrypt`—Is the encryption style used by *@password*; the possible values are:
  - None—The password is not encrypted (this is the default).
  - ODBC—The password is encrypted using the ODBC canonical encrypt function.

To activate the clerk application role you created earlier, issue this command:

```
exec sp_setapprole 'clerk', 'Clerk@ccess'
```

It is best practice to encrypt the password before sending it across the network to the SQL server. To activate the clerk role and encrypt the password, issue this command:

```
exec sp_setapprole 'clerk', {Encrypt N 'Clerk@ccess'}, 'ODBC'
```

Implementation

Again, as with the Oracle implementation, you utilize an authorization table to store the user names, passwords, and assigned application role names for your application. The difference is that your database user is explicitly denied access to all database objects except the authorization table. The user has select permissions to this table. Your application roles are then given the appropriate permissions, either directly or through database role membership.

To accomplish this implementation, you use the following pseudocode:

1. Connect to database as the proxy user (**PHARM_PROXY**).
2. Validate the user name and password entered by the end user against the authorization table and retrieve the application role name.
3. Activate the application role.

Now, you can simulate the pseudocode above using Query Analyzer and see the security context change. Set up the proxy user and test this out.

Here are the steps to set up a proxy user:

1. Create a SQL Server login called **PHARM_PROXY**.
2. Set the default database to **PHARMACY**.
3. Create the database user PHARM_PROXY for the login, and deny it all permissions on the table CUSTOMER.
4. Give the application role of clerk select permissions to the table CUSTOMER.

To test the proxy user setup, follow these steps:

1. Using Query Analyzer, connect to the pharmacy database as PHARM_PROXY. Execute the following command:

```
SELECT * FROM Customer
```

A permission denied error should be displayed as shown in Figure 5-12.

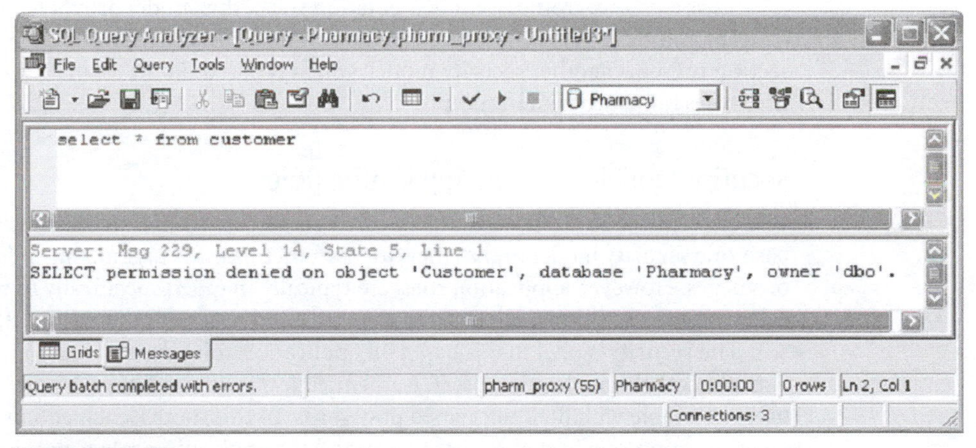

FIGURE 5-12 Permission denied error caused by lack of privileges

2. Now, activate the application role:

```
exec sp_setapprole 'clerk''Clerk@ccess'
```

3. Try executing the SELECT statement again. It now completes successfully as shown in Figure 5-13.

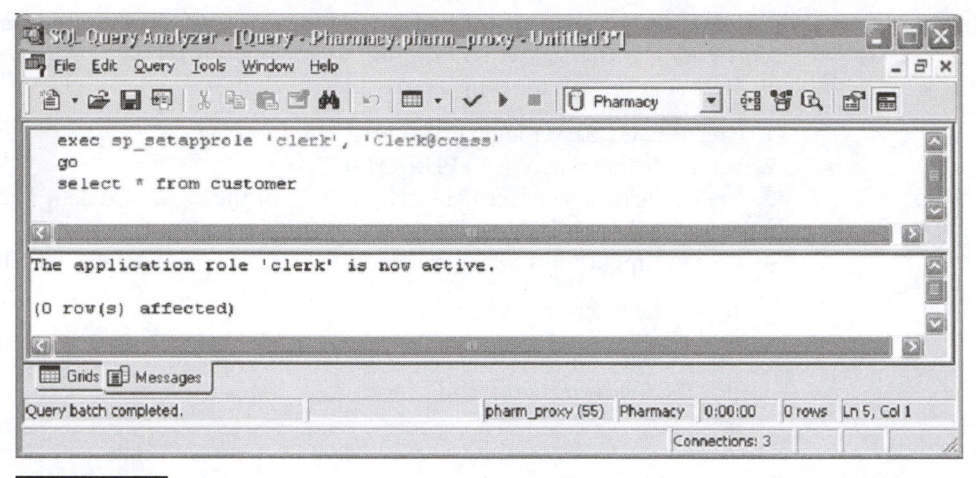

FIGURE 5-13 Application role activation

As you can see in Figure 5-13, once the application role is activated, you are now operating in the security context of the role, not as the user connected to the database.

As you can see from the implementation steps, this model proves to be simple in concept and development. In fact, it is too simple for many enterprise solutions. The next section presents another security model, which is based on the security model for database roles but extends the concepts.

Security Model Based on Application Roles

The concepts of an application role security model are similar to the concepts of a database role security model in that they are both methods for organizing and administering privileges. However, application roles are typically mapped specifically to real business roles (titles or positions).

The security model that is based on application roles depends on the application to authenticate the application users. Authentication is accomplished by maintaining all end users in a table with their encrypted passwords. In this model, each end user is assigned an application role (not a database role), and the application role is provided with application privileges to read/write specific modules of the application. Figure 5-14 shows the data model of this type of application. Table 5-3 contains description of tables used for this model. Figure 5-15 shows an architectural view of this model.

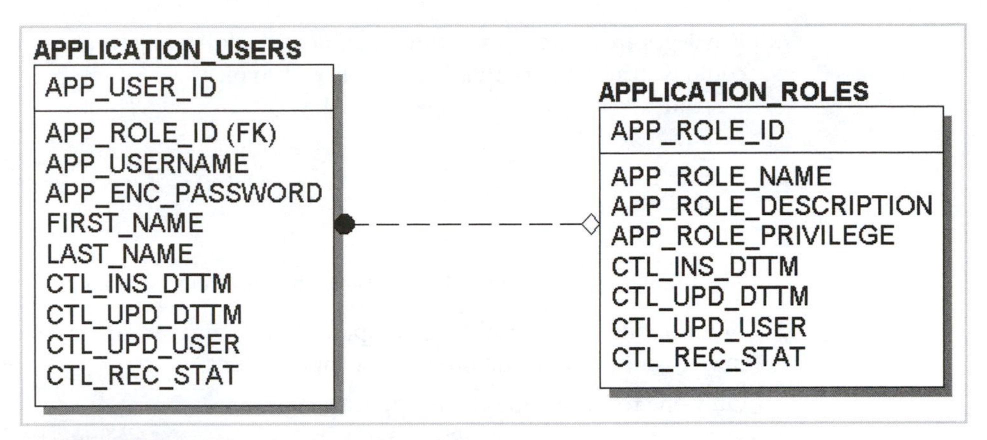

FIGURE 5-14 Security data model based on application roles

Table 5-3 Tables used in a security data model based on application roles

Table Name	Description
APPLICATION_USERS	This table is used to store and maintain all end users of the application with their encrypted passwords.
APPLICATION_ROLES	This table contains all roles defined by the application and for each role that a privilege is assigned. The privilege can be read, write, or read/write.

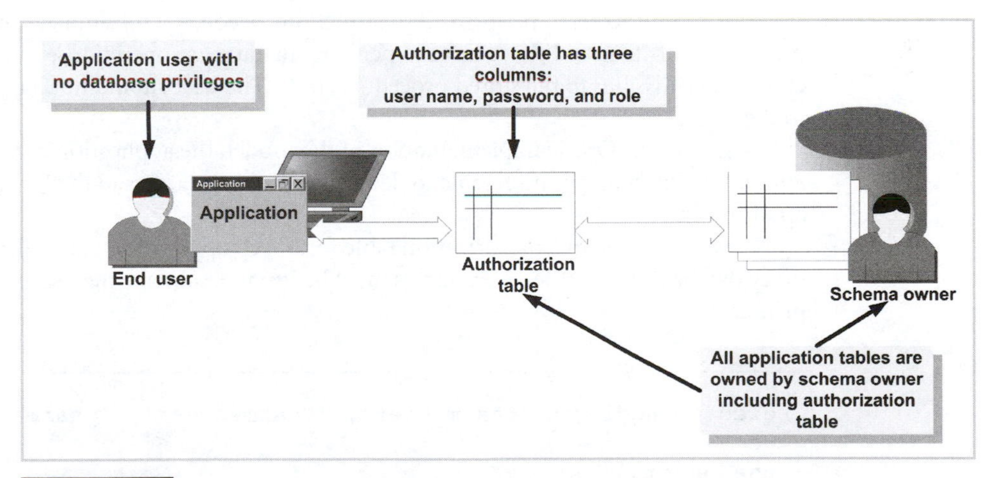

FIGURE 5-15 Architecture of a security data model based on application roles

When considering the security model, keep these points in mind:

- This model is primitive and does not allow the flexibility required to make changes necessary for security. For example, you may have a user called Scott who has a clerk role, and the clerk role has privileges to read, add, and modify. This means that Scott can perform these operations on all the modules of the application.

- Privileges are limited to any combination of the following (for example, privileges could be read, add, read/update/admin, and so on):
  - read
  - add
  - delete
  - update
  - admin

The following list presents characteristics of this security model:

- In this model you are isolating the application security from the database, which makes your implementation database independent.
- Only one role is assigned to an application user.
- Maintenance of the application security does not require specific database privileges. This lowers the risk of database violations.
- Passwords must be securely encrypted, preferably using private and public keys. In this case, you must modify the structure of the APPLICATION_USERS table by adding columns to store public and private keys.
- The application must use a real database user to log on and connect to the application database. The user name and password must be encrypted and stored in a configuration file.

Implementation in SQL Server

To implement this security model in SQL Server, you simply create a database user that the software developers will use to connect the application to the database. This user has the appropriate access to perform all the functions needed by the application. It is a best practice to create stored procedures to perform all database operations, and give this user execute permissions to the stored procedures only. The user should not have direct access to the tables.

Just as in the Oracle implementation of the model, the application makes use of an authentication/authorization table to determine end-user access to the front end of the application.

Create the user and authorization table for a Web application that utilizes your pharmacy database. Remember, user names must be unique, so make the user name column a primary key.

```
exec sp_addlogin 'pharm_user', 'Pharm@ccess', 'pharmacy'
go
use pharmacy
go
exec sp_grantdbaccess 'pharm_user'
go
create table auth_web (
     username   varchar(25) not null primary key,
     password   varchar(25)
)
go
```

```
grant all on auth_web to pharm_user
/* just use customer for demo purposes */
grant all on customer to pharm_user
go
```

A stored procedure used to authenticate the end user against the table would be:

```
Create procedure authenticate (@username varchar(25),
 @password varchar(25),
 output @success int
) as
select @success = count(*)
from auth_web
where username = @username and password = @password;
```

Security Model Based on Application Functions

The security model that is based on application functions depends on the application to authenticate the application users by maintaining all end users in a table with their encrypted passwords. In this model the application is divided into functions. For instance, if you were using an inventory application, you would have a function named CUSTOMERS that maintains customers and another function named PRODUCTS for maintaining products, and so on. Figure 5-16 represents a data model for this type of application showing the entities described in Table 5-4. Figure 5-17 shows the architectural diagram of this model.

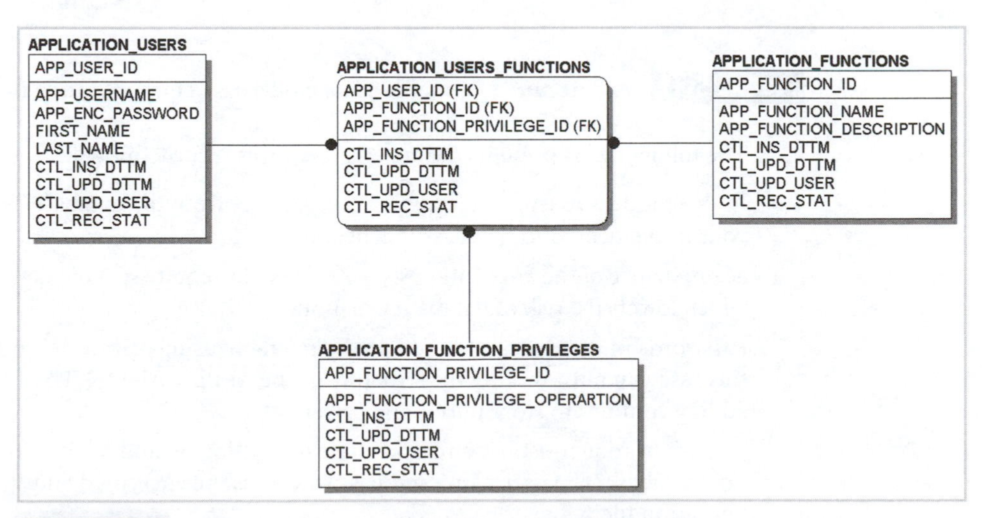

FIGURE 5-16 Security data model based on application functions

Table 5-4 Tables used in a security data model based on application functions

Table Name	Description
APPLICATION_USERS	This table is used to store and maintain all end users of the application with their encrypted passwords.
APPLICATION_FUNCTIONS	This table contains all logical functions of the application; a function is defined as a module or procedure or a set of tasks.
APPLICATION_FUNCTION_PRIVILEGES	This table stores all privileges for a function such as: read write read/write
APPLICATION_USERS_FUNCTIONS	This table contains all end users and their application function privileges.

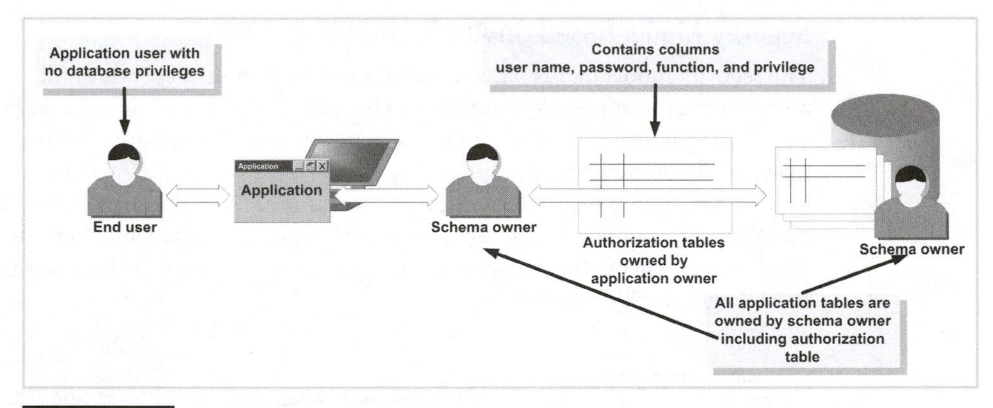

FIGURE 5-17 Architecture of a security data model based on application functions

The following list presents characteristics of this security model:

- In this model you isolate the application security from the database, which makes your implementation database independent.
- Maintenance of the application security does not require specific database privileges, which lowers the risk of database violations.
- Passwords must be securely encrypted, preferably using private and public keys. In this case you must modify the structure of the APPLICATION_USERS table by adding columns to store public and private keys.
- The application must use a real database user to log on and connect to the application database. The user name and password must be encrypted and stored in a configuration file.
- The application must be designed in a granular fashion. This means that the functions or modules of the application perform specific tasks and do not encompass a wide variety of tasks that are cross-functional. For example, if a function is ORDER_ENTRY, it should not include a task for reporting or adding products.

The more granular the privileges, the more effort is needed to implement them. For example, the WRITE privilege that allows the user to perform UPDATE, DELETE, and INSERT tasks on any table is easily implemented. In comparison, the WRITE privilege that allows the user to perform tasks on a specific column in a table is more complex to design and implement. But if you want to implement the update privilege only, your code and interface must work accordingly to prevent users from updating a value, but still allow the adding of new records. This requires a well-designed implementation.

Security Model Based on Application Roles and Functions

This security model is a combination of both the role and function security models. The application roles and functions security model depends on the application to authenticate the application users. The application authenticates users by maintaining all end users in a table with their encrypted passwords. In this model the application is divided into functions, and roles are assigned to functions that are in turn assigned to users. This model is highly flexible in implementing application security. Figure 5-18 represents a data model for this type of application showing the E-R (entity-relationship) diagram. See Table 5-5 for entity descriptions and Figure 5-19 for the architectural diagram of this model.

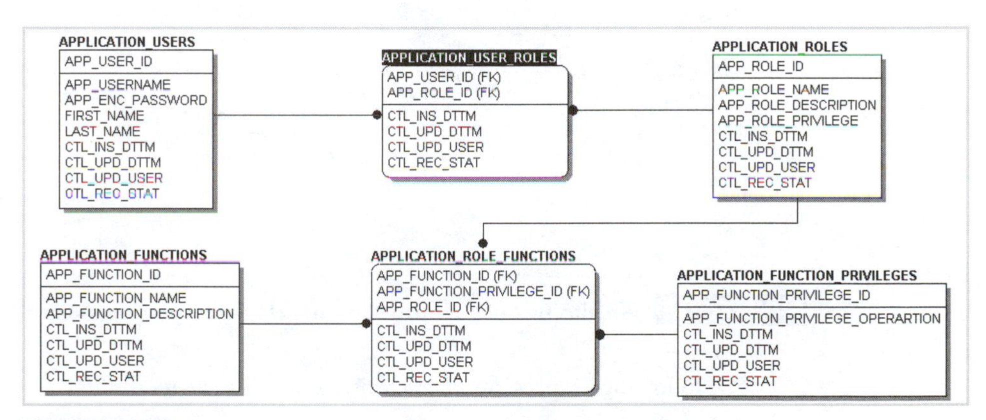

FIGURE 5-18 Security data model based on application roles and functions

Table 5-5 Tables used in a security data model based on application roles and functions

Table Name	Description
APPLICATION_USERS	Stores and maintains all end users of the application with their encrypted passwords
APPLICATION_USER_ROLES	Contains all roles assigned to users; a user can be assigned many roles
APPLICATION_ROLES	Contains all roles defined by the application and for each role a privilege is assigned; the privilege can be read, write, or read/write

Table 5-5 Tables used in a security data model based on application roles and functions (continued)

Table Name	Description
APPLICATION_FUNCTIONS	Contains all logical functions of the application; a function is defined as module or procedure or a set of tasks
APPLICATION_ROLE_FUNCTIONS	Contains all functions and privileges that are assigned to each role
APPLICATION_FUNCTION_PRIVILEGES	Stores all privileges for a function such as: read write read/write

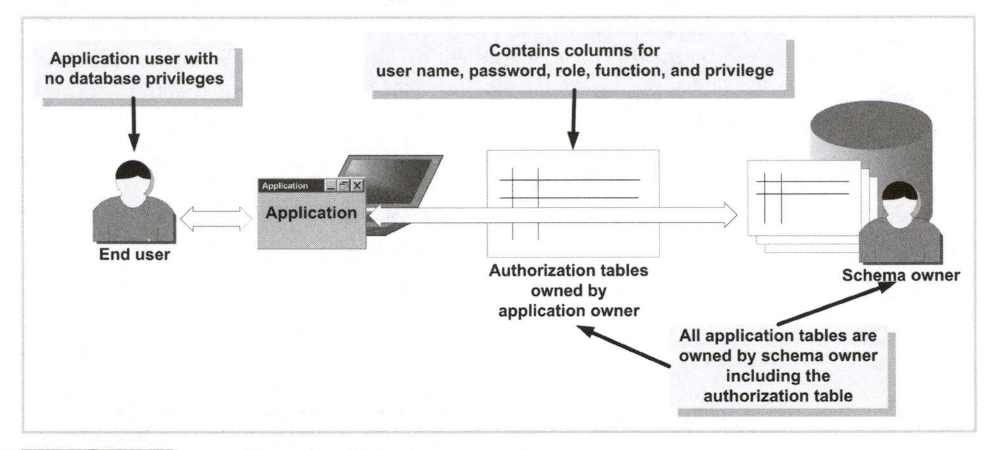

FIGURE 5-19 Architecture of a security data model based on application roles and functions

The following list presents characteristics of this security model:

- This model provides utmost flexibility for implementing application security.
- In this model you isolate the application security from the database, which makes your implementation database independent.
- Maintenance of the application security does not require specific database privileges, which lowers the risk of database violations.
- Passwords must be securely encrypted, preferably using private and public keys. In this case you must modify the structure of the APPLICATION_USERS table by adding columns to store public and private keys.
- The application must use a real database user to log on and connect to the application database. The user name and password must be encrypted and stored in a configuration file.
- The application must be designed in a very granular fashion, which means that the functions or modules of the application perform specific tasks and do not encompass a wide variety of tasks that are cross-functional. For example, if a function is ORDER_ENTRY, it should not include tasks for reporting or adding products.

- The more granular the privileges, the more effort needed to implement them. For example, this book recommends using the write privilege that allows the user to perform update, delete, and insert tasks on any table used by the function that is easily implemented by the code and interface. But if you want to implement the update privilege only, your code and interface must work accordingly to prevent users from updating a value, but allowing them to add a new record. This requires a well-designed implementation.

Security Model Based on Application Tables

This application security model depends on the application to authenticate users by maintaining all end users in a table with their encrypted passwords. Unlike previous models, the application provides privileges to the user based on tables, not on a role or a function. For example, an application user may be granted a read privilege on an application by adding an entry in APPLICATION_USER_TABLES table specifying (`user:SAM, table:Customers, Privilege:Read`). In this model the user is assigned access privilege to each table owned by the application owner. Figure 5-20 represents a data model for this type of application and shows the entities described in Table 5-6, and Figure 5-21 shows the architectural diagram of this model.

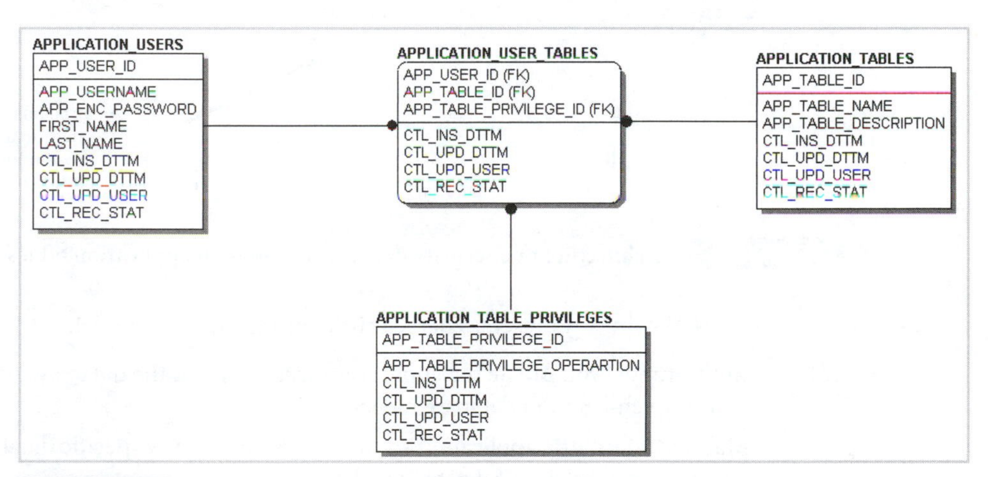

FIGURE 5-20 Security data model based on application tables

Table 5-6 Tables used in a security data model based on application tables

Table Name	Description
APPLICATION_USERS	Stores and maintains all end users of the application with their encrypted passwords
APPLICATION_USER_TABLES	Contains all tables assigned to users; user assignments can specify that he or she can access none or many tables
APPLICATION_TABLES	Contains all tables owned by the application

Table 5-6 Tables used in a security data model based on application tables (continued)

Table Name	Description
APPLICATION_TABLE_PRIVILEGES	Stores all privileges that specify access to a specific table such as: 0—No access 1—read only 2—read and add 3—read, add, and modify 4—read, add, modify, and delete 5—read, add, modify, delete, and admin

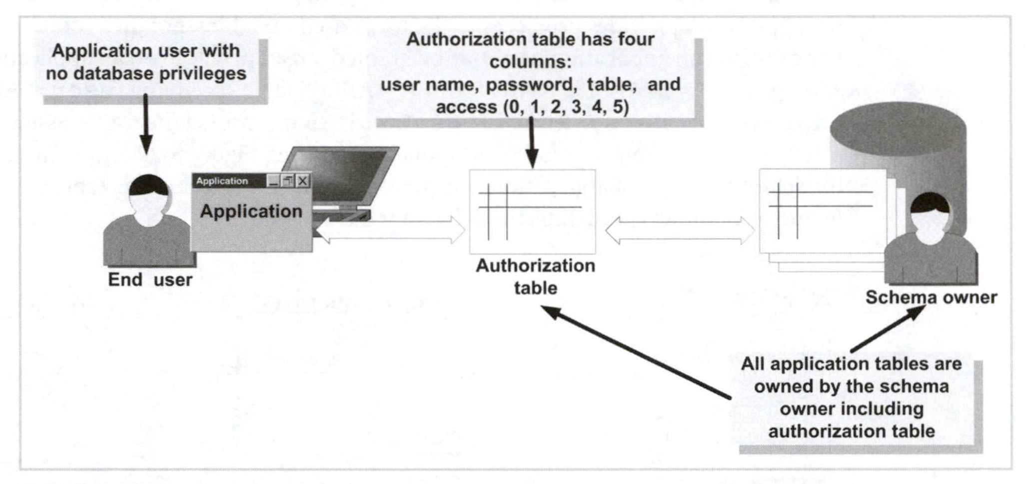

FIGURE 5-21 Architecture of a security data model based on application tables

The following list presents characteristics of this security model:

- In this model you isolate the application security from the database, which makes your implementation database independent.
- Maintenance of the application security does not require specific database privileges, which lowers the risk of database violations.
- Passwords must be securely encrypted, preferably using private and public keys. In this case you must modify the structure of the APPLICATION_USERS table by adding columns to store public and private keys.
- The application must use a real database user to log on and connect to the application database. The user name and password must be encrypted and stored in a configuration file.
- Security is implemented easily by using table access privileges that are assigned to each end user.

Implementation in SQL Server

As with the security model that is based on application roles you must take an extra step and grant authorization on application functions to the end user in order to use application. This must be done in the database tier instead of the application tier. Alter your authorization table from the security model based on database roles to incorporate the table and access columns required to support the model:

```
alter table auth_web
     add [table]     varchar(128),
     add access varchar(25)
```

A stored procedure to perform authentication and authorization looks like this:

```
create procedure authenticate (@username varchar(25),
 @password varchar(25)
) as

select [table], access
from auth_web
where username = @username and password = @password;
```

If the procedure returns a result set of access privileges, authentication was successful.

Table 5-7 presents a summary of all the security models presented in this chapter.

Table 5-7 Characteristics of security models

	Security Model				
Characteristics	Database Role Based	Application Role Based	Application Function Based	Application Role and Function Based	Application Table Based
Is flexible in implementing application security	No	No	No	Yes	No
Isolates application security from the database	Yes	Yes	Yes	Yes	Yes
Maintenance of application security does not require specific database privileges	No	No	No	Yes	No

Table 5-7 Characteristics of security models (continued)

Characteristics	Security Model Database Role Based	Application Role Based	Application Function Based	Application Role and Function Based	Application Table Based
Password must be securely encrypted	Yes	Yes	Yes	Yes	Yes
Uses real database user to log on	No	Yes	Yes	Yes	Yes
Is business-function specific	No	No	Yes	Yes	No

Data Encryption

Consider the following scenario: A customer signs up for a broadband Internet connection with a prominent Internet service provider. A few weeks later, this customer changes computers and as she is setting up her e-mail and news account, she forgets her password. She calls customer care and explains what she is trying to do. The customer care person is eagerly helpful and provides all the instructions that she needs. When the customer asks about her password, the customer care person recites it to her. Then she asks, "How do you know my password?" He answers that the password is recorded in the database. She replies, "Are you telling me that my password is recorded in the database unencrypted and can be easily seen by any customer care person?" She immediately closes her account with this provider.

The customer is justified in closing her account because her password is a piece of data that should be kept confidential and preferably encrypted.

When storing encrypted passwords in your authorization tables, it is a best practice never to decrypt the data, but rather to encrypt the password the user enters and compare the two hashes. (A hash is an algorithm that converts a varying text message to a fixed-length message.) Although the actual implementation is a concern for your developers, as a database administrator responsible for the protection of data, you should work closely with your development team to ensure that they are implementing best practices when accessing data.

This scenario leads into a discussion of data encryption, which should be part of your plan for securing or hiding important data. As you learned in previous chapters, confidentiality is one of the information security principles, which means that sensitive data must be protected from being accessed, disclosed, and exposed to unauthorized individuals. However, in many situations an authorized person has access to the data but should not read the actual unencrypted, sensitive data. In this case, sensitive data must be protected from being seen. To protect sensitive data, you must hide it through encryption.

Take an example of a database administrator who has access to the database and all the data residing in it. If sensitive data such as Social Security numbers is encrypted, the database administrator cannot read it, and therefore this confidential data is protected as long as the encryption method or key is not disclosed or broken. As mentioned earlier, Social Security numbers are considered private data and should be protected. Other examples of data that should be encrypted include passwords, salaries, and medical data.

This section presents one encryption method—base64 encryption—that you can use in your application. The implementation in this section uses the Java language. You can implement the algorithm presented in the code using any other computer language. Figure 5-22 shows a typical encryption method that uses private and public keys.

FIGURE 5-22 Encryption process using private and public keys

The following steps present the code to create a base64 Java class that can be used to encrypt and decrypt text messages. Following that, instructions are presented on how to test the encryption and decryption functionality of the base64 class.

The following code builds a base64 class using Java:

```
public class Base64 {
     /*
     This class implements a BASE64 character
     encoder and decoder as specified in RFC1521.
*/

     public Base64() {}

     //array for base64 character set.
     private char[] charSet =
          {
//        0   1   2   3   4   5   6   7
          'A','B','C','D','E','F','G','H', // 0
          'I','J','K','L','M','N','O','P', // 1
          'Q','R','S','T','U','V','W','X', // 2
          'Y','Z','a','b','c','d','e','f', // 3
          'g','h','i','j','k','l','m','n', // 4
          'o','p','q','r','s','t','u','v', // 5
          'w','x','y','z','0','1','2','3', // 6
          '4','5','6','7','8','9','+','/'  // 7
          };

     //define padding character
     private char padChar = '=';
```

```
 public String encrypt(String s1){
      //System.out.println("s1 length: " + s1.length());

      byte[] atom;
      StringBuffer sb = new StringBuffer();

      //loop through string taking blocks of 3
      for(int offset=0; offset<s1.length();offset += 3){

           //check offset length to avoid runtime error
           if( (s1.length() - 1) - offset >= 3 ){
                atom = s1.substring(offset, offset + 3).getBytes();
           } else {
                atom = s1.substring(offset).getBytes();
           }

           sb.append( encodeAtom(atom) );
      }

      return sb.toString();

 }//end encrypt

private char[] encodeAtom( byte[] atom ){

      byte a, b, c; //bytes to encode

      char[] retValue = new char[4]; //holding area for return value

      int len = atom.length; //get the length of the atom

      //System.out.println(len);

      //based on length, shift bytes creating 4 six bit bytes
      switch(len) {
           case 1:
               a = atom[0];
               b = 0;
               c = 0;

               retValue[0] = charSet[(a >>> 2) & 0x3F];
               retValue[1] = charSet[((a << 4) & 0x30) + ((b >>> 4) & 0xf)];
               retValue[2] = padChar;
               retValue[3] = padChar;

                break;
           case 2:
               a = atom[0];
               b = atom[1];
               c = 0;
```

```
                retValue[0] = charSet[(a >>> 2) & 0x3F];
                retValue[1] = charSet[((a << 4) & 0x30) + ((b >>> 4) & 0xf)];
                retValue[2] = charSet[((b << 2) & 0x3c) + ((c >>> 6) & 0x3)];
                retValue[3] = padChar;

                 break;
            case 3:
                a = atom[0];
                b = atom[1];
                c = atom[2];

                retValue[0] = charSet[(a >>> 2) & 0x3F];
                retValue[1] = charSet[((a << 4) & 0x30) + ((b >>> 4) & 0xf)];
                retValue[2] = charSet[((b << 2) & 0x3c) + ((c >>> 6) & 0x3)];
                retValue[3] = charSet[c & 0x3F];

                   break;
        }//end switch

        return retValue;

    }//end encodeAtom

     public String decrypt(String s1) {
        //System.out.println("s1 length: " + s1.length());

        byte[] atom;
        StringBuffer sb = new StringBuffer();

        //loop through taking the 4 byte encoded atom
        for(int offset=0; offset<s1.length();offset += 4){
            atom = s1.substring(offset, offset + 4).getBytes();

            sb.append( new String(decodeAtom(atom)) );
        }

        return sb.toString();

     }//end decrypt

    private byte[] decodeAtom(byte[] atom)
    {
        //initialize atom bytes
        byte a = (byte)-1;
        byte b = (byte)-1;
        byte c = (byte)-1;
        byte d = (byte)-1;

        int len = 3; //initial length of decoded byte array
```

```
        //get byte representation of each character in atom
        for (int i = 0; i < 64; i++) {
            if (atom[0] == charSet[i]) a = (byte) i;

            if (atom[1] == charSet[i]) b = (byte) i;

            if (atom[2] == charSet[i]) c = (byte) i;

            if (atom[3] == charSet[i]) d = (byte) i;

        }//end for(i)

        //get data length if atom was padded
        if ( atom[3] == '=') len = (atom[2] == '=') ? 1 : 2;

        byte[] retValue = new byte[len];

        //shift atom bytes to create 3 eight bit bytes
        switch (len) {
             case 1:
                 retValue[0] = (byte)(((a << 2) & 0xfc) | ((b >>> 4) & 3));
                 break;
             case 2:
                 retValue[0] = (byte)(((a << 2) & 0xfc) | ((b >>> 4) & 3));
                 retValue[1] = (byte)(((b << 4) & 0xf0) | ((c >>> 2) & 0xf));
                 break;
             case 3:
                 retValue[0] = (byte)(((a << 2) & 0xfc) | ((b >>> 4) & 3));
                 retValue[1] = (byte)(((b << 4) & 0xf0) | ((c >>> 2) & 0xf));
                 retValue[2] = (byte)(((c << 6) & 0xc0) | (d  & 0x3f));
                 break;
        }//end switch

        return retValue;
    }//end decodeAtom


}//end class
```

The following code tests the encryption:

```
import java.io.InputStreamReader;

public class TestEncryption {

     public static void main(String[] args){

          InputStreamReader rdr = new InputStreamReader(System.in);
          System.out.println("Base64 Encryption Test Container\n");
          System.out.print("Enter string to encrypt: ");
          StringBuffer sb = new StringBuffer();
          char c;
          try {
                while(true) {
                     c = (char)rdr.read();
                     if (c == '\n' || c == '\r') //break on CR/LF
                          break;
                     sb.append(c);

                }
          } catch (java.io.IOException e) {
                e.printStackTrace();
          }

          Base64 base64 = new Base64();

          String encryptedValue = base64.encrypt(sb.toString());

          System.out.println(" Encrypted String: " + encryptedValue );
          System.out.println("Original String: " + base64.decrypt(encrypted-
Value) );
     }
```

To run the code you should do the following:

```
Base64.class is a java class compiled on sdk 1.4.2
TestEncryption.class is a console application that accepts
input and uses Base64.class to encrypt the input.

To run:
java.exe TestEncryption.class
```

To recap, the main objective of this chapter is to teach you different security models that you can incorporate into any business application. You should note that the design of all security models enforces security at the application level rather than at the database level with the significant benefit of isolating security from the database and making security database independent. What does this mean to you? It actually means that you can build your security model on any database system such as Oracle10*g* and SQL Server. After working through the detailed tour of five security models, you are able to adopt any of these models as is or vary them depending on your business security needs.

Chapter Summary

- An application user is simply a record created for a user within the application schema. The record is used for authentication to the application.
- An application user (a record) usually does not have database privileges or roles assigned to the user.
- The application owner is a database user (schema owner) who owns the application tables and objects.
- The database user is a type of user account for accessing the database that has roles and/or privileges assigned to it.
- The proxy user is a database user that has specific roles and privileges assigned to it. The proxy user works on behalf of an application user.
- The access matrix model uses a matrix to represent two main entities that can be used for any security implementation.
- The access matrix model has columns that represent objects and rows that represent subjects.
- An object in the access matrix model can be a table, view, procedure, or any other database object.
- A subject in the access matrix model can be a user, role, privilege, or a module.
- The intersection of a row and column of the access matrix is an authorization cell representing the access details of the object to the subject.
- The access modes model uses both the subject and object as the main security entities and indicates what functions the subject is allowed to perform on the objects.
- Client/server architecture is composed of three main components that are typically found in a client/server application: user interface component, business logic component, and data access component.
- A Web application can be referred to as a Web-based application. A Web application is an application that uses the Web (HTTP protocol) to connect and communicate to the server.
- The Web browser is the front end of the Web application. It uses HTML pages embedded with other Web services, utilizing ActiveX, Java applets or beans, or VB scripts.
- A Web application typically consists of five layers: Web browser, Web server, application server, business logic, and database server.
- Data warehouses are used by decision-support applications to support executive management in their decision-making processes.

- The following list contains the five application security models, listed from most commonly used to least commonly used:
  - Database roles
  - Application roles
  - Application functions
  - Roles and functions in the application
  - Application tables
- Data encryption should be used to hide sensitive data.

Pharmacy Application

Figure 5-23 illustrates the data model for a small pharmacy store. This data model will be used for exercises presented in this chapter as well as in subsequent chapters in this book. Scripts to create the pharmacy schema can be found on the Student Resource CD provided with this book. Use create_pharmacy_sqlserver.sql to create the Pharmacy schema in SQL Server and use create_pharmacy_oracle.sql to create the Pharmacy schema in Oracle10*g*.

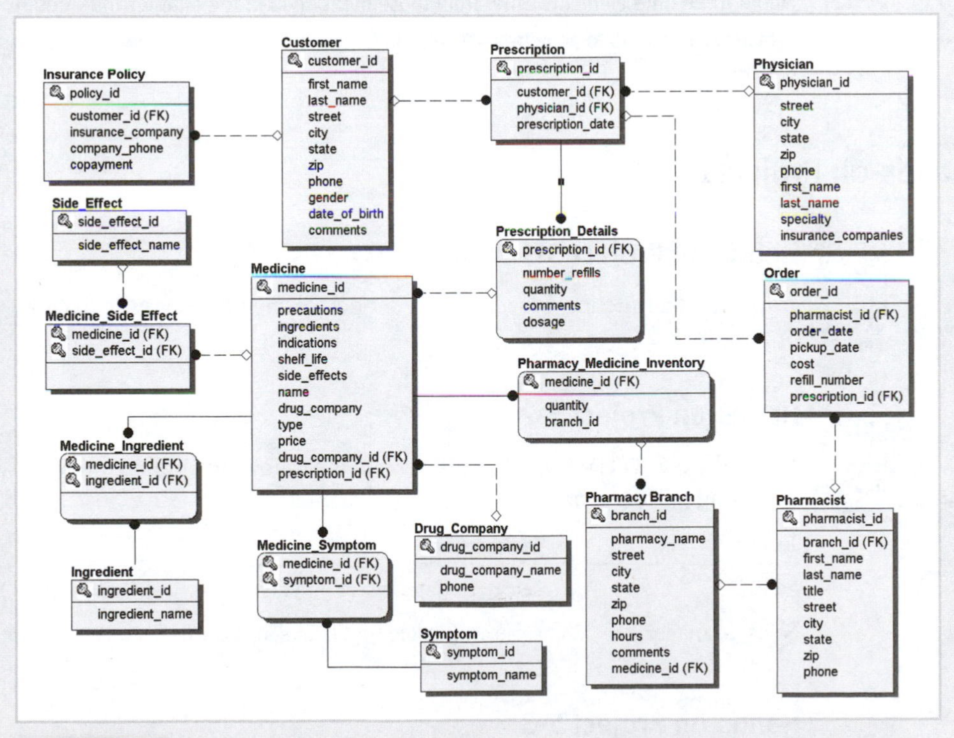

FIGURE 5-23 Pharmacy application data model

Review Questions

1. Explain the differences between a database user and an application user.
2. Provide an example of appropriate use of a proxy user.
3. Explain the difference between database roles and application roles.
4. Provide an example (no implementation is required) in Oracle of how to implement the dynamic modes of an access modes security model.
5. What are the differences between a client/server and a Web-based application?
6. Are there any security-related differences between client/server and Web-based applications?
7. Give an example why data encryption is needed and when you would use encryption.
8. Provide an example (no implementation is required) of when you would use the access matrix model.
9. Provide an example (no implementation is required) of when you would use the access modes model.
10. Which layer of the OSI model implements security in a Web application?
11. What type of users would you implement for a Web site that sells products and why?
12. Name three data elements other than those mentioned in the chapter that require data encryption.
13. Provide statements to activate user roles in Oracle and SQL Server. Explain why you would use role activation.

Hands-on Projects

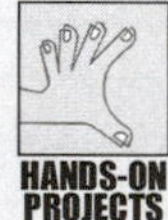

Hands-on Project 5-1

Implement a function-based security model for the PHARMACY database. See Figure 5-23.

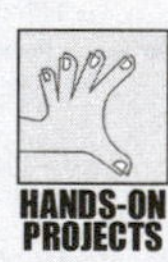

Hands-on Project 5-2

Using the NORTHWIND database, create the following application roles with the corresponding permissions:

Role	Permissions
Sales	select, update, insert on Customers and Orders and Order Details
Sales_Manager	sales and delete on Orders and Order Details

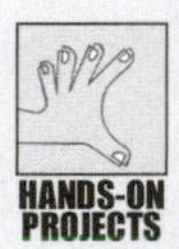

Hands-on Project 5-3

Using the NORTHWIND database, create a proxy user to activate the roles you created in Hands-on Project 5-2. Be sure to set the appropriate permissions for the user.

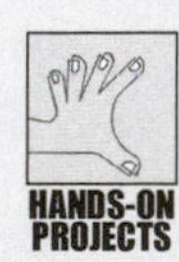

Hands-on Project 5-4

Using Query Analyzer, write a small script that selects a set of customers using the user and roles you created in Hands-on Projects 5-2 and 5-3.

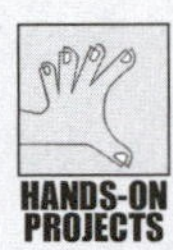

Hands-on Project 5-5

Using the concepts in the security data model based on application tables, set up the following permissions for the AUTHORS table in the PUBS database:

User	Access
John	select, update, insert
Jane	select, update, insert, delete
Sally	select

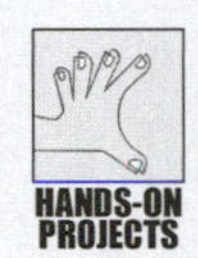

Hands-on Project 5-6

Drop the application roles created in Hands-on Project 5-2.

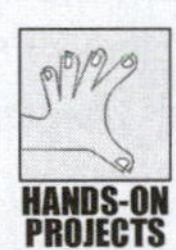

Hands-on Project 5-7

Create a report user in the PHARMACY database and give it access to a view that pulls data from the customer table.

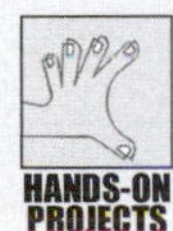

Hands-on Project 5-8

Using Oracle and the Pharmacy data model, devise an application security model that is based on access modes.

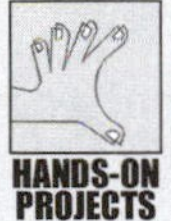

Hands-on Project 5-9

Using Oracle and the Pharmacy data model, devise an application security model that uses an access matrix.

Hands-on Project 5-10

Using Oracle and the Pharmacy data model, devise an application security model that is based on application functions.

Hands-on Project 5-11

Using Oracle and the Pharmacy data model, devise an application security model that is based on application tables.

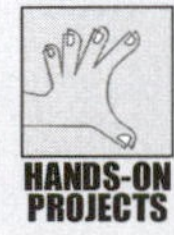

Hands-on Project 5-12

Using Oracle and the Pharmacy data model, devise an application security model that is based on application roles.

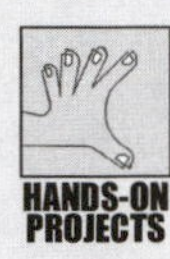

Hands-on Project 5-13

Using Oracle and the Pharmacy data model, devise an application security model that is based on database roles.

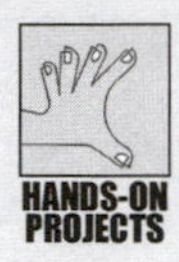

Hands-on Project 5-14

Using Oracle and the Pharmacy data model, devise an application security model that is based on application roles and functions.

Case Projects

Case 5-1 Secure Your Application with SQL Server

Acme Manufacturing is implementing a new Web-based ordering system for preferred customers and an internal client/server application so Acme sales people can manage the orders. Implement the following authorization table using any two suitable application models presented in this chapter.

Role	Permissions
clerk	select, update, and insert on all tables
super	clerk and delete
admin	super and all permissions on the authentication table

Case 5-2 Application Security with Oracle

You were hired by Acme Telecommunication Systems (ATS) to implement a number of security requirements. Provide the architecture of the application security model you select and the steps to implement it. The application has the following modules:

- Account maintenance
- Call records
- Transactions
- Call monitoring
- Backup and restore
- Security

The following ATS users will be accessing the system:

- Sammy Bright
- Tom Lord
- Linda Glass
- Joan Knight
- James Howell
- Dennis Wright
- Ray Stevenson
- Phil Magnet
- Kate Davis

ATS requested the following security implementation:

Roles

Role	Rank
Service	0
Operator	1
Supervisor	2
Administrator	3
CEO	4

Privileges

Privilege	Rank
query	0
update	1
delete	2
grant	3

Module—Role

Module	Service	Operator	Supervisor	Administrator	CEO
Account maintenance	query	update	delete	grant	query
Call records		update	delete	grant	query
Transactions		update	delete	grant	query
Call monitoring		query	query	grant	query
Backup and restore	delete			grant	
Security		query		grant	query

User—Role

User	Service	Operator	Supervisor	Administrator	CEO
Sammy Bright	X		X		
Tom Lord	X	X			
Linda Glass				X	
Joan Knight	X	X	X		
James Howell	X				
Dennis Wright			X		
Ray Stevenson		X			
Phil Magnet		X			
Kate Davis					X

Endnotes

1 Castano, Silvano (Editor), Fugini, Maria Grazia, Martella, Giancarlo. *Database Security*: Addison Wesley, 1994: 44-53.
2 Castano, Silvano (Editor), Fugini, Maria Grazia, Martella, Giancarlo. *Database Security*: Addison Wesley, 1994: 44-53.
3 This figure is adapted from *Management Information Systems for the Information Age* (1998) by Haag, Cummings, and Dawkins, Irwin/McGraw-Hill.

Virtual Private Databases

6

LEARNING OBJECTIVES

Upon completion of this material, you should be able to:

- Define the term "virtual private database" and explain its importance
- Implement a virtual private database by using the VIEW database object
- Implement a virtual private database by using Oracle's application context
- Implement the Oracle virtual private database feature
- Use a data dictionary to view an Oracle virtual private database
- Use Policy Manager to view an Oracle virtual private database
- Implement row-level and column-level security

Introduction

Chapter by chapter your database security schema has gradually enlarged, as illustrated in Figure 6-1. The figure shows your progress through the chapters as you have learned how to secure data to prevent data violations. As the last section in Part 1, Chapter 6 concludes the security portion of this book. In previous chapters, you learned database security concepts. You learned, in addition, methods and designs for developing security mechanisms at different levels of an information system: the database level, user level, application level, and table level. In this chapter, you advance one step further. You learn to enforce data security at its most granular level—rows and columns.

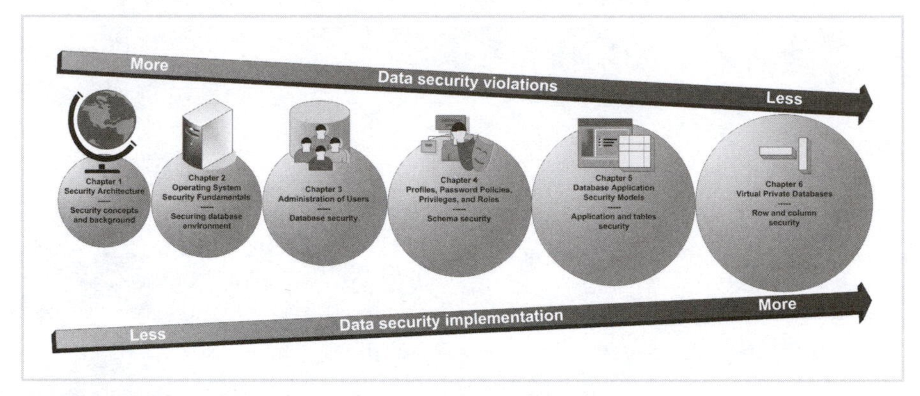

FIGURE 6-1 Topics covered by Chapters 1 through 6

One of the primary functions of any database application is to allow each department access to only its own data: data entered by Department A should be visible only to Department A. This is a typical situation for a virtual private database. One database model serves all departments, and all data for a specific entity from all departments resides in one table. Each department sees only its own data. This chapter illustrates this concept and demonstrates how to enable row-level and column-level security, an implementation referred to as a virtual private database.

Overview of Virtual Private Databases

Consider the following scenario. An educational software company sold its first enterprise application to a prominent university. The architecture of this application consists of customized Web pages connected to a database hosted by the vendor. It did not take long for the vendor to receive requests for the application from another university and then another. Every time a client was added, a new database was created for that specific client. Every time a client requested an application change or fix, it was replicated to all other clients. This process was manageable until the number of clients grew to a point at which it became unfeasible to roll out and track the changes for every client. The vendor reached a point at which existing architecture was not scalable and flexible enough to handle any new clients. A solution was needed to enable the vendor to grow and succeed.

The chief information officer (CIO) hired a database expert to recommend a new architectural approach to solve the database design problem. Over the next few weeks the database expert surveyed the existing database architecture for each client and conducted interviews with key engineers. He then drafted a report stating his findings and recommendations. The main recommendation was to consolidate all the clients' databases into one database and implement a virtual private database for each client. The main benefit was that database engineers would have one database architecture to work with and that would make maintenance easier.

After reading the scenario, you have probably guessed that a virtual private database deals with data access. More specifically, a VPD controls data access at the row or column level. With SQL Server 2000, you can implement row and column access by using the VIEW database object. However, Oracle10*g* implements the VPD concept differently. In fact, Oracle10*g* has a specific function that allows database administrators to implement the VPD concept. As shown in Figure 6-2, a **virtual private database (VPD)** is a shared database schema containing data that belongs to many different users, and each user can view or update only the data he or she owns.

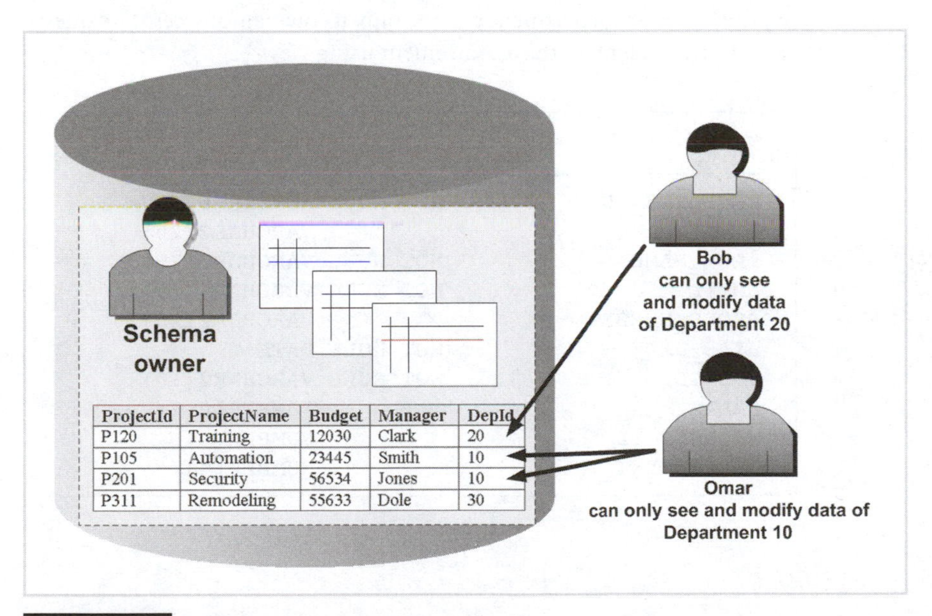

FIGURE 6-2 Virtual private database example

As mentioned earlier, not every database system offers a mechanism to implement this concept without VIEW objects, and Microsoft SQL Server is no exception. On the other hand, Oracle offered this feature in several versions before the release of 10*g*, but it had a different name. In fact, Oracle uses two other names to refer to VPDs: row-level security (RLS) and fine-grained access (FGA). At the same time, FGA is also an acronym for "fine-grained auditing," which has nothing to do with VPDs. Although many database experts cite numerous benefits of VPDs, there are really two main purposes for using them:

- The security requirements of a security policy necessitate that data access be restricted at the row or column level (fine-grained access).
- One database schema serves multiple unrelated groups or entities.

This section presents various methods of implementing a VPD and explains how it is implemented for both Oracle and SQL Server.

Implementing a VPD Using Views

Do you recall the first time you came across the VIEW object? Do you remember what you were told about the use of the VIEW object? You probably remember that one of the uses of a VIEW object is to limit what users can see and accomplish with the existing data in a table. In other words, you can construct a view that hides columns or rows from users, as illustrated in the following example. Consider the table EMPLOYEES. The table's data structure is shown in the following code segment. Suppose your business rules require that each department can see only its own employees. The question is, "How would you implement this requirement using views?"

```
Name                      Null?    Type
------------------------- -------- ------------
EMPLOYEE_ID               NOT NULL NUMBER(6)
FIRST_NAME                         VARCHAR2(20)
LAST_NAME                 NOT NULL VARCHAR2(25)
EMAIL                     NOT NULL VARCHAR2(25)
PHONE_NUMBER                       VARCHAR2(20)
HIRE_DATE                 NOT NULL DATE
JOB_ID                    NOT NULL VARCHAR2(10)
SALARY                             NUMBER(8,2)
MANAGER_ID                         NUMBER(6)
DEPARTMENT_ID                      NUMBER(4)
```

```
SQL> SELECT DISTINCT DEPARTMENT_ID FROM EMPLOYEES;

DEPARTMENT_ID
-------------
           10
           20
           30
           40
           50
           60
           70
           80
           90
          100
          110

12 rows selected.
```

Your first impulse might be to create a view for each department as follows:

```
SQL> CREATE VIEW  EMP_FOR_DEP_20 AS
  2     SELECT EMPLOYEE_ID, FIRST_NAME, LAST_NAME, EMAIL, PHONE_NUMBER, JOB_ID
  3       FROM EMPLOYEES
  4      WHERE DEPARTMENT_ID = 20
  5  /

View created.

SQL> SELECT * FROM EMP_FOR_DEP_20
/

EMPLOYEE_ID FIRST_NAME  LAST_NAME EMAIL    PHONE_NUMBER    JOB_ID
----------- ---------- --------- -------- ------------    ------
        201 Michael     Hartstein MHARTSTE 515.123.5555    MK_MAN
        202 Pat         Fay       PFAY     603.123.6666    MK_REP
```

Continuing with this example, you would create 11 different views, one for each department. This process might be manageable in some cases, but as the number of departments grows, it becomes a nightmare to administer and to maintain. Therefore, this method is not feasible. The solution you use depends on the database system you are using. However, you cannot implement this specific scenario with VIEW objects alone. Views are not the right solution for this type of a problem. The solution for this type of problem is a VPD.

Consider another example in which a VIEW object can be used to restrict access per user. Suppose you have the EMPLOYEES table, but with a modified data structure. Notice the new column, CTL_UPD_USER, in which CTL stands for Control and UPD stands for Update. Control indicates that it is a value generated and controlled by the application, not by the user. This column is used specifically to indicate the user who owns the row. For example, when a user inserts a new row, the user name value is automatically inserted in this column. Use the following steps to demonstrate this implementation:

1. Logon as DBSEC schema and display a column listing of the EMPLOYEES table. If this table does not exist, then create it according to the displayed structure:

```
SQ> DESC EMPLOYEES
Name                                       Null?    Type
------------------------------------------ -------- ------------
EMPLOYEE_ID                                NOT NULL NUMBER(6)
FIRST_NAME                                          VARCHAR2(20)
LAST_NAME                                  NOT NULL VARCHAR2(25)
EMAIL                                      NOT NULL VARCHAR2(25)
PHONE_NUMBER                                        VARCHAR2(20)
HIRE_DATE                                  NOT NULL DATE
JOB_ID                                     NOT NULL VARCHAR2(10)
SALARY                                              NUMBER(8,2)
MANAGER_ID                                          NUMBER(6)
DEPARTMENT_ID                                       NUMBER(4)
CTL_UPD_USER                                        VARCHAR2(30)
```

2. Create the table EMPLOYEES_VER1 as specified in the previous code:

```
SQL> CREATE TABLE EMPLOYEES_VER1
  2  (
  3     EMPLOYEE_ID     NUMBER(6),
  4     FIRST_NAME      VARCHAR2(20),
  5     LAST_NAME       VARCHAR2(25),
  6     EMAIL           VARCHAR2(25),
  7     PHONE_NUMBER    VARCHAR2(20),
  8     HIRE_DATE       DATE,
  9     JOB_ID          VARCHAR2(10),
 10     SALARY          NUMBER(8,2),
 11     MANAGER_ID      NUMBER(6),
 12     DEPARTMENT_ID   NUMBER(4),
 13     CTL_UPD_USER    VARCHAR2(30)
 14  )
 15  /

Table created.
```

3. Create a VIEW object to display rows that belong only to the logged on user. Note that USER is a function that returns the user name value of the person who is logged on. If HR is logged on, the returned value is HR.

```
SQL> CREATE VIEW EMPLOYEES_VIEW1 AS
  2    SELECT EMPLOYEE_ID, FIRST_NAME,
  3           LAST_NAME, EMAIL,
  4           PHONE_NUMBER, HIRE_DATE,
  5           JOB_ID, SALARY,
  6           MANAGER_ID, DEPARTMENT_ID,
  7           CTL_UPD_USER USER_NAME
  8      FROM EMPLOYEES_VER1
  9     WHERE CTL_UPD_USER = USER
 10  /

View created.
```

4. Grant SELECT and INSERT on this view to another user, such as SCOTT:

```
SQL> GRANT SELECT, INSERT ON EMPLOYEES_VIEW1 TO SCOTT
  2  /

Grant succeeded.
```

5. Insert a row using the EMPLOYEES_VIEW1 VIEW object just created:

```
SQL> INSERT INTO DBSEC. EMPLOYEES_VIEW1 (EMPLOYEE_ID, FIRST_NAME, LAST_NAME, EMAIL,
  2                                  PHONE_NUMBER, HIRE_DATE, JOB_ID, SALARY,
  3                                  MANAGER_ID, DEPARTMENT_ID, USER_NAME)
  4                           VALUES (100, 'Sam', 'Afyouni', 'safyouni',
  5                                  '123.234.3456', sysdate, 'WM_CLK', 1000,
  6                                  1000, 10, USER);

1 row created.

SQL> COMMIT;

Commit complete.
```

6. Now log on as SCOTT and insert a row:

```
SQL> INSERT INTO EMPLOYEES_VIEW1 (EMPLOYEE_ID, FIRST_NAME, LAST_NAME, EMAIL,
  2                               PHONE_NUMBER, HIRE_DATE, JOB_ID, SALARY,
  3                               MANAGER_ID, DEPARTMENT_ID, USER_NAME)
  4                        VALUES (101, 'Julia', 'Ronaldo', 'jronaldo',
  5                               '456.567.3678', sysdate, 'WM_ENG', 2000,
  6                               2000, 20, USER);

1 row created.

SQL> COMMIT;

Commit complete.
```

7. You know that there are two rows in the table EMPLOYEES_VER1.

```
SQL> SELECT EMPLOYEE_ID, FIRST_NAME, LAST_NAME, CTL_UPD_USER
  2    FROM EMPLOYEES_VER1
  3  /
EMPLOYEE_ID FIRST_NAME           LAST_NAME                  CTL_UPD_USER
----------- ----------           ---------                  ------------
        100 Sam                  Afyouni                    DBSEC
        101 Julia                Ronaldo                    SCOTT
```

8. As SCOTT, select the EMPLOYEES_VIEW1 VIEW object, and you see only the row that belongs to SCOTT.

```
SQL> SELECT DBSEC. EMPLOYEE_ID, FIRST_NAME, LAST_NAME, USER_NAME
  2    FROM EMPLOYEES_VIEW1
  3  /

EMPLOYEE_ID FIRST_NAME           LAST_NAME                  USER_NAME
----------- ----------           ---------                  ---------
        101 Julia                Ronaldo                    SCOTT
```

Now that you have implemented a VPD using a view, what happens when a user inserts a row? That user must enter the user name correctly for this to work. But you can add a trigger on insert to populate the user name automatically. As you can see, this implementation is limited and requires careful design and development.

```
SQL> CREATE OR REPLACE TRIGGER TRG_EMPLOYEES_VER1_BEFORE_INS
  2      BEFORE INSERT
  3      ON EMPLOYEES_VER1
  4      FOR EACH ROW
  5   BEGIN
  6      :NEW.CTL_UPD_USER := USER;
  7  END;
SQL> /

Trigger created.

SQL> INSERT INTO EMPLOYEES_VIEW1 (EMPLOYEE_ID, FIRST_NAME, LAST_NAME, EMAIL,
  2                                PHONE_NUMBER, HIRE_DATE, JOB_ID, SALARY,
  3                                MANAGER_ID, DEPARTMENT_ID)
  4                         VALUES (101, 'Aya', 'Afyouni', 'aafyouni',
  5                                '456.567.3678', sysdate, 'WM_DR', 2100,
  6                                1000, 10);

1 row created.

SQL> COMMIT;

Commit complete.

SQL> SELECT EMPLOYEE_ID, FIRST_NAME, LAST_NAME
  2     FROM EMPLOYEES_VIEW1
  3   /

EMPLOYEE_ID FIRST_NAME           LAST_NAME
----------- ----------           ---------
        100 Sam                  Afyouni
        101 Aya                  Afyouni
```

As you have noticed, the implementation of a VPD using VIEW is not difficult, but unfortunately using views does not lend itself to offer much scalability. For this specific reason, Oracle10*g* offers VPD functionality through which you can extend application security to any level. Before you look into the VPD feature, explore how VPDs are implemented using views in Microsoft SQL Server.

Hiding Rows Based on the Current User

SQL Server does not support the VPD function. However, you can implement a VPD using views. In this section you see how a VPD can be built in an application. You can determine which rows are returned based on the user logon. This technique works well for hosted applications in which several companies use the same database. Microsoft SQL Server 2000 has the system function USER, which returns the database user and thereby makes a connection similar to that used in Oracle, as illustrated in Figure 6-3.

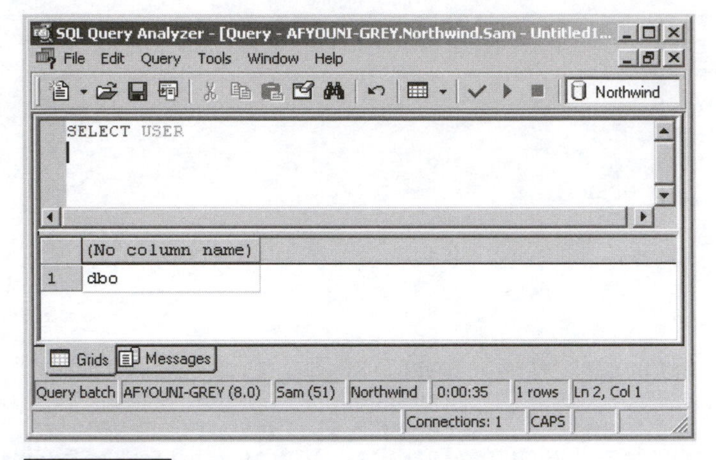

FIGURE 6-3 **Command to display current user**

Using this function, you can easily implement row-based security with views. To do this, you need to have a column in your tables to hold the user name of the row's owner. When adding this type of column to a table, it's a good practice to preface it with "CTL", indicating that this column is for system or application control and is not part of the entity's data.

1. Alter the Customer table in your PHARMACY database to contain the column CTL_UPD_USER. When defining the column, do not allow nulls and set a default value using the USER function; this function sets the user name for you on inserting new rows.

```
alter table customer
      add ctl_upd_user varchar(128) not null default user
go
```

2. Create database users in the PHARMACY database for the users sam and jason that you created in Chapter 3:

```
use pharmacy
go
exec sp_grantdbaccess 'sam'
exec sp_grantdbaccess 'jason'
go
grant all on customer to sam
grant all on customer to jason
go
```

3. Create the view that users will use to access the customer table:

```
create view vCustomer
as
      select customer_id, first_name, last_name, street, city,
 state, zip, phone,     gender, date_of_birth, comments
      from customer
      where ctl_upd_user = user
go
grant all on vCustomer to sam
grant all on vCustomer to jason
go
```

4. Test the code. Connect to the database as sam and insert some rows by referencing the view:

```
insert into vCustomer  (customer_id, first_name, last_name)
values (1000, 'John', 'Doe')
insert into vCustomer  (customer_id, first_name, last_name)
values (2000, 'Sarah', 'Jones')
go
```

5. Select from the view, then connect as jason, and select from the view as illustrated in Figures 6-4 and 6-5.

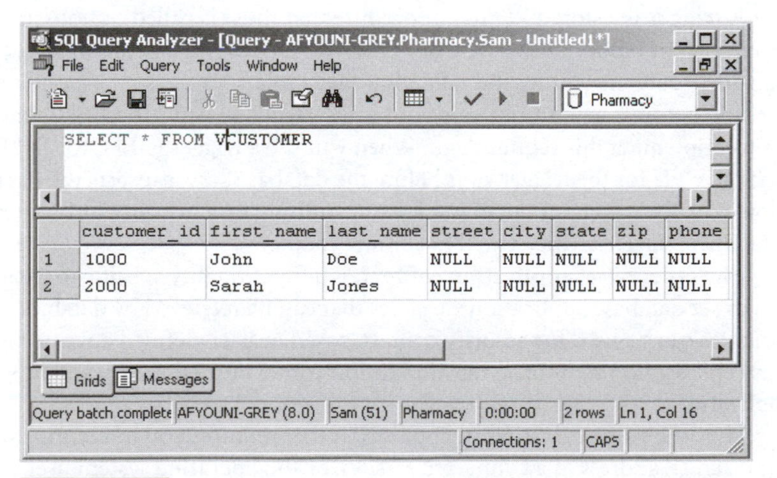

FIGURE 6-4 Query to show content of VCUSTOMER table before security enforcement

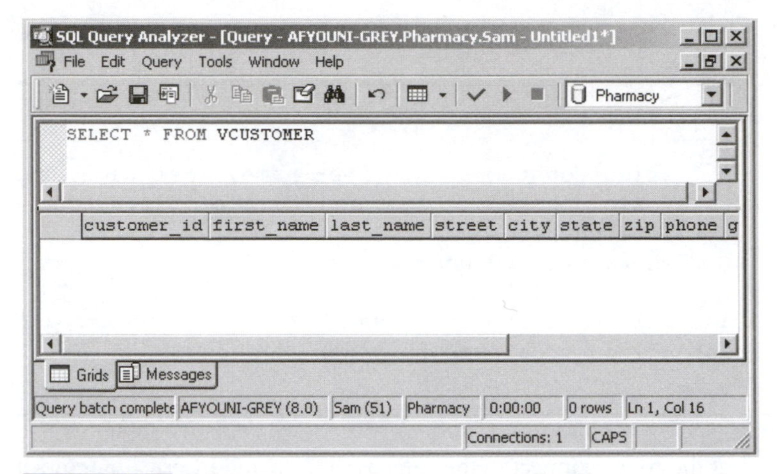

FIGURE 6-5 Query to show content of VCUSTOMER table after security enforcement

Notice that Sam can see the rows he inserted, but Jason cannot. You've just implemented a VPD in SQL Server.

Implementing a VPD Using Application Context in Oracle

You have not yet worked with the VPD feature in Oracle. To appreciate this feature, you need first to understand how it is implemented using views. In this section, you learn how to implement a VPD using views in combination with triggers and you'll learn about the application context functionality that is available in Oracle.

A **trigger** is a stored PL/SQL procedure that fires (is called) automatically when a specific event occurs, such as the BEFORE INSERT event. For example, suppose you have a business rule mandating the validation that a salary value is in a specific range for a position before a new employee record is inserted. You can use a BEFORE INSERT trigger to implement this requirement. When you use a trigger (BEFORE INSERT), the PL/SQL code for the trigger runs before the database server inserts the data. Triggers allow you, the programmer, to implement auditing functionality or business rules. More details on triggers are presented in coming chapters.

The other term, **application context,** is a functionality specific to Oracle that allows you to set database application variables that can be retrieved by database sessions. These variables can be used for security context-based or user-defined environmental attributes. This section shows you the application context functionality and how to use it in an application.

Suppose you are writing an application that requires you to identify a client host name, an IP address of a connected session, or the operating system user name of a connected session. You can get most of the session information from the dynamic performance view V$SESSION, if you have select privileges. You can also use the application context function SYS_CONTEXT in conjunction with predefined user-environment attributes, known as USERENV attributes, which are grouped as a namespace. Table 6-1 lists the most common USERENV attributes. Consult the *Oracle Database Security Guide* for a complete list of attributes. The following example illustrates how to get the IP address of the current connected session:

```
SQL> SELECT SYS_CONTEXT('USERENV', 'IP_ADDRESS') FROM DUAL
  2  /
SYS_CONTEXT('USERENV','IP_ADDRESS')
-----------------------------------
192.168.1.2

SQL> SELECT SYS_CONTEXT('USERENV', 'HOST') FROM DUAL
  2  /

SYS_CONTEXT('USERENV','HOST')
-----------------------------
AFYOUNI\AFYOUNI-BLACK

SQL> SELECT SYS_CONTEXT('USERENV', 'CURRENT_USER') FROM DUAL
  2  /

SYS_CONTEXT('USERENV','CURRENT_USER')
-------------------------------------
DBSEC
```

Table 6-1 Common USERENV namespaces

Attribute	Description of What the Attribute Returns
TERMINAL	Operating system terminal name for the current connected session
IP_ADDRESS	Network IP address for the current connected session
HOST	Name of the host machine for the current connected session
DB_NAME	Name of the database to which the current session is connected
CURRENT_USER	Database name for the current connected session
DB_DOMAIN	Network domain name for the database to which the current session is connected
OS_USER	Operating system user name for the current connected session
SERVER_HOST	Name of the host machine to which the current database session is connected
SESSIONID	Auditing session identifier for the current connected session
ISDBA	Information to indicate whether the connected session has DBA privileges or not; the returned value is a Boolean TRUE or FALSE

The information in this table is derived from the online documentation that Oracle provides at the Oracle Technology Network site: *www.otn.oracle.com.*

As you learned earlier, you can get the value for these namespaces (attributes) using a function called SYS_CONTEXT. Although Oracle provides a rich, full set of predefined attributes, you can actually set your own application context using the Oracle PL/SQL package called DBMS_SESSION. One of the procedures in the DBMS_SESSION package that is used to define your own application context is called SET_CONTEXT. In the following partial description of this package, only application context procedures are listed. You can get a full description of the package by issuing the DESC DBMS_SESSION command from SQL*Plus or from Oracle documentation.

```
PROCEDURE CLEAR_ALL_CONTEXT
 Argument Name                  Type            In/Out Default?
 ------------------------------ --------------- ------ --------
 NAMESPACE                      VARCHAR2        IN

PROCEDURE CLEAR_CONTEXT
 Argument Name                  Type            In/Out Default?
 ------------------------------ --------------- ------ --------
 NAMESPACE                      VARCHAR2        IN
 CLIENT_ID                      VARCHAR2        IN     DEFAULT
 ATTRIBUTE                      VARCHAR2        IN     DEFAULT

PROCEDURE LIST_CONTEXT
 Argument Name                  Type            In/Out Default?
 ------------------------------ --------------- ------ --------
 LIST                           TABLE OF RECORD OUT
 LSIZE                          NUMBER          OUT

PROCEDURE SET_CONTEXT
 Argument Name                  Type            In/Out Default?
 ------------------------------ --------------- ------ --------
 NAMESPACE                      VARCHAR2        IN
 ATTRIBUTE                      VARCHAR2        IN
 VALUE                          VARCHAR2        IN
 USERNAME                       VARCHAR2        IN     DEFAULT
 CLIENT_ID                      VARCHAR2        IN     DEFAULT
```

For example, to get a list of all the attributes for the current session, use the following PL/SQL block:

```
DECLARE
 AC_TBL DBMS_SESSION.AppCtxTabTyp;
 AC_SZE NUMBER;
BEGIN
 DBMS_SESSION.LIST_CONTEXT (AC_TBL, AC_SZE);
 FOR I IN 1..AC_TBL.COUNT LOOP
  DBMS_OUTPUT.PUT_LINE(AC_TBL(I).NAMESPACE||'.'||
           AC_TBL(I).ATTRIBUTE||'.'||
           AC_TBL(I).VALUE);
 END LOOP;
END;
/
```

Another important and useful procedure is SET_CONTEXT, and here is a typical example of how it is used. There is a need to allow different users with different levels to see specific rows in a table. For example, a clerk with level 2 authorization sees only rows that are marked "level 2" or below. This application has three levels, 1 through 3. When a user is logged on, the application obtains the security level from a table called APP_CONTEXT_USERS, and the application context for that user is set to that value. The following steps demonstrate the implementation of this scenario.

1. Using DBSEC that has privileges to create tables and other database objects, create the application context table APP_CONTEXT_USERS. If this table already exists you may either drop it or rename it:

```
SQL> CREATE TABLE APP_CONTEXT_USERS
 2 (
 3    APP_CONTEXT_ATTR  VARCHAR2(80),
 4    APP_CONTEXT_VALUE VARCHAR2(255),
 5    USER_NAME     VARCHAR2(30)
 6 )
 7 /

Table created.
```

2. As DBSEC, create the ORDERS table:

```
SQL> CREATE TABLE ORDERS
  2  (
  3     ORDER_ID     NUMBER,
  4     ORDER_DATE   DATE,
  5     CUSTOMER_ID  NUMBER,
  6     ORDER_AMOUNT NUMBER,
  7     CTL_REC_STAT NUMBER
  8  )
  9  /

Table created.
```

3. As DBSEC, insert rows into the ORDERS table:

```
SQL> INSERT INTO ORDERS VALUES(1, SYSDATE, 200, 1203.22, 1);

1 row created.

SQL> INSERT INTO ORDERS VALUES(2, SYSDATE, 210, 5431.23, 1);

1 row created.

SQL> INSERT INTO ORDERS VALUES(3, SYSDATE, 212, 100023.09, 2);

1 row created.

SQL> INSERT INTO ORDERS VALUES(4, SYSDATE, 210, 999210.55, 3);

1 row created.

SQL> COMMIT;

Commit complete.

SQL> SELECT * FROM ORDERS
  2  /

  ORDER_ID ORDER_DAT CUSTOMER_ID ORDER_AMOUNT CTL_REC_STAT
---------- --------- ----------- ------------ ------------
         1 07-JUN-04         200      1203.22            1
         2 07-JUN-04         210      5431.23            1
         3 07-JUN-04         212    100023.09            2
         4 07-JUN-04         210    999210.55            3
```

4. As DBSEC, insert the rows into the APP_CONTEXT_USERS table:

```
SQL> INSERT INTO APP_CONTEXT_USERS VALUES('SECURITY_LEVEL', '1', 'SCOTT');

1 row created.

SQL> INSERT INTO APP_CONTEXT_USERS VALUES('SECURITY_LEVEL', '2', 'HR');

1 row created.

SQL> INSERT INTO APP_CONTEXT_USERS VALUES('SECURITY_LEVEL', '3', 'SYSTEM');

1 row created.

SQL> COMMIT;

Commit complete.
```

5. As DBSEC, create a VIEW object to display rows based on the application context, SECURITY_LEVEL. After you create this view, if you try to view all rows from the view, you will not get any rows because the application context attribute SECURITY_LEVEL is not set.

```
SQL> CREATE VIEW ORDERS_VIEW AS
  2      SELECT ORDER_ID, ORDER_DATE, CUSTOMER_ID, ORDER_AMOUNT
  3        FROM ORDERS
  4       WHERE CTL_REC_STAT <= TO_NUMBER(SYS_CONTEXT('ORDERS_APP', 'SECURITY_LEVEL'))
  5  /

View created.

SQL> SELECT * FROM ORDERS_VIEW;

no rows selected
```

6. As DBSEC, create a context for ORDERS_APP to allow the user to set the context:

```
SQL> CREATE CONTEXT ORDERS_APP USING SYS.CONTEXT_PKG;

Context created.
```

7. In this step you will be creating a package. This package can be owned by SYS or SYSTEM, but if you prefer to use DBSEC you must provide EXECUTE grant privilege to DBSEC.

```
SQL> CREATE OR REPLACE PACKAGE CONTEXT_PKG AS
  2
  3      PROCEDURE SET_APP_CONTEXT(P_LEVEL VARCHAR2);
  4  END;
  5  /

Package created.

SQL> CREATE OR REPLACE PACKAGE BODY CONTEXT_PKG AS
  2
  3      PROCEDURE SET_APP_CONTEXT(P_LEVEL VARCHAR2) IS
  4      BEGIN
  5         DBMS_SESSION.SET_CONTEXT('ORDERS_APP', 'SECURITY_LEVEL', P_LEVEL);
  6      END;
  7  END;
  8  /

Package body created.
```

8. Grant the user CREATE ANY CONTEXT privilege and the execute privilege to DBSEC:

```
SQL> GRANT EXECUTE ON CONTEXT_PKG TO DBSEC;

Grant succeeded.

SQL> GRANT CREATE ANY CONTEXT TO DBSEC;

Grant succeeded.
```

9. Create a logon database trigger (you need to connect as SYSTEM or SYS):

```
SQL> CREATE OR REPLACE TRIGGER TRG_DB_LOGON
  2   AFTER LOGON ON DATABASE
  3  DECLARE
  4     V_LEVEL   VARCHAR2(255);
  5  BEGIN
  6     SELECT APP_CONTEXT_VALUE
  7       INTO V_LEVEL
  8       FROM HR.APP_CONTEXT_USERS
  9      WHERE USER_NAME = 'HR';
 10      CONTEXT_PKG.SET_APP_CONTEXT(V_LEVEL);
 11  END;
 12  /

Trigger created.
```

10. Connect as HR and select from the view. You will see only three rows out of a total of four rows:

```
SQL> CONN HR/HR@SEC
Connected.
SQL> SELECT * FROM ORDERS_VIEW
  2  /

  ORDER_ID ORDER_DAT CUSTOMER_ID ORDER_AMOUNT
---------- --------- ----------- ------------
         1 07-JUN-04         200      1203.22
         2 07-JUN-04         210      5431.23
         3 07-JUN-04         212    100023.09

SQL> SELECT SYS_CONTEXT('ORDERS_APP', 'SECURITY_LEVEL') FROM DUAL;

SYS_CONTEXT('ORDERS_APP','SECURITY_LEVEL')
------------------------------------------
2
```

This is a useful but not simple implementation; Oracle created the VPD feature to simplify it. In the next section you learn how to implement the same security more simply using VPDs in Oracle.

Implementing Oracle Virtual Private Databases

In the previous section, you implemented row-level security using a combination of VIEW objects, triggers, and application contexts. This process is not extremely complicated, but there is a simpler way. Oracle provides VPD features that are more direct. First, examine the architecture of a VPD illustrated in Figure 6-6 to understand its mechanics.

Notice that the PROJECT table has a registered policy that rewrites the query by using a policy function to add a WHERE clause predicate. However, before beginning the implementation, you need to create the necessary users and tables.

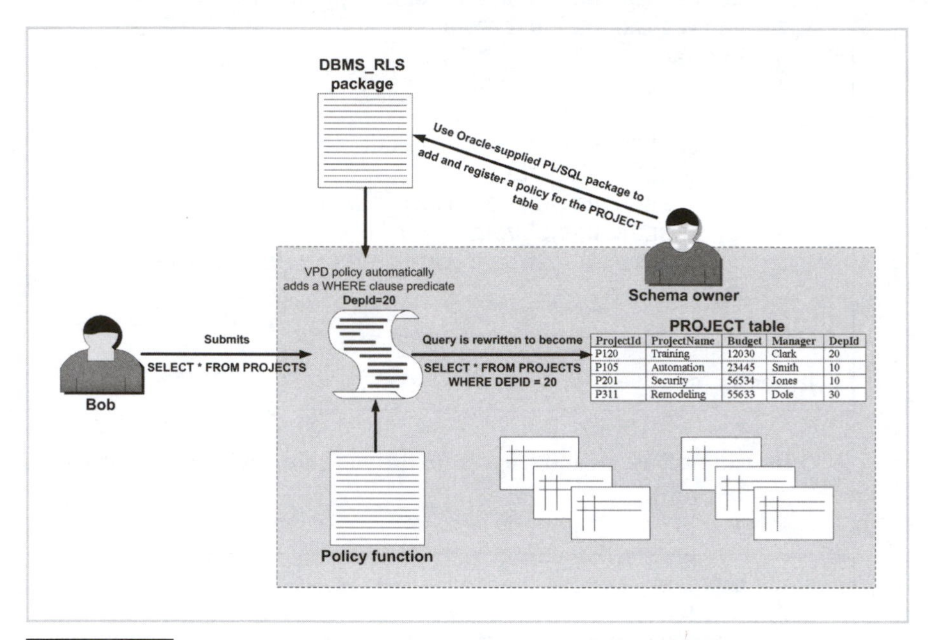

FIGURE 6-6 **Architecture of Oracle virtual private database feature**

Before creating a USER table, you need to understand the functions of the "users." The following list outlines the user functions as well as the entities and components involved in the scenarios contained in this chapter.

- *DBSEC user*—The application schema owner that owns all tables and the tables' policies
- *CUSTOMERS table*—Used to demonstrate VPDs, this table contains additional control columns (CTL) used by each scenario. In addition the table will be populated with 100 rows.
- *VPD_CLERK1, VPD_CLERK2, and VPD_CLERK3 users*—The database users that are used to test VPDs

Create the table for customer users through the following steps:

1. Create the CUSTOMERS table:

```
SQL> CREATE TABLE CUSTOMERS
  2  (
  3     SALES_REP_ID   NUMBER(4),
  4     CUSTOMER_ID    NUMBER(8),
  5     CUSTOMER_SSN   VARCHAR2(9),
  6     FIRST_NAME     VARCHAR2(20),
  7     LAST_NAME      VARCHAR2(20),
  8     ADDR_LINE      VARCHAR2(80),
  9     CITY           VARCHAR2(30),
 10     STATE          VARCHAR2(30),
 11     ZIP_CODE       VARCHAR2(9),
 12     PHONE          VARCHAR2(15),
 13     EMAIL          VARCHAR2(80),
 14     CC_NUMBER      VARCHAR2(20),
 15     CREDIT_LIMIT   NUMBER,
 16     GENDER         CHAR(1),
 17     STATUS         CHAR(1),
 18     COMMENTS       VARCHAR2(1024),
 19     CTL_UPD_DTTM   DATE,
 20     CTL_UPD_USER   VARCHAR2(30),
 21     CTL_REC_STAT   CHAR(1)
 22  );
Table created.
```

2. Insert rows into the CUSTOMERS table, using the script `dbsec_vpd_script_06.sql` which should populate 100 rows. The columns used in each scenario categorize the row count statistics.Here are three definitions that may help:
   - *CTL_UPD_USER*—Contains the name of the user who inserted the row and is allowed to update it
   - *CTL_REC_STAT*—Contains the security level of the row (1–5)
   - *SALES_REP_ID*—Represents the identification number of the sales representative assigned to a customer

```
-- USED FOR SCENARIO #1
SQL> SELECT CTL_UPD_USER, COUNT(*)
  2    FROM CUSTOMERS
  3   GROUP BY CTL_UPD_USER
  4  /

CTL_UPD_USER                        COUNT(*)
------------                        --------
VPD_CLERK1                                21
VPD_CLERK2                                37
VPD_CLERK3                                42

-- USED FOR SCENARIO #2
SQL> SELECT SALES_REP_ID, COUNT(*)
  2    FROM CUSTOMERS
  3   GROUP BY SALES_REP_ID
  4  /

SALES_REP_ID   COUNT(*)
------------   --------
        2200         18
        2336         15
        4587          8
        4710          9
        5605         10
        6415         18
        7719         15
        9644          7

-- USED FOR SCENARIO #3
SQL> SELECT CTL_REC_STAT, COUNT(*)
  2    FROM CUSTOMERS
  3   GROUP BY CTL_REC_STAT
  4  /

CTL_REC_STAT   COUNT(*)
------------   --------
1                    12
2                    34
3                    26
4                    15
5                    13
```

3. Create three users for testing, VPD_CLERK1, VPD_CLERK2, and VPD_CLERK3, and grant the necessary privileges. Use SYSTEM or SYS to perform this step:

```
--VPD_CLERK1
SQL> CREATE USER VPD_CLERK1 IDENTIFIED BY VPD_CLERK1
  2  /

User created.

SQL> GRANT CREATE SESSION TO VPD_CLERK1
  2  /

Grant succeeded.

--VPD_CLERK2
SQL> CREATE USER VPD_CLERK2 IDENTIFIED BY VPD_CLERK2
  2  /

User created.

SQL> GRANT CREATE SESSION TO VPD_CLERK2
  2  /

Grant succeeded.


-- VPD_CLERK3
SQL> CREATE USER VPD_CLERK3 IDENTIFIED BY VPD_CLERK3
  2  /

User created.

SQL> GRANT CREATE SESSION TO VPD_CLERK3
  2  /

Grant succeeded.
```

4. Grant the necessary privileges on the CUSTOMERS table to use each test:

```
SQL> GRANT SELECT, INSERT, UPDATE, DELETE ON CUSTOMERS TO VPD_CLERK1
  2  /

Grant succeeded.

SQL> GRANT SELECT, INSERT, UPDATE, DELETE ON CUSTOMERS TO VPD_CLERK2
  2  /

Grant succeeded.

SQL> GRANT SELECT, INSERT, UPDATE, DELETE ON CUSTOMERS TO VPD_CLERK3
  2  /

Grant succeeded.
```

Now you are ready to implement each scenario. The first scenario, ROW-OWNER security, is an application that enforces row-level security based on the user that owns the row. For example, if VPD_CLERK1 inserts a row, CTL_UPD_USER is populated with the name of the user (VPD_CLERK1) to identify the user that owns the row. Each user will only see and update his own rows. Explore this concept by following these steps:

1. Create a policy function to add a predicate to the WHERE clause. As illustrated earlier, the VPD feature adds a policy that uses a function to add a WHERE clause predicate. In this case, the predicate filters out the rows that are owned by the current user.
2. Using DBMS_RLS, which is an Oracle-supplied package, add the VPD policy used to enforce VPDs: RLS (row-level security). DBMS_RLS contains several functions and procedures. Table 6-2 lists those most commonly used. For full details on this package, consult Oracle documentation.

Table 6-2 Most commonly used procedure of DBMS_RLS

Procedure	Description
PROCEDURE ADD_POLICY	Adds and registers a VPD policy for a table
PROCEDURE ADD_POLICY_CONTEXT	Adds an application context to a policy
PROCEDURE DROP_POLICY	Removes a VPD policy from a table
PROCEDURE ENABLE_POLICY	Enables or disables a policy

The information in this table is derived from the online documentation that Oracle provides at the Oracle Technology Network site: *www.otn.oracle.com.*

```
SQL> EXEC DBMS_RLS.ADD_POLICY(OBJECT_SCHEMA   => 'DBSEC', -
>                            OBJECT_NAME     => 'CUSTOMERS', -
>                            POLICY_NAME     => 'DBSEC_ROW_OWNER_POLICY', -
>                            FUNCTION_SCHEMA => 'DBSEC', -
>                            POLICY_FUNCTION => 'DBSEC_ROW_OWNER_WHERE', -
>                            STATEMENT_TYPES => 'SELECT, UPDATE, INSERT, DELETE', -
>                            ENABLE => TRUE)

PL/SQL procedure successfully completed.
```

3. Log in as VPD_CLERK1 and display the number of records that this user can see:

```
SQL> CONN VPD_CLERK1/VPD_CLERK1@SEC
Connected.
SQL> SELECT COUNT(*) FROM DBSEC.CUSTOMERS
  2  /

  COUNT(*)
----------
        21
```

As you can see from this data, VPD_CLERK1 owns 21 rows, which is exactly the same as in the previous results. Disable this policy using DBMS_RLS:

```
SQL> EXEC DBMS_RLS.ENABLE_POLICY(OBJECT_SCHEMA=>'DBSEC', -
>                                OBJECT_NAME  =>'CUSTOMERS', -
>                                POLICY_NAME  =>'DBSEC_ROW_OWNER_POLICY', -
>                                ENABLE       =>FALSE)

PL/SQL procedure successfully completed.
```

In the second scenario APPLICATION-CONTEXT, you use an application context to allow specific users to see only rows for a specific sales representative. For example, VPD_CLERK1 is assigned to see customers that have sales representative 2336. Follow these steps for a complete implementation of this scenario.

1. Create the DBSEC_CUSTOMERS_APP_CONTEXT table that contains a row for each user with the appropriate sales representative ID assigned to the user:

```
SQL> CREATE TABLE DBSEC_CUSTOMERS_APP_CONTEXT
  2  (
  3     SALES_REP_ID      NUMBER,
  4     USER_NAME         VARCHAR2(30)
  5  )
  6  /
Table created.
SQL> GRANT SELECT ON DBSEC_CUSTOMERS_APP_CONTEXT
  2     TO VPD_CLERK1, VPD_CLERK2, VPD_CLERK3
  3  /
Grant succeeded.
```

2. Insert rows into the DBSEC_CUSTOMERS_APP_CONTEXT table:

```
SQL> INSERT INTO DBSEC_CUSTOMERS_APP_CONTEXT ( SALES_REP_ID, USER_NAME )
  2 VALUES ( 2336, 'VPD_CLERK1');
1 row created.
SQL> INSERT INTO DBSEC_CUSTOMERS_APP_CONTEXT ( SALES_REP_ID, USER_NAME )
  2 VALUES ( 9644, 'VPD_CLERK2');
1 row created.
SQL> INSERT INTO DBSEC_CUSTOMERS_APP_CONTEXT ( SALES_REP_ID, USER_NAME )
  2 VALUES ( 4587, 'VPD_CLERK3');
1 row created.
SQL> COMMIT;
Commit complete.
```

3. Create a trusted package that allows DBSEC to execute DBMS_SESSION. You must create this package as SYS:

```
SQL> CREATE OR REPLACE PACKAGE PKG_DBSEC_CUST_SALES_REP AS
  2
  3      PROCEDURE SET_CONTEXT;
  4  END;
  5  /
Package created.
SQL> CREATE OR REPLACE PACKAGE BODY PKG_DBSEC_CUST_SALES_REP IS
  2     PROCEDURE SET_CONTEXT IS
  3        V_SALES_REP_ID  NUMBER;
  4     BEGIN
  5
  6        SELECT SALES_REP_ID
  7          INTO V_SALES_REP_ID
  8          FROM DBSEC.DBSEC_CUSTOMERS_APP_CONTEXT
  9         WHERE USER_NAME = USER;
 10
 11        DBMS_SESSION.SET_CONTEXT('DBSEC_CUSTOMERS_SALESREP',
 'SALES_REPID', V_SALES_REP_ID);
 12     END;
 13  END;
 14  /
Package body created.
```

4. Create an application context for this policy:

```
SQL> CREATE OR REPLACE CONTEXT DBSEC_CUSTOMERS_SALESREP
  2    USING PKG_DBSEC_CUST_SALES_REP
  3  /

Context created.
```

5. Create a new VPD function policy to add a WHERE clause predicate:

```
SQL> CREATE OR REPLACE FUNCTION
  2     DBSEC_CUST_SALESREP_WHERE(P_SCHEMA_NAME IN VARCHAR2,
  3                              P_OBJECT_NAME IN VARCHAR2) RETURN VARCHAR2 IS
  4   V_WHERE   VARCHAR2(4000);
  5 BEGIN
  6
  7     V_WHERE := 'SALES_REP_ID = ' ||
SYS_CONTEXT('DBSEC_CUSTOMERS_APP_CONTEXT', 'SALES_REPID');
  8     RETURN V_WHERE;
  9  END;
 10  /

Function created.
```

6. Add a VPD policy for the CUSTOMERS table:

```
SQL> EXEC DBMS_RLS.ADD_POLICY(OBJECT_SCHEMA   => 'DBSEC', -
>                            OBJECT_NAME     => 'CUSTOMERS', -
>                            POLICY_NAME     => 'DBSEC_CUST_SALESREP_POLICY', -
>                            FUNCTION_SCHEMA => 'DBSEC', -
>                            POLICY_FUNCTION => 'DBSEC_CUST_SALESREP_WHERE', -
>                            STATEMENT_TYPES => 'SELECT, UPDATE, INSERT, DELETE', -
>                            ENABLE => TRUE)

PL/SQL procedure successfully completed.
```

7. Create an after-logon trigger to set the application context:

```
CREATE OR REPLACE TRIGGER TRG_AFTER_LOGON
 AFTER LOGON
 ON DATABASE
BEGIN
 PKG_DBSEC_CUST_SALES_REP.SET_CONTEXT;
END;
/
```

8. Now log on as VPD_CLERK2. This user should only see rows for SALES_REP_ID 9644 based on the DBSEC_CUSTOMERS_APP_CONTEXT table. This has been fairly easy to implement.

This scenario is called ROLE SECURITY LEVEL. In this scenario the application detects the role of the user. For example, if the role is clerk, the security level for the user assigned to that role is set to 2. See Table 6-3 for more details. After the application detects the security level, a predicate is used to filter the rows that can be seen by each user. For example, a security level 3 role can see any row that is assigned to CTL_REC_STAT equal to 3 or below.

Table 6-3 Setup for using roles to determine security levels

User	Assigned Role	Security Level
VPD_CLERK1	CLERK_ROLE	2
VPD_CLERK2	MANAGER_ROLE	3
VPD_CLERK3	ADMIN_ROLE	4

Follow these steps for a demonstration:

1. Make sure to disable any policies on the CUSTOMERS table as well as disabling the AFTER LOGON database trigger. Then create three roles as specified in Table 6-3 and assign each role to a corresponding user:

```
SQL> CREATE ROLE CLERK_ROLE;

Role created.

SQL> CREATE ROLE MANAGER_ROLE;

Role created.

SQL> CREATE ROLE ADMIN_ROLE;

Role created.

SQL> GRANT CLERK_ROLE TO VPD_CLERK1;

Grant succeeded.

SQL> GRANT MANAGER_ROLE TO VPD_CLERK2;

Grant succeeded.

SQL> GRANT ADMIN_ROLE TO VPD_CLERK3;

Grant succeeded.

SQL> GRANT SELECT ON DBA_ROLE_PRIVS TO DBSEC
  2  /

Grant succeeded.
```

2. Create an application context for the security level:

```
SQL> CREATE OR REPLACE CONTEXT DBSEC_ROLE_SECURITY_LEVEL
  2    USING PKG_DBSEC_ROLE_SECURITY_LEVEL
  3  /

Context created.
```

3. Create an application context package to set the application context:

```
SQL> CREATE OR REPLACE PACKAGE PKG_DBSEC_ROLE_SECURITY_LEVEL IS
  2     PROCEDURE SET_CONTEXT(P_ROLE VARCHAR2 DEFAULT NULL);
  3  END;
  4  /

Package created.

SQL> CREATE OR REPLACE PACKAGE BODY Pkg_Dbsec_Role_Security_Level IS
  2
  3     PROCEDURE SET_CONTEXT(P_ROLE VARCHAR2 DEFAULT NULL) IS
  4
  5     V_USER      VARCHAR2(30);
  6        V_ROLE      VARCHAR2(30);
  7        V_LEVEL     NUMBER;
  8
  9     BEGIN
 10     V_USER := USER;
 11     V_ROLE := P_ROLE;
 12        IF P_ROLE IS NULL THEN
 13           SELECT GRANTED_ROLE
 14             INTO V_ROLE
 15             FROM DBA_ROLE_PRIVS
 16      WHERE GRANTEE = V_USER;
 17        END IF;
 18
 19        IF     V_ROLE = 'CLERK_ROLE' THEN
 20           V_LEVEL := 2;
 21        ELSIF V_ROLE = 'MANAGER_ROLE' THEN
 22           V_LEVEL := 3;
 23        ELSIF V_ROLE = 'ADMIN_ROLE' THEN
 24           V_LEVEL := 4;
 25        ELSE
 26           V_LEVEL := 0;
 27        END IF;
 28
 29        DBMS_SESSION.SET_CONTEXT('DBSEC_ROLE_SECURITY_LEVEL',
'SECURITY_LEVEL', V_LEVEL);
 30
```

```
 31      EXCEPTION WHEN OTHERS THEN
 32         DBMS_OUTPUT.PUT_LINE(SQLERRM);
 33         RETURN;
 34      END;
 35
 36   END;
 37   /

Package body created.

SQL> GRANT EXECUTE ON PKG_DBSEC_ROLE_SECURITY_LEVEL TO
VPD_CLERK1, VPD_CLERK2, VPD_CLERK3
  2  /

Grant succeeded.
```

4. Create a policy function to implement row-level security (VPD):

```
SQL> CREATE OR REPLACE FUNCTION
  2              DBSEC_ROLE_SECURITY_LEVEL(P_SCHEMA_NAME IN VARCHAR2,
  3                           P_OBJECT_NAME IN VARCHAR2) RETURN VARCHAR2 IS
  4
  5      V_WHERE   VARCHAR2(4000);
  6   BEGIN
  7
  8      V_WHERE := 'TO_NUMBER(CTL_REC_STAT) <= ' ||
  9                nvl(SYS_CONTEXT('DBSEC_ROLE_SECURITY_LEVEL',
'SECURITY_LEVEL'),0);
 10      RETURN V_WHERE;
 11   END;
 12   /

Function created.
```

5. Create a policy to enforce a WHERE clause predicate. (A policy name cannot exceed 30 characters.) Also, you can use a package for a policy function.

```
SQL> EXEC DBMS_RLS.ADD_POLICY(OBJECT_SCHEMA   => 'DBSEC', -
>                            OBJECT_NAME     => 'CUSTOMERS', -
>                            POLICY_NAME     => 'DBSEC_ROLE_SECURITY_POLICY', -
>                            FUNCTION_SCHEMA => 'DBSEC', -
>                            POLICY_FUNCTION => 'DBSEC_ROLE_SECURITY_LEVEL', -
>                            STATEMENT_TYPES => 'SELECT, UPDATE, INSERT, DELETE', -
>                            ENABLE          => TRUE)

PL/SQL procedure successfully completed.
```

6. Now the application logs on as VPD_CLERK3 and executes a PKG_DBSEC_ROLE_SECURITY_LEVEL package. You should notice in the code that follows that before you set the application context for a security level, the number of rows returned is 0, but after setting security level to 4 for VPD_CLERK3 role, the user is able to see 87 rows:

```
SQL> CONN VPD_CLERK3/VPD_CLERK3@SEC
Connected.
SQL> SELECT SYS_CONTEXT('DBSEC_ROLE_SECURITY_LEVEL', 'SECURITY_LEVEL')
  2    FROM DUAL
  3  /

SYS_CONTEXT('DBSEC_ROLE_SECURITY_LEVEL','SECURITY_LEVEL')
---------------------------------------------------------
SQL> SELECT COUNT(*) FROM DBSEC.CUSTOMERS
  2  /

  COUNT(*)
  --------
         0

SQL> EXEC DBSEC.PKG_DBSEC_ROLE_SECURITY_LEVEL.SET_CONTEXT

PL/SQL procedure successfully completed.

SQL> SELECT SYS_CONTEXT('DBSEC_ROLE_SECURITY_LEVEL', 'SECURITY_LEVEL')
  2    FROM DUAL
  3  /

SYS_CONTEXT('DBSEC_ROLE_SECURITY_LEVEL','SECURITY_LEVEL')
---------------------------------------------------------
4

SQL> SELECT COUNT(*) FROM DBSEC.CUSTOMERS
  2  /

  COUNT(*)
  --------
        87
```

7. Repeat Step 6, but this time use VPD_CLERK1:

```
SQL> CONN VPD_CLERK1/VPD_CLERK1@SEC
Connected.
SQL> SELECT SYS_CONTEXT('DBSEC_ROLE_SECURITY_LEVEL', 'SECURITY_LEVEL')
  2    FROM DUAL
  3  /

SYS_CONTEXT('DBSEC_ROLE_SECURITY_LEVEL','SECURITY_LEVEL')
---------------------------------------------------------


SQL> SELECT COUNT(*) FROM DBSEC.CUSTOMERS
  2  /

  COUNT(*)
  --------
         0

SQL> EXEC DBSEC.PKG_DBSEC_ROLE_SECURITY_LEVEL.SET_CONTEXT

PL/SQL procedure successfully completed.

SQL> SELECT SYS_CONTEXT('DBSEC_ROLE_SECURITY_LEVEL', 'SECURITY_LEVEL')
  2    FROM DUAL
  3  /

SYS_CONTEXT('DBSEC_ROLE_SECURITY_LEVEL','SECURITY_LEVEL')
---------------------------------------------------------
2

SQL> SELECT COUNT(*) FROM DBSEC.CUSTOMERS
  2  /

  COUNT(*)
  --------
        46
```

You should know that VPD policies can be grouped for organizational purposes. For example, suppose you have several policies for each table in the application schema. If so, it would be logical to group them together in a policy group. One final note: In the scenarios presented in this section, you worked through row-level security enforcement using the SELECT statement. Oracle also enforces row-level security using all the DML statements: UPDATE, INSERT, and DELETE.

At this point, you should understand why the implementation of row-level security is better organized and simpler with the Oracle VPD feature. Now, suppose you want to see what VPD policies and their related information exist in your database. To view them, you use the data dictionary or the Policy Manager tool provided by Oracle Enterprise Manager. The next two parts of this chapter cover VPD policies.

Viewing VPD Policies and Application Context Using the Data Dictionary

Oracle is rich with data dictionary views that enable you to see everything created and stored in the database. VPDs are no exception. Oracle provides a set of data dictionary views for all application contexts. Oracle also provides data dictionary views for VPD policies. Table 6-4 lists these views.

Table 6-4 Oracle policy and application context data dictionary views

View Name	Description
DBA_POLICIES	Contains all policies that are created in the database and their attributes
ALL_POLICIES	Contains all policies that the current user owns and has access to and their attributes
USER_POLICIES	Contains all policies owned by the current user and their attributes
V$VPD_POLICY	Lists all policies that have been used and executed and are still cached in memory
ALL_CONTEXT	Lists all contexts that the current user owns or has privileges to view
SESSION_CONTEXT	Lists all active contexts for the current session

The information in this table is derived from the online documentation that Oracle provides at the Oracle Technology Network site: *www.otn.oracle.com.*

The following queries show the contents of some of the data dictionary views listed in Table 6-4:

```
SQL> SELECT OBJECT_NAME, POLICY_NAME, ENABLE
  2    FROM USER_POLICIES
  3  /

OBJECT_NAME                     POLICY_NAME                    ENABLE
-----------                     --------------------------     ------
CUSTOMERS                       DBSEC_CUST_SALESREP_POLICY     NO
CUSTOMERS                       DBSEC_ROW_OWNER_POLICY         NO
CUSTOMERS                       DBSEC_ROLE_SECURITY_POLICY     YES
```

```
SQL> SELECT SQL_HASH, SQL_ID, OBJECT_OWNER OWNER,
  2         OBJECT_NAME OBJECT, POLICY, PREDICATE
  3    FROM V$VPD_POLICY
  4  /

SQL_HASH   SQL_ID              OWNER     OBJECT    POLICY                   PREDICATE
---------- ------------------- --------- --------- ------                   ---------
4193346388 djn29uvwz2sun DBSEC CUSTOMERS DBSEC_ROLE_SECURITY_POLICY
TO_NUMBER(CTL_REC_STAT) <= 2
4193346388 djn29uvwz2sun DBSEC CUSTOMERS DBSEC_ROLE_SECURITY_POLICY
TO_NUMBER(CTL_REC_STAT) <= 4
4193346388 djn29uvwz2sun DBSEC CUSTOMERS DBSEC_ROLE_SECURITY_POLICY
TO_NUMBER(CTL_REC_STAT) <= 0
```

```
SQL> EXEC DBSEC.PKG_DBSEC_ROLE_SECURITY_LEVEL.SET_CONTEXT('CLERK_ROLE');

PL/SQL procedure successfully completed.

SQL> SELECT * FROM all_CONTEXT
  2  /

NAMESPACE                  SCHEMA  PACKAGE
---------                  ------  -------
DBSEC_ROLE_SECURITY_LEVEL DBSEC   PKG_DBSEC_ROLE_SECURITY_LEVEL

SQL> SELECT * FROM SESSION_CONTEXT
  2  /

NAMESPACE                  ATTRIBUTE      VALUE
---------                  ---------      -----
DBSEC_ROLE_SECURITY_LEVEL SECURITY_LEVEL 2
```

Viewing VPD Policies and Application Contexts Using Policy Manager

Everyone appreciates interface tools that are easy to use. Oracle provides such a tool for its VPD policies and application contexts. This tool is called Policy Manager. You can run this tool from the Oracle10*g* Enterprise Edition as illustrated in Figure 6-7.

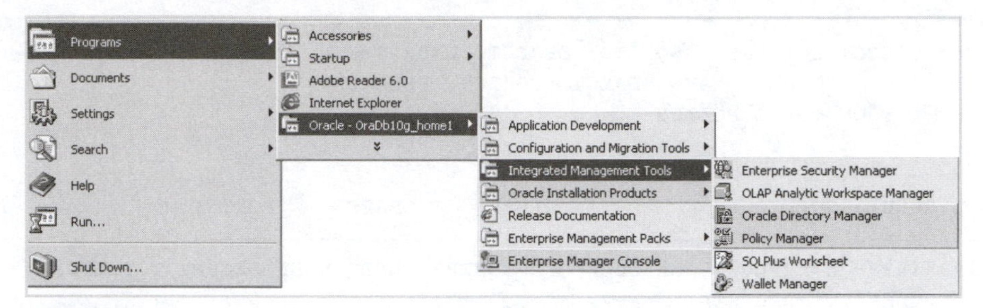

FIGURE 6-7 Policy Manager shortcut in the Start menu

After you click the shortcut shown in Figure 6-7, a login screen is displayed. (Note that although this screen logs you into Policy Manager, Oracle displays only Enterprise Manager on the screen.) You can use SYSTEM credentials to log in, as shown in Figure 6-8. After you are authenticated by the database, an Oracle Enterprise Manager Console type window is displayed, as illustrated in Figure 6-9. In the figure, you see that the left side shows that the FGA control policies are divided into two parts: Policy groups and Application contexts. This tool allows the administration of VPD policies and application contexts.

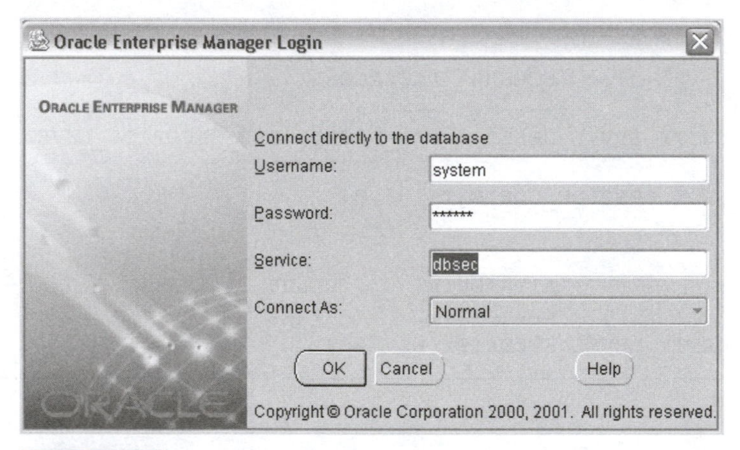

FIGURE 6-8 Logging into Oracle Policy Manager

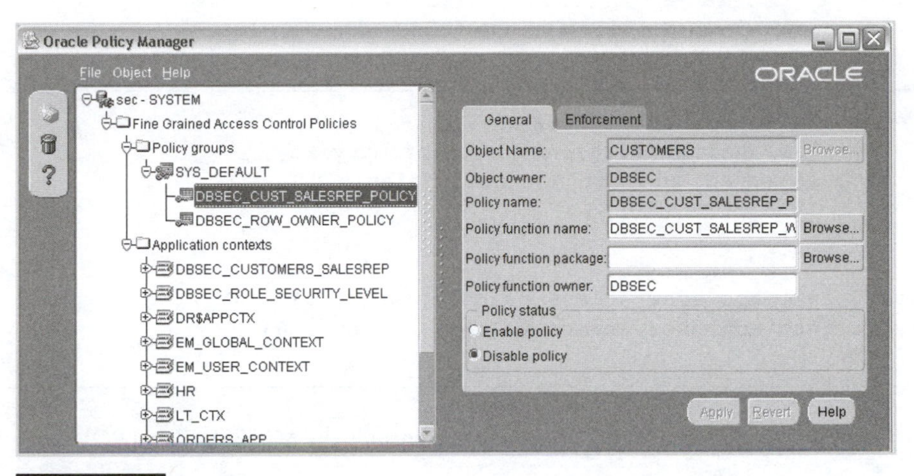

FIGURE 6-9 Oracle Policy Manager

Implementing Row- and Column-level Security with SQL Server

Although Microsoft SQL Server 2000 does not directly support VPDs, you can mimic their functionality. You might recall that in Oracle the DBMS automatically appends a WHERE clause to your queries and then pulls only records that belong to you. Using views and expanding on some of the security models that you learned about in Chapter 5, you can create your own VPDs, complete with access security policies.

Row-based Security Using Access Levels

To increase the level of detail of row-based security, you can also implement a variation of both the application table-based security model and the application function-based security model.

Recall the access-level control model from Chapter 5. In this model, you use access levels from 0 to 5 where:

0 = No access
1 = select
2 = select, insert
3 = select, insert, update
4 = select, insert, update, delete
5 = administrator access

To implement this at the row level, use a table that contains users and their access codes. A column is added to each table to hold the access level of that row.

1. Create the APPLICATION USERS table:

```
create table app_user_access (
    username varchar(128) not null primary key,
    access_level  tinyint not null default 0
)
go

insert into app_user_access values ('sam', 4)
insert into app_user_access values ('jason', 0)
go
```

2. Alter the CUSTOMER table to include the ACCESS CONTROL column:

```
alter table customer
    add access_level  integer
go
```

3. With the security structure in place, you need a way to retrieve the data. One way to retrieve data is by using a view:

```
create view vCustomer_Secure
as
      select customer_id, first_name, last_name, street, city, state, zip,
 phone,
       gender, date_of_birth, comments
      from customer
      where access_level > 0
and access_level <= (select isnull(access_level, 0)
                              from app_user_access where username = user)
go
```

4. Apply privileges:

```
grant select on vCustomer_Secure to sam
grant select on vCustomer_Secure to jason
go
```

The drawback of this view is that it allows insertion, update, and deletion of records in the underlying table, regardless of your application's security. A preferred method is to use a stored procedure to return the data and revoke privileges on the tables. Follow these steps to demonstrate this method:

1. Create a procedure:

```
create procedure Customer_Sel
as
      select customer_id, first_name, last_name, street, city, state, zip,
 phone,
       gender, date_of_birth, comments
      from customer
      where access_level > 0
and access_level <= (select isnull(access_level, 0)
from app_user_access where username = user)
go
```

2. Apply privileges:

```
grant execute on Customer_Sel to sam
grant execute on Customer_Sel to jason
go
```

In either case, you must use stored procedures to perform insert, update, and delete operations, so that your access level can be verified. The following steps show you how this is done:

1. Create the procedure:

```
create procedure Customer_Del (@id int)
as
      declare @level tinyint;
select @level = (select isnull(access_level, 0)
  from app_user_access where username = user);
    if @level >= 4 begin
          delete from customer where customer_id = @id and access_level
>= @level;
    end
go
```

2. Apply privileges:

```
grant execute on Customer_Del to sam
grant execute on Customer_Del to jason
go
```

These models can be much more complex, but this gives you a good idea of how you can implement access-level security at the row level.

Row-based Security Using Application Functions

An alternative to using access-level control for row-based security is to use application functions. This is implemented in the same way, except that your access table lists a function instead of an access level. Here is an example of the application function:

```
create table app_user_function (
    username varchar(128) not null primary key,
    function_name varchar(50) not null default 'NONE'
)
Go
```

Column-based Security

You have learned how to enforce security on the row level. There is one more important element at this level of security implementation, and that is column-level security. This section presents a step-by-step demonstration of implementing column-level security in Oracle and SQL Server. It is important to note that VPD column-level security is only available in Oracle10*g*.

VPD and Column Access Using Oracle

Oracle extended the VPD feature from rows to columns. You already know how to create a policy to restrict rows, as you did in the last VPD scenario, which covered role security levels. You now use a similar process to control security through column access. Follow these steps:

1. Log in as VPD_CLERK2 and view rows and columns in the CUSTOMERS table. VPD_CLERK2 must have EXECUTE privilege on PKG_DBSEC_ROLE_SECURITY_LEVEL granted by DBSEC:

```
SQL> CONN VPD_CLERK2/VPD_CLERK2@SEC
Connected.
SQL> EXEC DBSEC.PKG_DBSEC_ROLE_SECURITY_LEVEL.SET_CONTEXT

PL/SQL procedure successfully completed.

SQL> SELECT COUNT(*) FROM DBSEC. CUSTOMERS
/


  COUNT(*)
  --------
        72
```

```
SQL> SELECT * FROM DBSEC.CUSTOMERS
  2   WHERE ROWNUM < 2
  3  /

Note: Output is not shown here because of the excessive amount of data that
results from the  query. However you should see all the columns that exist
 in the CUSTOMERS table.
```

2. Log in as the DBSEC user and recreate the policy on customers to restrict columns PHONE and EMAIL. Two parameters are added to the call of the procedure that create a new policy. They are:
   - *SEC_RELEVANT_COLS*—Contains all columns whose values are to be hidden from the user. Column names must be separated by a comma or a space.
   - *SEC_RELEVANT_COLS_OPT*—Notifies Oracle to hide or reveal column values. Use DBMS_RLS.ALL_ROWS for every row in the table.

DBSEC must have EXECUTE privilege on DBMS_RLS in order to do this step.

```
SQL> EXEC DBMS_RLS.DROP_POLICY (OBJECT_SCHEMA => 'DBSEC', -
>                 OBJECT_NAME    => 'CUSTOMERS', -
>                 POLICY_NAME    => 'DBSEC_ROLE_SECURITY_POLICY')

PL/SQL procedure successfully completed.

SQL> EXEC DBMS_RLS.ADD_POLICY(OBJECT_SCHEMA           => 'DBSEC', -
>                 OBJECT_NAME            => 'CUSTOMERS', -
>                 POLICY_NAME            => 'DBSEC_ROLE_SECURITY_POLICY', -
>                 FUNCTION_SCHEMA        => 'DBSEC', -
>                 POLICY_FUNCTION        => 'DBSEC_ROLE_SECURITY_LEVEL', -
>                 STATEMENT_TYPES        => 'SELECT, UPDATE, INSERT, DELETE', -
>                 SEC_RELEVANT_COLS      => 'PHONE,EMAIL', -
>                 POLICY_TYPE            => DBMS_RLS.CONTEXT_SENSITIVE, -
>                 SEC_RELEVANT_COLS_OPT  => DBMS_RLS.ALL_ROWS, -
>                 ENABLE                 => TRUE)
PL/SQL procedure successfully completed.
```

3. Log in as VPD_CLERK2 and query the CUSTOMERS table:

```
SQL> CONN VPD_CLERK2/VPD_CLERK2@SEC
Connected.
SQL> EXEC DBSEC.PKG_DBSEC_ROLE_SECURITY_LEVEL.SET_CONTEXT

PL/SQL procedure successfully completed.

SQL> SELECT FIRST_NAME, LAST_NAME, PHONE, EMAIL
  2    FROM DBSEC.CUSTOMERS
  3   WHERE ROWNUM < 2
  4  /

FIRST_NAME LAST_NAME PHONE      EMAIL
---------- --------- -----      -----
Otto       Huff
```

VPD_CLERK2 must have EXECUTE privilege on PKG_DBSEC_ROLE_SECURITY_LEVEL granted by DBSEC.

Column Privileges in Oracle

There is one final comment on column-level security for Oracle. Oracle has the capability to restrict updates or inserts at the column level, using GRANT UPDATE(column) and INSERT(column). Follow these steps for a demonstration.

1. Log in as DBSEC and create a table called TEST:

```
SQL> CONN DBSEC/DBSEC@SEC
Connected.
SQL> CREATE TABLE TEST
  2  (
  3    NUM  NUMBER,
  4    TEXT VARCHAR2(20)
  5  )
  6  /

Table created.
```

2. Grant SELECT on the TEST table to SCOTT:

```
SQL> GRANT SELECT ON TEST TO SCOTT
  2  /

Grant succeeded.
```

3. Grant UPDATE only on the column TEXT in the TEST table to SCOTT:

```
SQL> GRANT UPDATE(TEXT) ON TEST TO SCOTT
  2  /

Grant succeeded.
```

4. Insert a row into the TEST table and save it:

```
SQL> INSERT INTO TEST VALUES(1, 'LINE 1')
  2  /

1 row created.

SQL> COMMIT
  2  /

Commit complete.
```

5. Log in as SCOTT and query the TEST table owned by DBSEC:

```
SQL> CONN SCOTT/TIGER@SEC
Connected.
SQL> SELECT * FROM DBSEC.TEST
  2  /

       NUM TEXT
---------- ------
         1 LINE 1
```

6. Update the TEXT column in the TEST table:

```
SQL> UPDATE DBSEC.TEST SET TEXT = 'LINE 1 MOD BY SCOTT';

1 row updated.

SQL> COMMIT;

Commit complete.

SQL> SELECT * FROM DBSEC.TEST
  2  /

       NUM TEXT
---------- -------------------
         1 LINE 1 MOD BY SCOTT
```

7. Now try to update the NUM column in the TEST table; you will see that you cannot:

```
SQL> UPDATE DBSEC.TEST SET NUM = 3;
UPDATE DBSEC.TEST SET NUM = 3
             *
ERROR at line 1:
ORA-01031: insufficient privileges
```

The next section shows how access-level security can be implemented at the column level with SQL Server.

Access-level Control with SQL Server

The access-level security model can also be implemented at the column level. In addition to the APP_USER_ACCESS table, you need two other tables. One table maintains a list of tables (APP_TABLES), and another maintains a list of each column within each table and the related access level for each column (APP_COLUMNS).

1. Create the APP_TABLES table:

```
create table app_tables (
    table_id int not null primary key,
    table_name varchar(128) not null
)
go
```

2. Create the APP_COLUMNS columns:

```
create table app_columns (
    column_id int not null primary key,
    table_id int not null references app_tables(table_id),
    column_name varchar(128) not null,
    access_level tinyint not null default 0
)
Go
```

3. Because the access control must be verified on each column, you cannot use a simple view to return the data. All access to the tables must be performed with stored procedures.

```
create procedure Customer_Sel
as
      declare @qry varchar(4000), @level tinyint, @column varchar(128),
@counter int
      select @level = (select access_level app_user_access where username
= user);
      declare @columns cursor forward_only for
                  select column_name
                  from app_columns c
                  inner join app_tables t on c.table_id = t.table_id
                  where t.table_name = 'customer'
                        and c.access_level <= @level;

     select @qry = 'select ';
     select @counter = 0;
     open @columns;
     fetch next from @columns into @column
     while @@fetch_status = 0 begin
           if counter = 0 begin
                 select @qry = @qry + @column;
           end
           else begin
                 select @qry = @qry + ', ' + @column;
           end
           select @counter = @counter + 1;
           fetch next from @columns into @column;
    end
    select @qry = @qry + ' from customer';
    execute(@qry);
go
```

Column Privileges with SQL Server

In Chapter 4, you learned that Microsoft SQL Server 2000 supports privileges at the column level. This section presents steps for enforcing security on the column level.

1. Set update permissions for sam on the column phone in the Customer table:

```
grant update on customer(phone) to sam
go
```

Why would you use your own, customized, column-access control when you can simply set privileges on the columns? It all depends on your application. If you are implementing an access-level control model at either the table or row level, it may be better to implement column security in the same way. This facilitates management, because all security operations for an application are maintained within the application itself.

You just finished the database security part of the book. To tie concepts together, think over the applications that you have worked with and try to connect them back to the materials in the security part of the book. The second part of this book covers database auditing concepts, design, and implementation. Database auditing is another tool that you will use to validate security measures you put in place to protect data.

Chapter Summary

- A virtual private database allows or prevents data access at the row or column level.
- Row and column access can be implemented by using the VIEW database object.
- VPDs are also referred to as row-level security (RLS) or fine-grained access (FGA).
- One of the uses of a VIEW object is to limit what users can see and do with the existing data in a table.
- SQL Server does not support VPDs.
- The Microsoft SQL Server 2000 system function of USER returns the name of the connected database user.
- A trigger is a stored PL/SQL procedure that fires (calls) automatically when a specific event occurs.
- Application context is functionality specific to Oracle that allows the setting of database application variables that can be retrieved by database sessions.
- The SYS_CONTEXT function is used to get values for all predefined attributes in Oracle.
- In Oracle, you can set your own application context using the procedure supplied in the Oracle PL/SQL package called DBMS_SESSION.
- Use the SET_CONTEXT procedure within DBMS_SESSION to set a user's defined application context.

- Use the Oracle-supplied package DBMS_RLS to add the VPD policy used to enforce row- and column-level security.
- Oracle data dictionary views enable you to see everything created and stored in the database.
- Oracle Policy Manager is a graphical tool used to administer VPD policies.
- Oracle has the capability to restrict updates or inserts on columns, using GRANT UPDATE(column) and INSERT(column).

Review Questions

1. In this chapter, FGA stands for fine-grained auditing. True or false?
2. SQL Server does not provide the specific feature that Oracle does for implementing a VPD. True or false?
3. Oracle and SQL Server provide a function called USER to display the name of the connected user. True or false?
4. There is no drawback to using VIEW in SQL Server to implement a VPD. True or false?
5. Use SQL Server Policy Manager to view all policies created in the database. True or false?
6. Explain the key differences between SQL Server and Oracle in implementing VPDs.
7. Using Oracle SQL*Plus or iSQLPlus, display three predefined system context attributes listed in Table 6-1.
8. Using Oracle SQL*Plus or iSQLPlus, display three predefined system context attributes not listed in Table 6-1.
9. Create two application context attributes and verify the creation of the attributes by viewing the data dictionary views.
10. Create an application context using Policy Manager.
11. List two key benefits of VPDs.

Hands-on Projects

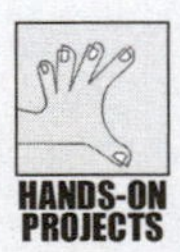

Hands-on Project 6-1

Using SQL Server views, set up row-based security for the Medicine table in the Pharmacy database.

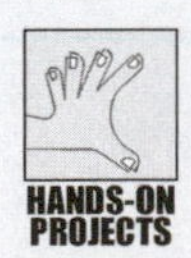

Hands-on Project 6-2

Using access-level control, set up row-level security on the PHARMACY database. Include the functionality for select, insert, update, and delete in SQL Server.

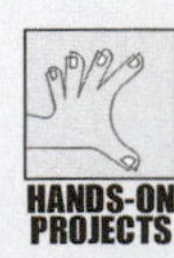

Hands-on Project 6-3

Using access-level control, set up column-level security on the Physician table. Include the functionality for select, insert, update, and delete in SQL Server.

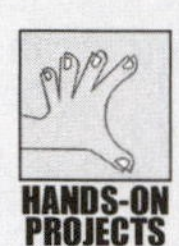

Hands-on Project 6-4

In Oracle, create a policy on a table that uses a user name to enforce row-level security.

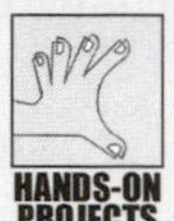

Hands-on Project 6-5

Using SQL Server 2000 column privileges, set up column-level security on the Prescription table.

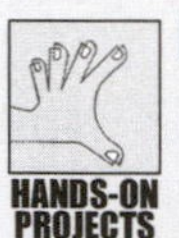

Hands-on Project 6-6

In Oracle, display all policies existing in the database.

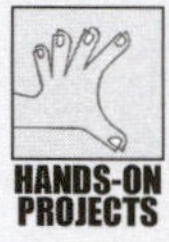

Hands-on Project 6-7

Using access-level control, set up row- and column-level security on the Drug Company table. Clerks can only edit phone numbers and add new rows, but managers can perform all DML operations in SQL Server.

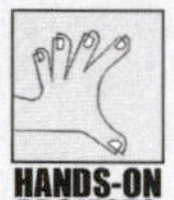

Hands-on Project 6-8

Using views, set up row-based security on the SIDE_EFFECTS table in SQL Server.

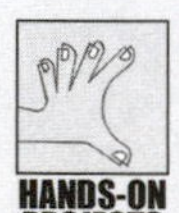

Hands-on Project 6-9

In Oracle, demonstrate how to set and use application contexts using any DML or database trigger.

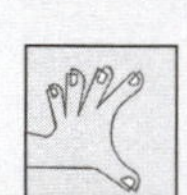

Hands-on Project 6-10

Using views, set up row-security on the Ingredient table in SQL Server.

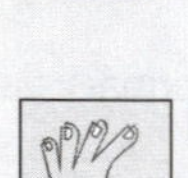

Hands-on Project 6-11

Using Oracle, demonstrate how INSERT(column) privileges work.

Hands-on Project 6-12

Using Oracle Policy Manager, create a policy on a table and then disable it, using the DBMS_RLS package.

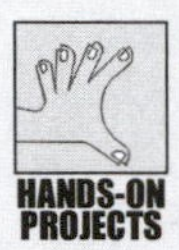

Hands-on Project 6-13

In Oracle, demonstrate how to implement the VIRTUAL PRIVATE view using VIEW and TRIGGER objects.

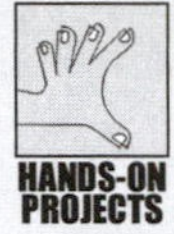

Hands-on Project 6-14

Draw a flowchart for all three methods used to implement VPDs. *Hint*: Methods are as follows:

- Using view and trigger database objects
- Using views and application context
- Using the VPD feature

Case Projects

Case 6-1 Working Virtually with SQL Server

You are contracted by an application service provider to configure a VPD so that the company can support multiple customers with a single instance of an application. Using the ORDERS database, configure a VPD for the following users: acme_user, penco_user, and mfg_user. The solution should support all DML operations.

Case 6-2 Working Privately with Oracle

You were promoted to a project leader position at a software company that specializes in security application development. Your manager assigned you to design and develop a prototype (template) for a client. The client will use this prototype to implement the application. Your job is to provide this prototype according to these requirements:

- The application must use the following security access levels: 0 = none, 1 = select, 2 = update, 3 = delete
- A user should only see rows that are created by the user.
- A user with security access of level 6 can use SELECT, UPDATE, and DELETE functions (1+2+3), even if the row does not belong to the current user.
- Each row is assigned a security level.
- Each user is assigned a security level.

PART TWO

7

Database Auditing Models

LEARNING OBJECTIVES

Upon completion of this material, you should be able to:

- Gain an overview of auditing fundamentals
- Understand the database auditing environment
- Create a flowchart of the auditing process
- List the basic objectives of an audit
- Define the differences between auditing classifications and types
- List the benefits and side effects of an audit
- Create your own auditing models

Introduction

A software engineer retired at age 65 and decided to invest some of his retirement money in a convenience store down the road from his house. He purchased the store to keep himself busy and generate income in addition to his retirement pension. He hired several teens from his community to help him operate the store. On the recommendation of the previous owner, he installed security cameras to monitor and record store activities 24 hours a day. Three months later he noticed that the store was not generating the sales income it should have and the inventory was shrinking. A friend advised him to keep an eye on his employees and told him that most shoplifting is done by insiders. The storeowner replied, "But I have cameras. They wouldn't dare." "Did you ever watch the tapes?" the friend asked. The storeowner was surprised by the question. "Of course not! How can I watch a 24-hour tape every day?" His friend asked, "You didn't even randomly watch portions of the tapes? That's your problem!" The moral of the story is that no matter what security measures you put in place, you must plan ahead with an auditing procedure to understand if your security measures are working or not. Otherwise, your effort and investment in security measures could be futile.

Security is the buzzword of this decade. It's on everyone's mind. In the past, the word crime connoted burglary or assault. Today, crime brings to mind a whole new set of risks to privacy and confidentiality. Security requires action. Many public and private institutions are taking serious action against security risks, old and new. These actions encompass not only the establishment and enforcement of new security measures, but also the reinforcement of those measures through tough audit controls.

Previous chapters introduced you to the building blocks necessary for developing a security model to suit most business requirements. In this chapter another key aspect of data confidentiality, integrity, and accessibility is presented—database auditing, thus providing the full complement to database security. The two concepts are inseparable. Together they ensure that your data is protected. You guard your data by enforcing database security, and you ensure that data is well guarded through database auditing. Database security is not effective without database auditing and vice versa. Figure 7-1 illustrates that when you decouple database auditing from database security, and vice versa, data violations emerge. To prevent data violations, the best practice is to have both database security and auditing in place.

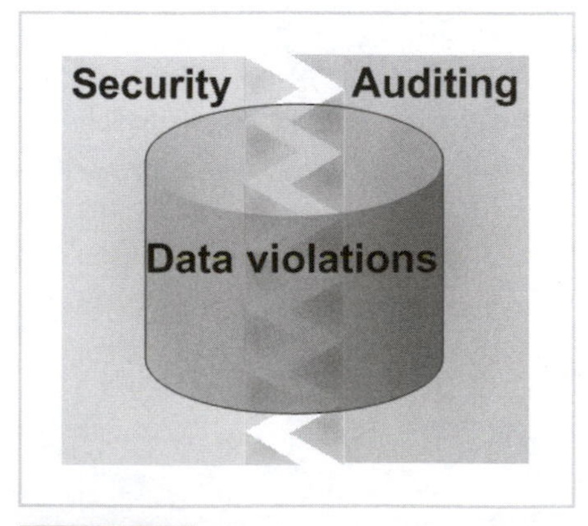

FIGURE 7-1 Database security and auditing

Auditing is the responsibility of developers, database administrators, and business managers. As a developer, you might be asked to provide an auditing mechanism for a new application. This auditing mechanism would enable users to trace changes to sensitive data. As a database administrator, you might be summoned to your manager's office, where the business managers are waiting to question you about a mischievous incident that left the database unavailable for hours. As a business manager for a financial company, you might be involved in supplying business requirements to a systems analyst, who is responsible for implementing an application's auditing functionality.

These are but a few of the endless examples of the need to keep your database security inseparable from audit functions. In this chapter you learn about database auditing, the importance of auditing, auditing types, and database auditing models.

Auditing Overview

To understand database auditing, you first need to know some fundamentals about auditing itself. In this section, you learn the definitions of basic auditing terms, you read a real-life auditing story, and you learn how security measures alone are not effective without good auditing processes.

Definitions

There are always stories in the news about companies being audited by the government for suspicious financial activities. Perhaps you know of someone who was audited by the Internal Revenue Service. Audits are conducted for many reasons. However, the goals of an audit are usually consistent. In general, an audit examines the documentation that reflects the actions, practices, and conduct of a business or individual. Then it measures their compliance to policies, procedure, processes, and law. Database auditing follows this general definition. Therefore the list that follows contains both general auditing and database auditing definitions.

- ***Audit/Auditing***—The process of examining and validating documents, data, processes, procedures, systems, or other activities to ensure that the audited entity complies with its objectives
- ***Audit log***—A document that contains all activities that are being audited ordered in a chronological manner. Usually, an automated system generates the log. For example, an audit log could be maintained to track visitors to a company, as well as their times of entrance and exit.
- ***Audit objectives***—A set of business rules, system controls, government regulations, or security policies against which the audited entity is measured to determine compliance
- ***Auditor***—A person with the proper qualifications and ethics, who is authorized to examine, verify, and validate documents, data, processes, procedures, systems, or activities and to produce an audit report
- ***Audit procedure***—A set of step-by-step instructions for performing the auditing process
- ***Audit report***—A document that contains the audit findings and is generated by an individual(s) conducting the audit
- ***Audit trail***—A chronological record of document changes, data changes, system activities, or operational events

- ***Data audit***—A chronological record of data changes stored in a log file or a database table object
- ***Database auditing***—A chronological record of database activities, such as shutdown, startup, logons, and data structure changes of database objects
- ***Internal auditing***—An examination, verification, and validation of documents, processes, procedures, sytems, or activities conducted by staff members of the organization being audited
- ***External auditing***—An examination, verification, and validation of documents, processes, procedures, systems, or activities conducted by staff members outside of the organization being audited

Auditing Activities

A large retail company based in New York was under the microscope after its name was in the national news. One of the company's customer care representatives, who had access to all the customer information, struck a deal with a prison inmate to provide customer information, including financial data for a sum of money. The inmate used this information for a variety of scams, including identity theft. Even though security was very tight at the company, a failure to perform a complete and thorough security audit allowed this incident to happen. Tight security cannot prevent or even minimize risk without tight auditing. Auditing should not be taken lightly. It must be thorough, and it must apply to every document, to every person, and of course to every system.

This section covers **auditing activities**. Auditing activities are performed as a part of an audit, audit process, or audit plan. Some of these activities can be thought of as the auditor's responsibilities or they can be incorporated into an organization's audit policies. The purpose of listing these activities is to broaden your understanding of the internal auditor's job. These activities are not listed in any specific order.

- Evaluate and appraise the effectiveness and adequacy of the audited entity according to the auditing objectives and procedures.
- Ascertain and review the reliability and integrity of the audited entity (the organization being audited).
- Ensure the organization being audited is in compliance with the policies, procedures, regulations, laws, and standards of the government and the industry.
- Establish plans, policies, and procedures for conducting audits.
- Keep abreast of all changes to the audited entity, such as design, process, or structure.
- Keep abreast of updates and new audit regulations, laws, standards, and policies set by industry, government, or the company itself.
- Provide all audit details to all company employees involved in the audit. These details include: resource requirements, audit plans, and audit schedules.

- Publish audit guidelines and procedures to the company itself and its partners and clients when appropriate.
- Act as liaison between the company and the external audit team.
- Act as a consultant to architects, developers, and business analysts to ensure that the company being audited is structured in accordance with the audited objectives.
- Organize and conduct internal audits.
- Ensure all contractual items are met by the organization being audited.
- Identify the audit types that will be used (audit types are explained later in this chapter).
- Work jointly with the Security Department to identify security issues that must be addressed.
- Provide consultation to the Legal Department to identify regulations and laws with which the company must comply.

The importance of these responsibilities is exemplified in the strict guidelines of the financial industry. Their standards and regulations dictate that all financial transactions must be stored for a specific time and all changes to an account must be recorded, including the name of the person who performed the change, the time of the change, and the previous state before it was changed. The purpose of this regulation is to uncover any discrepancies or unlawful activities with the account when it is audited and to assure that all bank account activities can be traced back to the moment of its inception.

Auditing Environment

The auditing practice that is most often publicized is the review of an organization's documents such as the financial statements to make sure that every change to the account is recorded and is accurate. An audit also assures that all company transactions comply with government regulations and laws. As mentioned previously, an audit can be a process of reviewing a system application, such as an inventory control system, to validate that all transactions are processed and stored according to the system objectives. Most importantly, an audit can be conducted as a review of the enforcement of security policies and procedures. These are only a few of the many reasons for audits. However, as diverse as these audits are, they all take place in an **auditing environment**. In this section you learn about the general auditing environment and its components (see Figure 7-2) and how it compares to the **database auditing environment**.

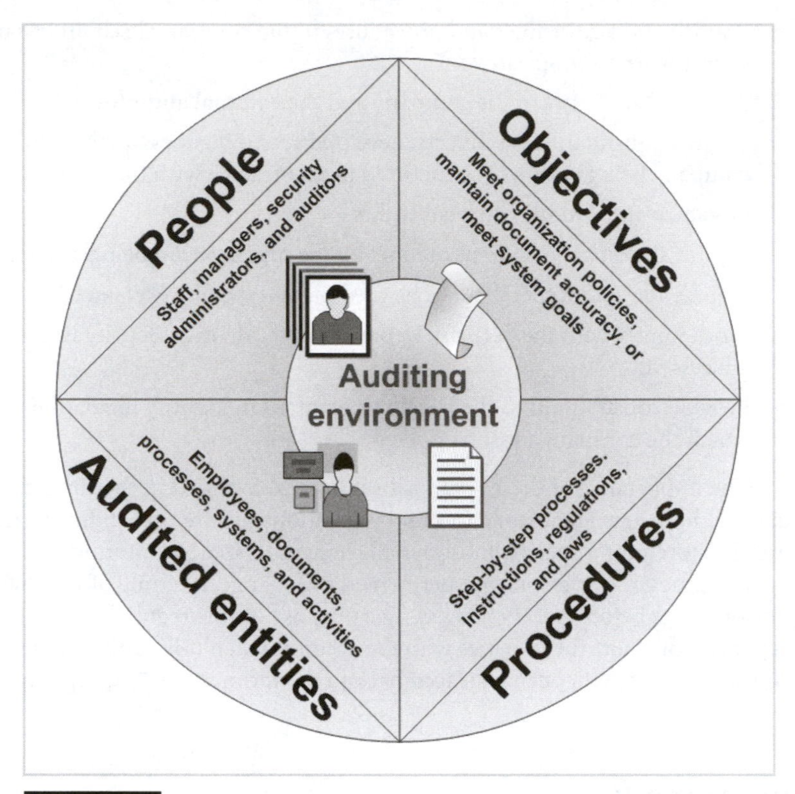

FIGURE 7-2 **Auditing environment components**

Figure 7-2 shows the four major components of the auditing environment:

- *Objectives*—An audit without a set of objectives is useless. To conduct an audit you must know what the audited entity is to be measured against. Usually the objectives are set by the organization, industry standards, or government regulations and laws.
- *Procedures*—To conduct an audit, step-by-step instructions and tasks must be documented ahead of time. In the case of a government-conducted audit, all instructions are available to the public. In the case of an organizational audit, specialized personnel document the procedures to be used not only for the business itself, but also for the audit.
- *People*—Every auditing environment must have an auditor, even in the case of an automatic audit, which is explained later in this chapter. Other people involved in the audit are employees, managers, and anyone being audited.
- *Audited entities*—This includes people, documents, processes, systems, activities, or any operations that are being audited.

The auditing environment presented in Figure 7-2 is applicable to all aspects of the business including the database. The auditing environment for database auditing differs slightly from the generic auditing environment in that it distinguishes the database from the other auditable entities, as shown in Figure 7-3.

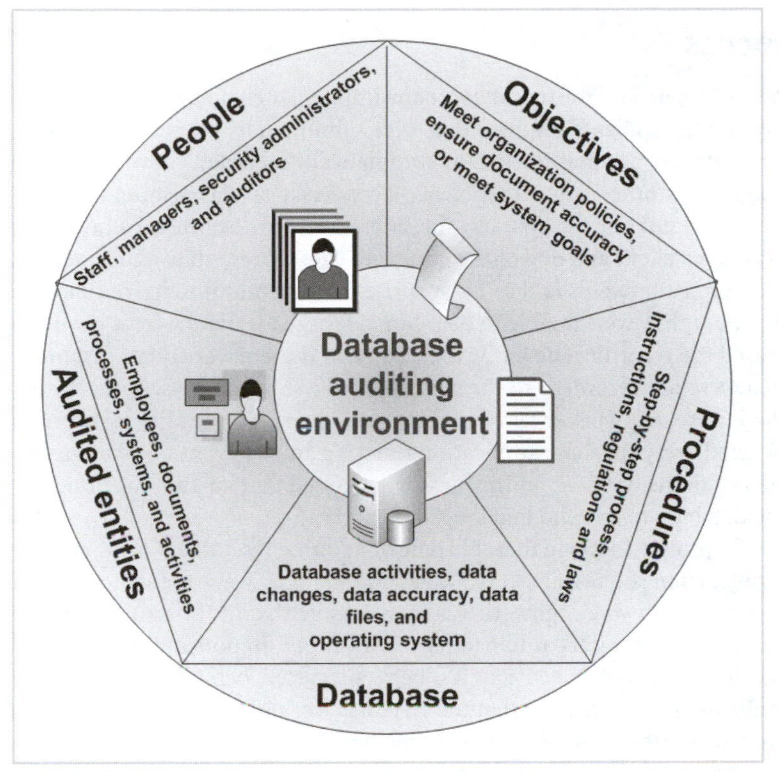

FIGURE 7-3 Components of a database auditing environment

You will find Figure 7-3 helpful when designing an auditing model for your database application. You need to account for all the components shown in the figure to ensure your auditing model is complete and effective. When you design and develop an application system or business process that requires auditing, you must think about every objective related to these components. Now that you know the components of the auditing environment, it's time to learn about the auditing process.

Recall the concept that security measures are inseparable from auditing. To illustrate the concept, consider this scenario of the database application architect who works for a small software vendor. The vendor designs auditing systems to track orders used by their customers' Purchasing Departments. The application was built with every bell and whistle that could be imagined, and sales soared. The system had the capability to generate a trail of every change for every piece of data. Customers praised the product, except one small company whose owner complained. He said that the software was a great help for tracking all orders with historical changes, but because of the lengthy auditing reports, it was not feasible for a person to go through the reports to identify any suspicious behavior. Indeed, that was the shortfall of the system. The application architect had not put enough thought into auditing and how it would be conducted. If she had considered auditing more thoroughly, she would have designed the application with intelligence that would alert the system administrator when an exceptional or deviant incident occurred. As you now know, your risks increase if you develop an audit process that won't be used.

Auditing Process

A large financial investment company implemented a new database application system to log and track clients' requests that were submitted to the customer care representatives. The database application system automatically recorded conversations between the client and the customer care person. Once a conversation was recorded, the customer care person could not erase it. One day the shift supervisor overheard a heated conversation between a client and one of the customer care representatives. He decided to listen to the recording afterwards and to his surprise he could not find it. He called application support, but they were unable to help him. He asked if there were a way to find out who erased the recording, how it was erased, and if there were other instances of client calls that were not recorded. To the surprise of the shift supervisor and the development team there were many instances of missing recordings. The **quality assurance (QA)** team retested every database application function and could not find a bug. Everyone was puzzled until the database administrator suggested that an audit be done on the system and the database to see who had logged on and off.

If you are thinking that this type of auditing resembles QA or even **performance monitoring**, you're right. It is similar, but the goal is not the same. The purpose of the QA process in software engineering is to make sure that the system is bug free and that the system is functioning according to its specifications. In nonsoftware fields, QA ensures that the product is not defective as it is being produced, before it is sold to the customer. The **auditing process** ensures that the system is working and complies with the policies, standards, regulations, or laws set forth by the organization, industry, or government.

Another way to distinguish between QA and auditing processes is by examining the timing of each. The QA process is active during the development phase of the product and before the implementation of the system. The auditing process is usually active after the system is implemented and in production.

Auditing is also not the same as performance monitoring. Their objectives are totally different. The main objective of performance monitoring is to observe if there is degradation in performance at various operation times, which is totally different from auditing. Auditing validates compliance to policy, not performance. See Table 7-1 for comparisons of these three processes.

Table 7-1 Differences in QA, auditing, and performance monitoring processes

Process	Timing	Objectives
QA process	During development and before the product is commissioned into production	Test the product to make sure it is working properly and is not defective
Auditing process	After the product is commissioned into production	Verify that the product or system complies with policies, laws, and standards
Performance monitoring process	After the product is commissioned into production	Monitor response time

Now you understand the distinctions among all three processes. Before returning to the auditing process, remember that you were previously shown that the difference between the auditing and database auditing environments is the addition of the database component in the latter. Again it should be emphasized that the auditing process for a nonsoftware system is exactly the same as it is for the database application system.

Figure 7-4 illustrates the auditing process flow. The first block on the left is the full **system development life cycle**, known as SDLC. In this block the system goes through a structured process of development. One of the phases in this block is QA testing, which is designed to identify bugs in the code and make sure the system fulfills the requirements of the functional specifications.

After the system is in production, the auditing process starts. The first phase of the process is to understand the objectives of the audit and plan a procedure to execute the auditing process. In this phase you are identifying what should happen. The second phase is to review, verify, and validate the system according to the objectives set in the previous phase. Divergence from the auditing objectives is documented. In this phase you are observing what is happening. The last phase is to document the results of the audit and recommend changes to the system. In this phase you are identifying the changes that are needed.

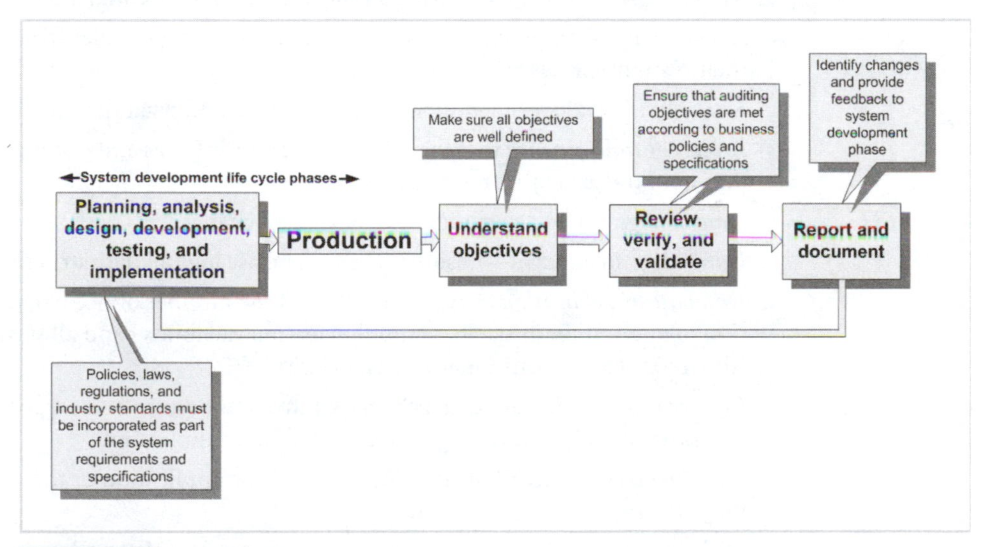

FIGURE 7-4 Auditing process phases

Of course, you should not forget the fact that auditing objectives include security goals. Making sure that users are not logging onto the system at odd times outside of work hours is an example of a security goal that should be auditable.

Auditing Objectives

It is important to remember that **auditing objectives** are established as part of the development process of the entity to be audited. For example, when a software application is being coded, the developers include in their software development design objectives the capability to audit the application. Auditing objectives are established and documented for the following reasons:

- *Complying*—Identify all company policies, government regulations, laws, and the industry standards with which your company must comply.
- *Informing*—All policies, regulations, laws, and standards must be published and communicated to all parties involved in the development and operation of the audited entity. Informing relevant personnel of these details instills awareness and increases their ability to incorporate the necessary controls to ensure that the audited entity is compliant.
- *Planning*—Knowing all the objectives enables the auditor to plan and document procedures to assess the audited entity.
- *Executing*—Without auditing objectives, the person conducting the audit cannot evaluate, verify, or review the audited entity and cannot determine if the auditing objectives have been met.

Here are the top ten database auditing objectives:

- *Data integrity*—Ensure that data is valid and in full referential integrity.
- *Application users and roles*—Ensure that users are assigned roles that correspond to their responsibilities and duties.
- *Data confidentiality*—Identify who can read data and what data can be read.
- *Access control*—Ensure that the application records times and duration when a user logs onto the database or application.
- *Data changes*—Create an audit trail of all data changes.
- *Data structure changes*—Ensure that the database logs all data structure changes.
- *Database or application availability*—Record the number of occurrences and duration of application or database shutdowns (unavailability) and all startup times. Also, record all reasons for any unavailability.
- *Change control*—Ensure that a change control mechanism is incorporated to track necessary and planned changes to the database or application.
- *Physical access*—Record the physical access to the application or the database where the software and hardware resides.
- *Auditing reports*—Ensure that reports are generated on demand or automatically, showing all auditable activities.

Auditing Classifications and Types

The Information System (IS) Department of a small engineering company installed sophisticated software to track all software licenses purchased and used within the company. On a daily basis, the software checked every machine connected to the network to identify all installed software on that machine and determined if any pirated software or software that had not been registered by the IS Department was installed. The application generated a daily audit log, which was mailed to the system administrator. One day the system administrator noticed that 12 machines in the Engineering Department had installed software for which the company had not purchased licenses. The system administrator talked to the department manager inquiring about the new software that was causing concern. The department manager dismissed the system administrator, saying that the software was being used only temporarily and would be uninstalled later when it was not needed. A month later the company received a letter from the Legal Department of the software vendor whose software had not been registered with IS. The letter demanded that all machines in the Engineering Department be seized and not tampered with because a software audit was pending.

Every industry and business sector uses different classifications of audits. In addition, the definition of each classification can differ from business to business. In this section you learn the most generic definitions of audit classifications. Generally, these categories are applicable to all industries. You also learn about the types of audits that are performed. Note that some texts refer to audit classifications as types and audit types as purposes.

Audit Classifications

This section outlines the different audit classifications.

Internal Audit

An **internal audit** is an audit that is conducted by a staff member of the company being audited. The purpose and intention of an internal audit is to:

- Verify that all auditing objectives are met by conducting a well-planned and scheduled audit.
- Investigate a situation that was prompted by an internal event or incident. This audit is random, not planned or scheduled.
- Investigate a situation that was prompted by an external request. This audit is random, not planned or scheduled.

External Audit

An **external audit** is conducted by a party outside the company that is being audited. The purpose and intention of this audit is to:

- Investigate the financial or operational state of the company. This audit is initiated at will by the government or is prompted by suspicious activities or accusations. The person conducting this audit is usually employed and appointed by the government.
- Verify that all auditing objectives are met. This audit typically is planned and scheduled. To ensure objectivity and accuracy, the individual or team conducting the audit is not employed by the company and usually specializes in the field. This audit is typically performed to certify that the company is complying with standards and regulations.

Automatic Audit

An **automatic audit** is prompted and performed automatically (without human intervention). Automatic audits are used mainly for systems and database systems. Some systems that employ this type of audit generate reports and logs. The administrator reads and interprets these reports to evaluate the integrity of the system and identify abnormal activities. Other advanced and sophisticated systems employ inference engine or artificial intelligence to interpret suspicious activities and alert administrators of these activities, and some even more advanced systems can take preventive and corrective action.

Manual Audit

A manual audit is performed completely by humans. The audit team uses various methods to collect audit data, including interviews, document reviews, and observation. The auditors may even perform the operational tasks of the audited entity.

Hybrid Audit

A hybrid audit is simply a combination of automatic and manual audits. Most audits fall into this classification.

Audit Types

The following types of audits are conducted by various business sectors:

- **Financial audit**—Ensures that all financial transactions are accounted for and comply with the law. For example, all financial companies save all trading transaction records for a period of time to comply with government regulations.
- **Security audit**—Evaluates if the system is as secure as it should be. This audit identifies security gaps and vulnerabilities. For example, your company might ask a hacker to break into the company's network system to determine how secure or vulnerable the network is.
- **Compliance audit**—Verifies that the system complies with industry standards, government regulations, or partner and client policies. For example, all pharmaceutical companies must keep paper trails of all research activities to comply with the industry standards as well as government regulations.
- **Operational audit**—Verifies if an operation is working according to the policies of the company. For example, when a new hire starts work, the Human Resources Department provides employment identification card, sign disclosure, confidentiality papers, tax forms, and so on.
- **Investigative audit**—Performed in response to an event, request, threat, or incident to verify the integrity of the system. For example, an employee might have committed a fraudulent activity by which he or she diverted some of the financial transactions to a fake account. In this case, your company conducts an investigative audit to discover how this incident happened and how to prevent it.
- **Product audit**—Performed to ensure that the product complies with industry standards. This audit is sometimes confused with testing; it should not be. For instance, a product audit does not include auditing of its functionality but entails how it was produced and who worked on its development.
- **Preventive audit**—Performed to identify problems before they occur. For example, your company should conduct both random and routine audits to verify that business operations are being performed according to specifications.

Benefits and Side Effects of Auditing

This list of audit benefits should enhance your understanding of the importance of audits, as well as provide a review of the previous sections:

- Enforces company policies and government regulations and laws
- Lowers the incidence of security violations
- Identifies security gaps and vulnerabilities
- Provides an audit trail of activities
- Provides another means to observe and evaluate operations of the audited entity
- Provides a sense or state of security and confidence in the audited entity
- Identifies or removes doubts
- Makes the organization being audited more accountable
- Develops controls that can be used for purposes other than auditing

Auditing Side Effects

Nothing is perfect. Conducting audits more frequently than necessary causes side effects. Moderation is the key. Frequent audits can cause the following:

- Performance problems due to preoccupation with the audit instead of the normal work activities
- Generation of many reports and documents that may not be easily or quickly disseminated
- Disruption to the operations of the audited entity
- Consumption of resources, and added costs from downtime
- Friction between operators and auditor
- From a database perspective, could degrade the performance of the system; might also generate a massive number of logs, reports, and data that require a system purge

Auditing Models

As you now know, security and auditing are the two inseparable elements of data and database integrity. Database auditing can be implemented by utilizing built-in features of the database or by creating your own mechanism. Both methods serve the same purpose. If you decide to build your own auditing utility, this section will help you in designing an audit model to fit your business requirements.

The section includes several auditing models. Though some of these models are simple and others are quite sophisticated, they are all easy to implement. Before you learn about these models, it is important that you understand how auditing is processed for data and database activities. Figure 7-5 presents a flowchart of data auditing. The flowchart shows what happens when a user performs an action on a database object. Specific checks occur to verify if the action, the user, or the object are registered in the auditing repository. If they are registered, the following are recorded:

- State of the object before the action was taken along with the time of the action
- Description of the action that was performed
- Name of the user who performed the action

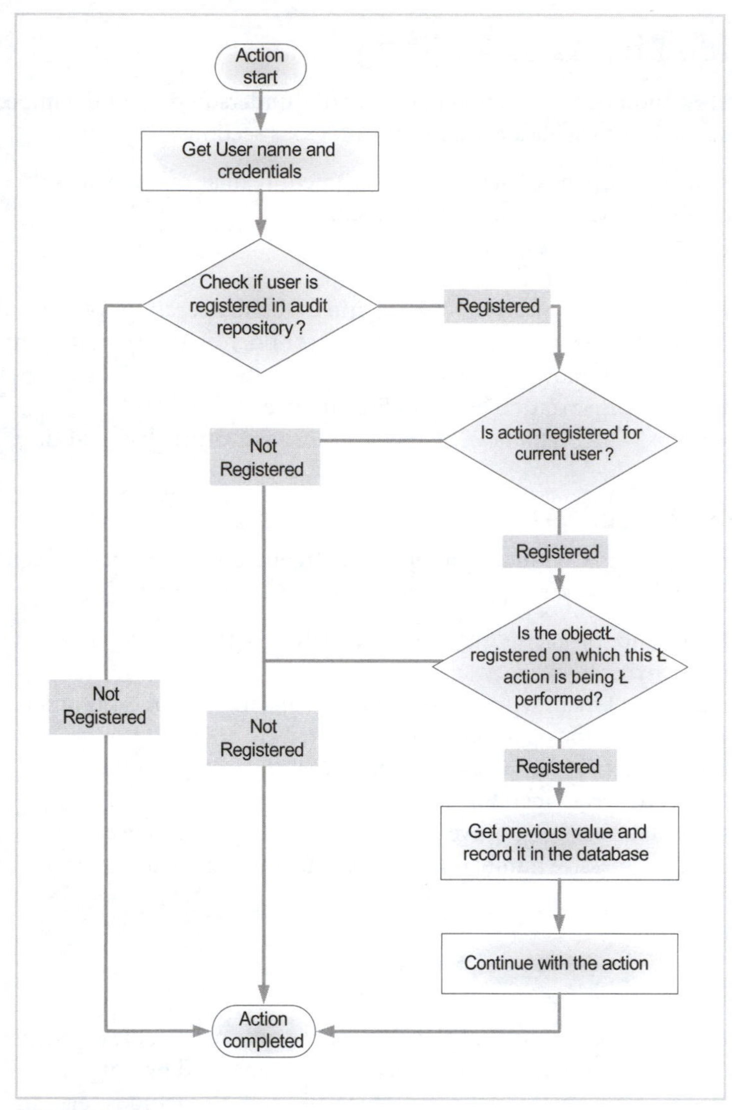

FIGURE 7-5 Data auditing flowchart

NOTE

"Not registered" indicates that the audited entity (user, table, column, or action) is not found in the auditing repository.

For example, suppose that MAAN is the user performing an UPDATE statement on the EMPLOYEE table. The auditing process checks whether MAAN is being audited. If so, the process continues to check whether MAAN is being audited for updates. Again, if MAAN is being audited for updates, the process checks whether the EMPLOYEE object is being audited. If it is, the state of the row(s) being modified is stored in another table or the action is recorded with the data being stored. What is stored depends on the auditing requirements.

Simple Auditing Model 1

The first auditing model is called "simple" because it is easy to understand and develop. This model registers audited entities in the audit model repository to chronologically track activities performed on or by these entities. An entity can be a user, table, or column, and an activity can be a DML transaction or logon and off times. The repository is used by the auditing process to check if the user, action, and object are to be audited. Figure 7-6 illustrates this model.

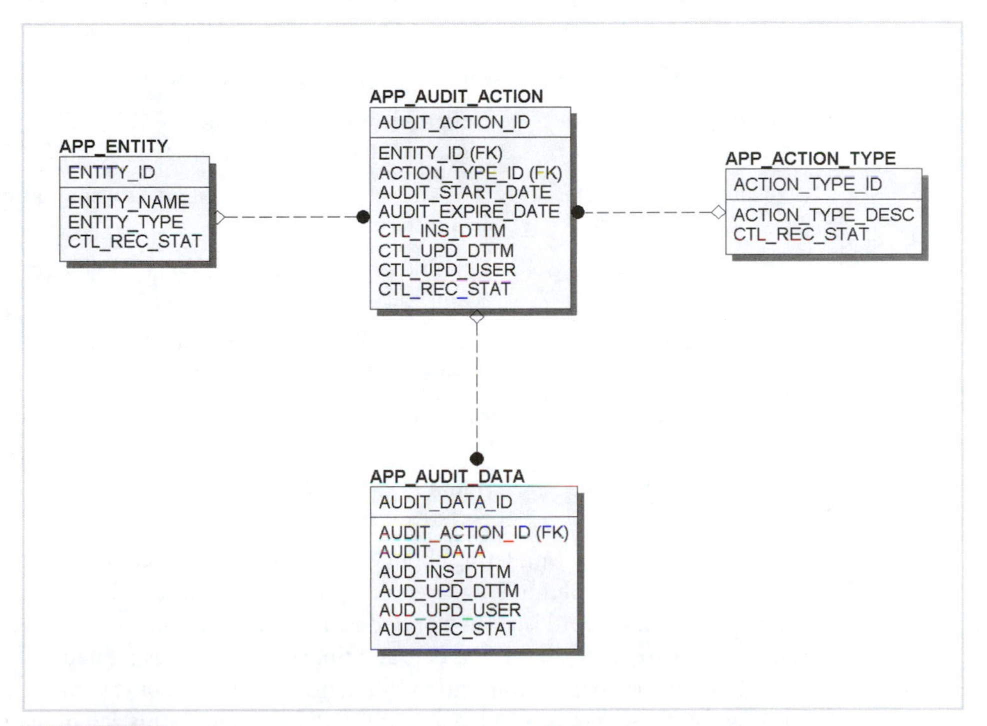

FIGURE 7-6 Data model of a repository for simple auditing model 1

Before you proceed with the illustration of simple auditing model 1, you should look at Tables 7-2 and 7-3 for brief descriptions of the tables and columns presented in this model.

Table 7-2 Description of tables presented in simple auditing model 1

Table	Description
APP_ENTITY	Holds the name of the entity to be audited; an entity can be a name of a user, name of a table, or name of the column
APP_AUDIT_ACTION	Holds entities and the actions that are audited
APP_ACTION_TYPE	Holds the actions to be audited; an action can be UPDATE, DELETE, INSERT, LOGIN, or LOGOUT
APP_AUDIT_DATA	Holds audit trail data generated by the auditing process

Table 7-3 Description of columns presented in simple auditing model 1

Column	Description
ACTION_TYPE_DESC	Name or description of the audited action such as UPDATE, INSERT, DELETE, LOGIN, or LOGOUT
ACTION_TYPE_ID	Unique identification number of APP_ACTION_TYPE table generated automatically by the application
AUDIT_ACTION_ID	Unique identification number of APP_AUDIT_ACTION table generated automatically by the application
AUDIT_DATA	Audit data trail generated by the auditing process
AUDIT_DATA_ID	Unique identification number of APP_AUDIT_DATA table generated automatically by the application
AUDIT_START_DATE	Date and time when the audit on a specific entity and action starts
AUDIT_EXPIRE_DATE	Data and time when the audit on a specific entity and action ends
ENTITY_ID	Unique identification number of APP_ENTITY table generated automatically by the application
ENTITY_NAME	Name of the user, table, or column to be audited
ENTITY_TYPE	Type of the entity to be audited; a type can be user, table, or column

You are probably wondering about control columns—what they are and how they are used. A control column is a placeholder for data that the application inserts automatically when a record is created or updated. Typically, a control column stores data about the current record, such as date and time the record was created and updated. There are several naming conventions for control columns. Some companies do not distinguish these columns with a prefix (CTL). For example, instead of naming the column CTL_INS_DTTM, they may use DATE_CREATED or other similar name. This book highly recommends using a prefix to distinguish these control columns from the other data columns; a prefix could be CTL for control or AUD for audit, and so on. Table 7-4 lists control columns that you may have seen or might see in the future.

Table 7-4 Description of control columns

Column	Stands for	Description of the Control Column
CTL_ARC_FLAG	CONTROL ARCHIVE FLAG	Indicates whether current record can be archived or not; possible values are Yes and No
CTL_AUD_END	CONTROL AUDIT END	Stores audit end date and time for current record
CTL_AUD_FLAG	CONTROL AUDIT FLAG	Indicates whether current record is audited or not; possible values are Yes and No

Table 7-4 Description of control columns (continued)

Column	Stands for	Description of the Control Column
CTL_AUD_START	CONTROL AUDIT START	Stores audit start date and time for current record
CTL_INS_DTTM	CONTROL INSERT DATE TIME	Stores the date and time the record is created
CTL_INS_USER	CONTROL INSERT USER	Stores the user name that created the record
CTL_PUR_FLAG	CONTROL PURGE FLAG	Indicates whether current record can be purged or not; possible values are Yes and No
CTL_REC_STAT	CONTROL RECORD STATUS	Stores the status of the current record; record status could be A for active, D for deleted, or I for inactive
CTL_SEC_LEVEL	CONTROL SECURITY LEVEL	Is used to define security access level for current record
CTL_UPD_DTTM	CONTROL UPDATE DATE TIME	Stores the date and time of the most recent update on the current record
CTL_UPD_USER	CONTROL UPDATE USER	Stores the user name that created or performed the last update on the record

Consider the following to illustrate how simple auditing model 1 works. Suppose your business requirements dictate that all updates on the SALARY table and all DML actions by the user SAM must be audited. In this case you must insert records to the repository as outlined in Table 7-5.

Table 7-5 Sample data for simple auditing model 1

Table	New Records
APP_ENTITY	10, SAM, USER, A 11, SALARY, TABLE, A
APP_ACTION_TYPE	1, UPDATE, A 2, INSERT, A 3, DELETE, A
APP_AUDIT_ACTION	1, 10, 1, 10-MAY-2005, 10-JUN-2005, 15-APR-2005, NULL, DBSEC, A 2, 10, 2, 10-MAY-2005, 10-JUN-2005, 15-APR-2005, NULL, DBSEC, A 3, 10, 3, 10-MAY-2005, 10-JUN-2005, 15-APR-2005, NULL, DBSEC, A 4, 11, 1, 10-MAY-2005, 10-JUN-2005, 15-APR-2005, NULL, DBSEC, A

Simple Auditing Model 2

In this model, only the column value changes are stored for audit purposes. The audit data table APP_AUDIT_DATA contains chronological data on all changes on columns that are registered in APP_AUDIT_TABLE. Note, in this model there is a purging and archiving mechanism to help reduce the amount of data stored in the database. Figure 7-7 illustrates this auditing model. Also note that this model does not register an action that was performed on the data. In addition, the AUDIT_DATA column in the APP_AUDIT_DATA table must be Large Object data type in order to fit all column values being audited. This model is ideal for auditing a column or two of a table, as opposed to all columns. Tables 7-6 and 7-7 describe simple auditing model 2 tables and columns respectively.

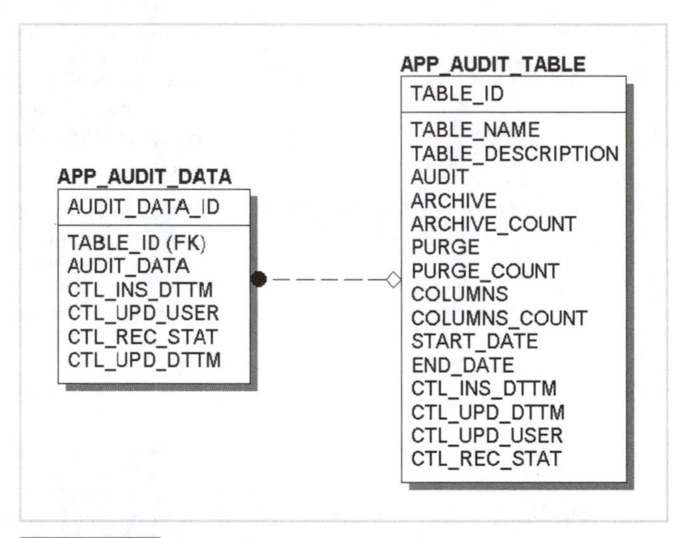

FIGURE 7-7 Data model of a repository for simple auditing model 2

Table 7-6 Description of tables presented in simple auditing model 2

Table	Description
APP_AUDIT_DATA	Contains all audit data generated by the auditing process
APP_AUDIT_TABLE	Contains table name and columns to be audited as well as audit start and end time, purging indicator, and archiving indicator

Table 7-7 Description of columns presented in simple auditing model 2

Column	Description
ARCHIVE	Indicates whether records stored in APP_AUDIT_DATA for the audited table are archived or not
ARCHIVE_COUNT	Contains the number of records to be archived from APP_AUDIT_DATA for audited table

Table 7-7 Description of columns presented in simple auditing model 2 (continued)

Column	Description
AUDIT	Indicates whether table contained in TABLE_NAME column is audited or not
AUDIT_DATA	Value of the columns before the table row (record) was modified
AUDIT_DATA_ID	Unique identifier for APP_AUDIT_DATA table generated automatically by the application
COLUMNS_COUNT	Number of columns contained in the COLUMNS value
COLUMNS	A list of columns to be audited
END_DATE	Auditing end date
PURGE	Indicates whether the audited table records are to be purged or not
PURGE_COUNT	Number of records to be purged from APP_AUDIT_TABLE for audited table
START_DATE	Auditing start date
TABLE_DESCRIPTION	Contains description of the audited table
TABLE_ID	Unique identifier of APP_AUDIT_TABLE generated automatically by the application
TABLE_NAME	Name of the audited table

Advanced Auditing Model

This model is called "advanced" because of its flexibility. It is more flexible than the simple models because it can be used as an auditing application with a user interface. Figure 7-8 presents the process flow of the user interface.

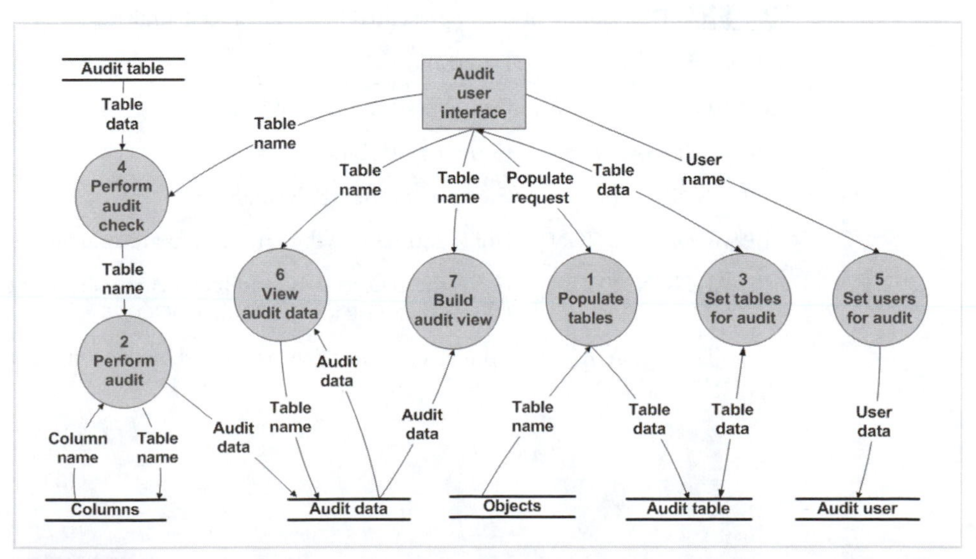

FIGURE 7-8 Process flow for an audit user interface

Of course the repository for this model is more complex than previous models. It contains data stores to register all entities (users, actions, tables, and columns) that can be audited. The auditing level of the model is referred to as fine grained (column level). The model can handle auditing users, actions, tables, and column entities, a feature that few database systems incorporate. See Figure 7-9 for an illustration. Tables 7-8 and 7-9 describe the advanced auditing model tables and columns, respectively.

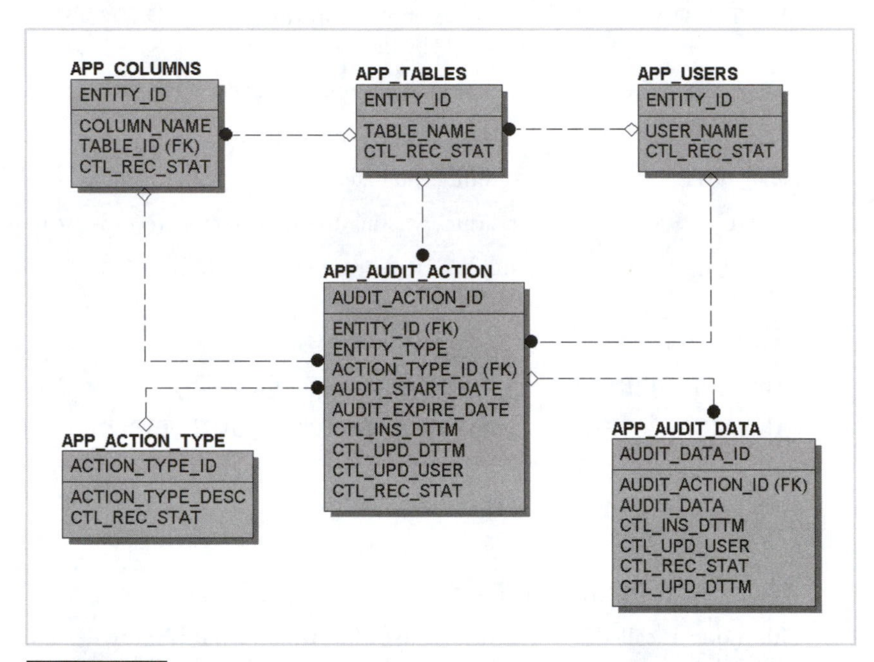

FIGURE 7-9 Data model of the repository for an advanced auditing model

Table 7-8 Description of tables presented in the advanced auditing model

Table	Description
APP_COLUMNS	Contains all columns to be audited
APP_TABLES	Contains all tables to be audited
APP_USERS	Contains all users to be audited
APP_AUDIT_ACTION	Holds entities and the actions that are audited
APP_ACTION_TYPE	Holds the actions to be audited; an action can be UPDATE, DELETE, INSERT, LOGIN, or LOGOUT
APP_AUDIT_DATA	Contains all audit data generated by the auditing process

Table 7-9 Description of columns presented in the advanced auditing model

Column	Description
ACTION_TYPE_DESC	Name or description of the audited action such as UPDATE, INSERT, DELETE, LOGIN, or LOGOUT
ACTION_TYPE_ID	Unique identification number of APP_ACTION_TYPE table generated automatically by the application
AUDIT_ACTION_ID	Unique identification number of APP_AUDIT_ACTION table generated automatically by the application
AUDIT_DATA	Value of the columns before the table row (record) was modified
AUDIT_DATA_ID	Unique identifier for APP_AUDIT_DATA table generated automatically by the application
AUDIT_EXPIRE_DATE	Audit end date
AUDIT_START_DATE	Audit start date
COLUMN_NAME	Name of the audited column
ENTITY_ID	Unique identifier for APP_TABLES, APP_COLUMNS, or APP_USERS generated automatically by the application; note that all values for ENTITY_ID in this model must be unique across all tables previously listed in this table
ENTITY_TYPE	Type of the entity to be audited; a type can be user, table, or column
TABLE_ID	Table identification number, this column references ENTITY_ID in APP_TABLES
TABLE_NAME	Name of the audited table
USER_NAME	Name of the audit application user

Historical Data Model

This model is used for applications that require a record of the whole row when a DML transaction is performed on the table. This model is typically used in most financial applications. With this model, the whole row is stored in the HISTORY table, before it is changed or deleted. See Figure 7-10 for an illustration of this model.

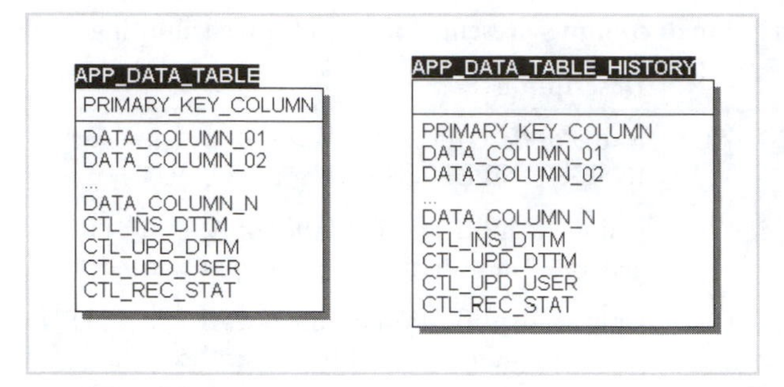

FIGURE 7-10 Data model of a repository for a historical auditing model

Auditing Application Actions Model

There may be a requirement for an application to audit specific operations or actions, as illustrated in Figure 7-11. For example, you may want to audit a credit to an invoice, the reason for it being credited, the person who credited it, and the time it was credited. The following model allows you to do just that.

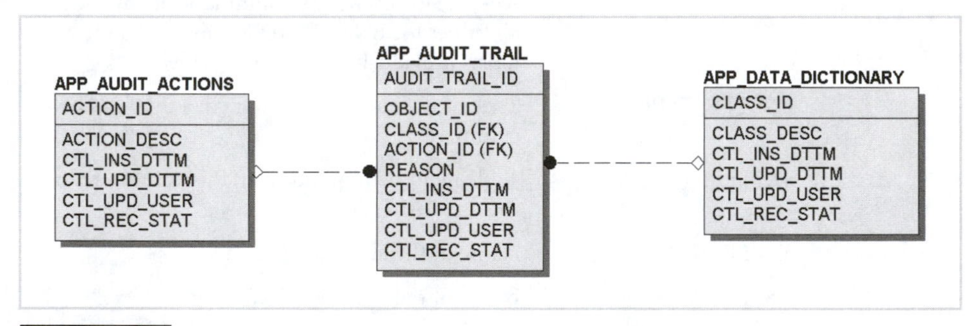

FIGURE 7-11 Data model of a repository for auditing application actions

The sample data in these tables would look like the data presented in Tables 7-10, 7-11, and 7-12.

Table 7-10 APP_AUDIT_ACTIONS data

Column Name	Column Value
ACTION_ID	1
ACTION_DESC	Invoice Credit
CTL_INS_DTTM	2004-05-10 12:30:21
CTL_UPD_DTTM	NULL
CTL_UPD_USER	ADMIN
CTL_RECT_STAT	ACTIVE

Table 7-11 DATA_DICTIONARY data

Column Name	Column Value
CLASS_ID	1
CLASS_DESC	Invoice Business Object
CTL_INS_DTTM	2004-05-10 12:30:21
CTL_UPD_DTTM	NULL
CTL_UPD_USER	ADMIN
CTL_REC_STAT	ACTIVE

Table 7-12 APP_AUDIT_TRAIL data

Column	Column Value
AUDIT_TRAIL_ID	100001
OBJECT_ID	135432
CLASS_ID	1
ACTION_ID	1
REASON	Customer-presented coupon
CTL_INS_DTTM	2004-06-12 12:30:21
CTL_UPD_DTTM	NULL
CTL_UPD_USER	JDOE
CTL_REC_STAT	ACTIVE

The OBJECT_ID column of the APP_AUDIT_TRAIL table is the ID of the invoice record. As you can see, the model doesn't track changes to data in specific columns or rows, but rather identifies an operation performed within the application itself.

C2 Security

The National Security Administration has given a C2 security rating to Microsoft SQL Server 2000. This means that the server passes requirements set by the Department of Defense and is typically implemented in military and government applications. When configured as a C2 system, SQL Server utilizes **DACLs** (**discretionary access control lists**) to manage security and audit activity. Requirements for enabling C2 auditing in SQL Server include the following:

- The Microsoft Windows Server must be configured as a C2 system.
- Windows Integrated Authentication is supported, but SQL native security is not supported.
- Only transactional replication is supported.
- The following SQL Server services are not included in a C2 evaluation:
  - SQL Mail, Full Text Search, English Query, DTC, Meta Data Services, Analysis Services (OLAP)

Chapter Summary

- An audit examines, verifies, and validates that documents, procedures, processes, operations, systems, and other activities comply with the original objectives.
- Auditing activities are performed as part of an audit. The activities can be considered an auditor's responsibilities and incorporated into audit policies.
- The auditing environment consists of objectives, procedures, people, and audited entities.
- The database-auditing environment consists of objectives, procedures, people, audited entities, and databases.
- The purpose of an auditing process is to make sure that the system is working and complies with the policies, standards, regulations, and laws set forth by the organization, the industry, or the government.
- Auditing objectives are established and incorporated as part of the development process of the audited entity, regardless of the nature of the audited entity (the audited entity could be a document, process, procedure, system, data, or activities).
- Auditing has four main objectives—compliance, informing, planning, and executing.
- A staff member of a company conducts an internal audit.
- A party outside the company conducts an external audit.
- An automatic audit is prompted and performed automatically through a system process without human intervention. Automatic audits are mainly used for systems and database systems.
- A manual audit is performed completely by humans. The audit team uses various methods to obtain auditable information, including interviews, documents, and observations. They may even perform the tasks of the audited entity.
- A hybrid audit is simply a combination of automatic and manual audits.
- Excessive use of auditing can degrade the performance of the system.
- Simple auditing model 1 is an auditing repository for audited entities such as user, table, column, and action.
- A control column is a placeholder for data that the application inserts automatically when a record is created or updated.
- Simple auditing model 2 is an auditing repository for column changes only.
- The advanced auditing model contains data stores to register all entities (users, actions, tables, and columns) that can be audited.
- The advanced auditing model is referred to as fine grained (column level).
- The historical data model stores the entire row in a HISTORY table, before it is changed or deleted.
- The historical data model is used mainly by financial institutions.
- The auditing application actions model is a repository for audited operations or actions performed by an application.

Review Questions

1. The purpose of the auditing process is to make sure that the system is working and functional. True or false?
2. A security audit verifies that the system complies with industry standards, government regulations, or partner and client policies. True or false?
3. An audit trail is a document generated by an individual conducting the audit; it contains auditing findings. True or false?
4. Database auditing is the process of examining and validating documents, data, processes, procedures, systems, or activities to ensure that the audited entity complies with its original objectives. True or false?
5. An external audit is conducted to investigate the financial or operational state of a company and is brought about randomly by the government because of suspicious activities or accusations. True or false?
6. List three side effects of excessive database auditing.
7. Give an example of an application in which auditing is implemented.
8. State the differences between internal and external audits and provide an example of each.
9. Using a business situation, list the activities that are executed for each phase of the auditing process.
10. List the four purposes of having auditing objectives.
11. List five auditing activities, and for each activity provide a real-life example.
12. List two reasons for an internal audit and provide a real-life example.
13. List two reasons for an external audit and provide a real-life example.
14. Define the term "auditing environment" and include the components of the auditing environment.

Hands-on Projects

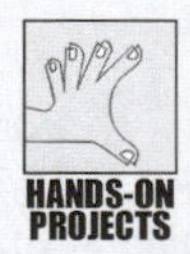

Hands-on Project 7-1

Your colleague at work was browsing through this chapter when she came across the CTL columns. She asked you what the columns were and how to use them. Briefly describe the definition of three of the CTL columns and outline how they can be used.

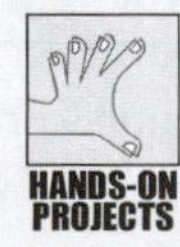

Hands-on Project 7-2

Provide an example of the historical auditing model, showing the structure of the table that is being audited, as well as its history table.

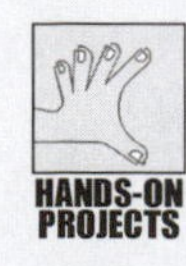

Hands-on Project 7-3

Using simple auditing model 1, provide the sample data with an explanation of how it would be used.

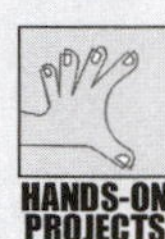

Hands-on Project 7-4

Translate Figure 7-5 into pseudocode (English-like instructions), and provide an example of executing these instructions.

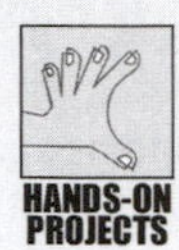

Hands-on Project 7-5

Using simple auditing model 2, provide the sample data with an explanation of how it would be used.

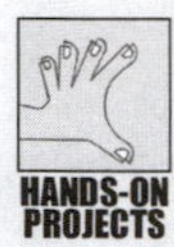

Hands-on Project 7-6

Using the historical auditing model, give an example of how this model is implemented. (No code is necessary for this problem.)

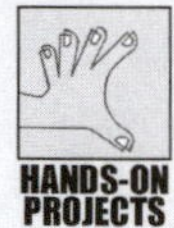

Hands-on Project 7-7

Using simple auditing model 2, provide a flowchart for the purging or archiving process.

Hands-on Project 7-8

Translate Table 7-1 into a chart with time lines, showing the differences among the QA, the auditing, and the performance monitoring processes.

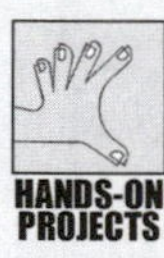

Hands-on Project 7-9

Define three auditing terms that were presented in the definition section of this chapter.

Case Projects

Case 7-1 Audit Process Flow Case

Your company received a request from its most important partner to comply with industry standards, which requires an audit trail of all financial data. Your manager summoned you to her office, explained this new requirement, and assigned you to design the architecture and a high-level auditing model. Compliance requires that all data changes be recorded in the history tables and that the auditing model employ a mechanism to audit any user on demand. Use a data model or a combination of the auditing models presented in this chapter to produce the following:

- The process flow diagram or flowchart of your auditing model
- The data model of your auditing model
- Identification of all components that will be involved in the auditing model
- A list of any concerns you have about your auditing model

Application Data Auditing 8

LEARNING OBJECTIVES

Upon completion of this material, you should be able to:

- Understand the difference between the auditing architecture of DML Action Auditing Architecture and DML changes
- Create and implement Oracle triggers
- Create and implement SQL Server triggers
- Define and implement Oracle fine-grained auditing
- Create a DML statement audit trail for Oracle and SQL Server
- Generate a data manipulation history
- Implement a DML statement auditing using a repository
- Understand the importance and the implementation of application errors auditing in Oracle
- Implement Oracle PL/SQL procedure authorization

Introduction

You are an information security manager responsible for all data security policies. You've just read an article in a reputable magazine about a fraud scheme that is becoming epidemic in many companies. The theme of the article is that the very people who are guarding a company's data are likely to be those who threaten the data. Even if this is not a major issue in your company at this time, you should realize that you must guard against every type of security risk and that you must use all resources available to assure that you are well versed in all security and audit issues.

Chapter 7 covered the theory of database auditing and a number of models for auditing. This chapter complements Chapter 7 by presenting the practical side of database auditing and the implementations of those models. This chapter also covers Oracle10*g* fundamentals, built-in Oracle auditing capabilities, and SQL Server triggers. After you have mastered these topics, the chapter moves on to Oracle10*g* fine-grained auditing and procedures for invoking method features. In addition, you will learn about data manipulation auditing and application error auditing.

DML Action Auditing Architecture

One of the leading cellular phone service providers hired a customer service representative. The representative was very excited about his new job, especially because he could create accounts for customers, as well as change and assign their plans. He soon told his friends about the new job and asked them to subscribe to his company. However, he did not stop there. Without authorization, he offered to give them more minutes for a lower price than the standard plan. Then, he pushed his scheme to a higher level by charging his friends directly and giving them a price lower than any the company offered. This made him rich for two years, until a business associate made a call to the vice president of Customer Services. "Bad news," he said. "one of your customer service reps is selling minutes to your customers under the table. Yesterday at a party I overheard a guy talking about his cheap subscription and nearly unlimited minutes. Immediately, I asked him how he got such a deal. I was appalled at his answer and decided to get his name and number." The vice president was lost for words.

This situation actually occurred. What is more amazing is that it took so long for the scam to be uncovered. Would it be possible for this scam to continue for a long time, even if audit reports were being published? The answer is "yes", especially if audit reports aren't being reviewed.

In this section, you are introduced to auditing **Data Manipulation Language (DML)** statements from two approaches. With the first approach, you create an audit trail for DML activities that are occurring on a table, as shown in Figure 8-1. As you can see, the action is recorded before the statement is applied to the table. The other approach is to register all column values either before or after the DML statement is applied to the table, as illustrated in Figure 8-2.

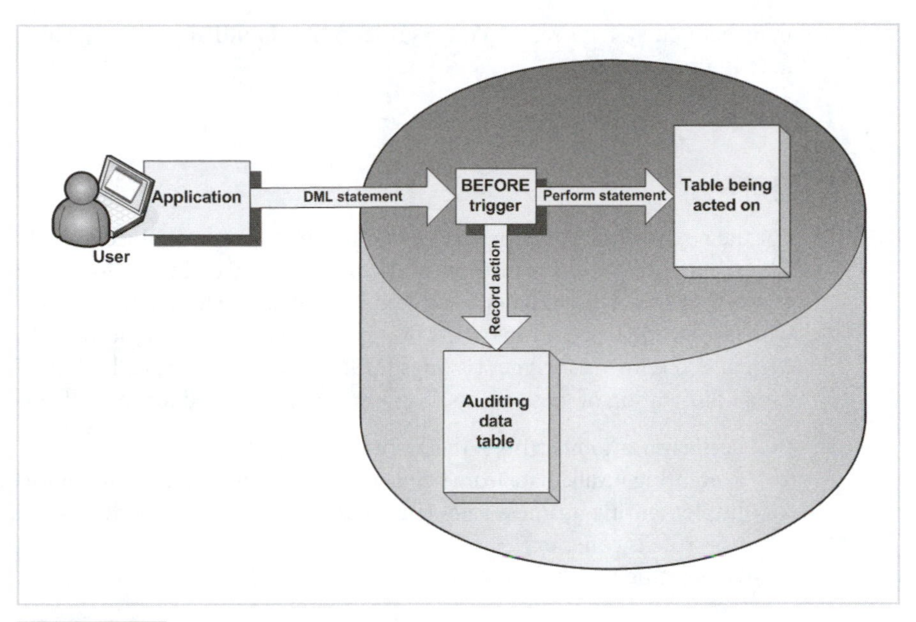

FIGURE 8-1 Auditing architecture for DML action

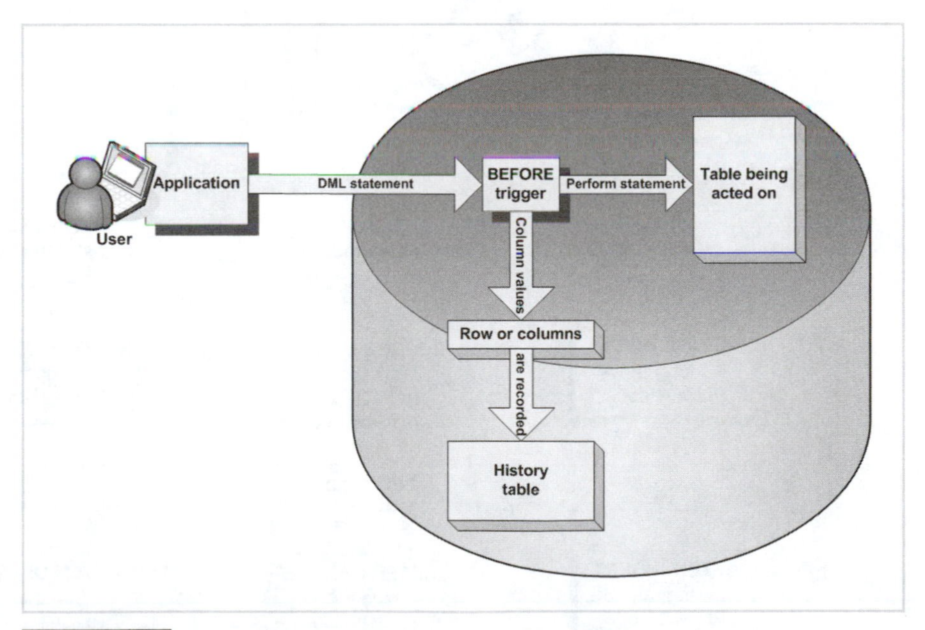

FIGURE 8-2 Auditing architecture for DML changes

Companies with sensitive data often use auditing architecture for DML changes. DML changes can be performed on two levels: row level and column level. For example, when an UPDATE statement that modifies only one column is applied to a table, the DML auditing mechanism can record all values for every column in the table. This is called row-level auditing. However, if you are interested in recording only the before value of the modified column(s), use column-level auditing. Oracle and other database management systems refer to this as **fine-grained auditing (FGA)**, which is discussed later in this chapter.

Oracle Triggers

For the next auditing scheme, notice that the main database objects are database triggers. This section discusses the purpose and syntax of triggers. First, a formal definition of a trigger is warranted: a **trigger** is a stored PL/SQL procedure that is executed automatically whenever a DML operation occurs or a specific database event occurs. Oracle has six DML events, also known as **trigger timings** for INSERT, UPDATE, and DELETE. See Figure 8-3 for an illustration of these events. Triggers are mainly used for the following purposes:

- Performing audits (this is their primary use)
- Preventing invalid data from being inserted into the tables (but not for data validations)
- Implementing business rules (this is not highly recommended especially if the business rule is complex)
- Generating values for columns

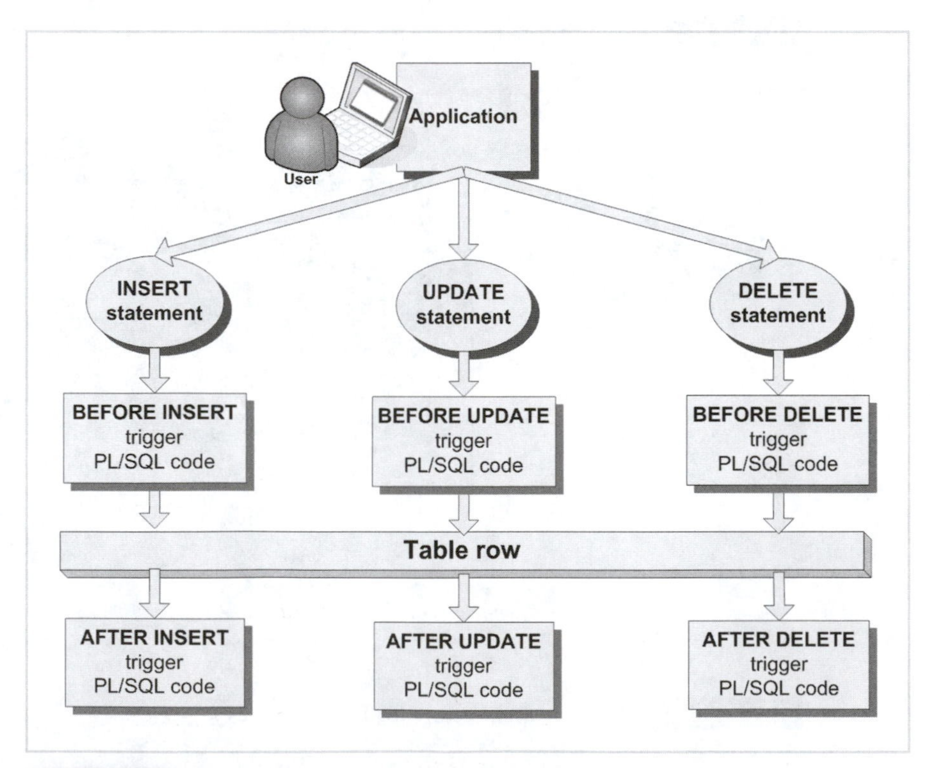

FIGURE 8-3 Trigger timings for DML statements

Now look at trigger syntax. You need to learn the grammar for creating DML statement triggers. You can use the CREATE TRIGGER statement on a table by following the syntax presented in the following example. DDL statements are presented in the next chapter. If you need more details, refer to the Oracle documentation found at *otn.oracle.com*.

```
CREATE [OR REPLACE ] TRIGGER trigger_name
       {BEFORE | AFTER | INSTEAD OF }
       {INSERT [OR] | UPDATE [OR] | DELETE}
       [OF col_name]
       ON table_name
       [FOR EACH ROW]
       [REFERENCING {OLD [AS] old | NEW [AS] new | PARENT [AS] parent}]
       WHEN (condition)
   { pl/sql_block | call_procedure_statement } ;
```

Where:

BEFORE: This indicates that the trigger executes just before the DML statement is applied on the affected rows.

AFTER: This indicates that the trigger executes just after the DML statement is applied on the affected rows.

INSTEAD OF: This is only applicable to DML statements. It indicates that the trigger should be executed, instead of the DML statement.

OF: This specifies the columns that you want the trigger to fire when the column mentioned in this OF clause is affected. If you omit the OF clause, the trigger fires when any column in the table is affected by the statement.

ON: This specifies the table that the trigger affects.

FOR EACH ROW: This indicates that the trigger fires for each row affected by the DML statement. For example, suppose you issue an UPDATE statement that affects 10 rows. If you issue that statement, the trigger fires 10 times, once for each row. This is known as the ROW LEVEL trigger. If you omit this clause, the trigger fires only once, regardless of the number of rows affected by the statement. This is known as the STATEMENT LEVEL trigger or TABLE LEVEL trigger.

REFERENCING: When a trigger contains the FOR EACH ROW clause, Oracle offers two pseudorows, NEW and OLD. These are virtual rows that contain the new values for each column in the row and the old values for each column in the row, respectively. For example, if you want to know the new value for a column when it is updated, you can reference it as NEW.ColumnName. However, if you don't want to use NEW or OLD, you can change them to any desired keywords.

WHEN: This specifies the criterion or condition that the trigger must meet to fire.

It is worth noting that in Oracle triggers are always executed in a specific order. The STATEMENT LEVEL triggers are executed before the ROW LEVEL triggers, and the BEFORE triggers are executed before the AFTER triggers. See Figure 8-4 for an illustration.

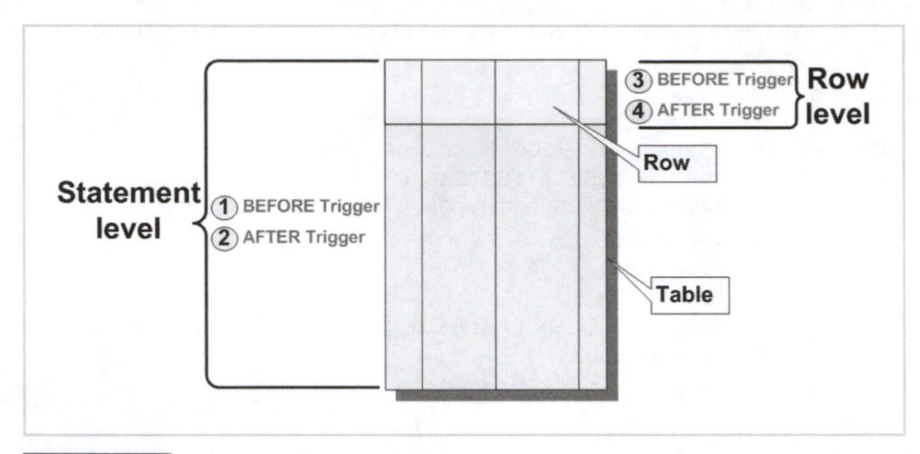

FIGURE 8-4 Order of trigger execution

Because this syntax is complex, you can benefit from examining some examples of trigger object creation. Here are two examples:

```
--Example 1: Row-Level Trigger
CREATE OR REPLACE TRIGGER TRG_EMPLOYEES_BUR
  BEFORE UPDATE ON APP_TBL2
  FOR EACH ROW
DECLARE

   V_OPERATION  VARCHAR2(20);

BEGIN

   -- INSERTING, UPDATING and DELETING
   -- variables that are set to TRUE automatically
   -- by Oracle, based on the action DML operation
   -- that fired the trigger. If an INSERT fires
   -- the trigger, then INSERTING is set TRUE and
   -- so forth for the UPDATE and DELETE.
   IF INSERTING THEN
      V_OPERATION := 'INSERT';
   ELSIF UPDATING THEN
      V_OPERATION := 'UPDATE';
   ELSE
      V_OPERATION := 'DELETE';
   END IF;

   ...

END;
/
```

```
--Example 2: Statement-Level Trigger
CREATE OR REPLACE TRIGGER TRG_EMPLOYEES_AUDR
  AFTER UPDATE OR DELETE ON EMPLOYEES
DECLARE

   V_OPERATION  VARCHAR2(20);

BEGIN

   IF INSERTING THEN
      V_OPERATION := 'INSERT';
   ELSIF UPDATING THEN
      V_OPERATION := 'UPDATE';
   ELSE
      V_OPERATION := 'DELETE';
   END IF;

   ...

END;
/
```

You can view all triggers created on a table by using the USER_TRIGGERS data dictionary view. The structure of this view is listed in the example that follows. The body of the trigger is contained in the TRIGGER_BODY column. A table can have unlimited triggers, but overuse is not recommended because it can lead to database performance degradation.

```
SQL> DESC USER_TRIGGERS
 Name                    Null?    Type
 ----------------------- -------- --------------
 TRIGGER_NAME                     VARCHAR2(30)
 TRIGGER_TYPE                     VARCHAR2(16)
 TRIGGERING_EVENT                 VARCHAR2(227)
 TABLE_OWNER                      VARCHAR2(30)
 BASE_OBJECT_TYPE                 VARCHAR2(16)
 TABLE_NAME                       VARCHAR2(30)
 COLUMN_NAME                      VARCHAR2(4000)
 REFERENCING_NAMES                VARCHAR2(128)
 WHEN_CLAUSE                      VARCHAR2(4000)
 STATUS                           VARCHAR2(8)
 DESCRIPTION                      VARCHAR2(4000)
 ACTION_TYPE                      VARCHAR2(11)
 TRIGGER_BODY                     LONG
```

SQL Server Triggers

Similar to Oracle10g, SQL Server provides a trigger mechanism that fires automatically when a DML statement occurs. In this section you will learn the syntax of creating triggers, examine trigger examples, and learn how to implement both an historical and an application actions model with SQL Server.

Creating Triggers

As with Oracle, the heart of any auditing implementation is the trigger. This section covers trigger syntax, trigger conditions, logical tables, and restrictions on triggers.

Trigger Syntax

To create a trigger in Microsoft SQL Server 2000, you use the CREATE TRIGGER DDL statement. The following is the full syntax for creating a trigger:

```
CREATE TRIGGER trigger_name
ON { table | view }
[ WITH ENCRYPTION ]
{
    { { FOR | AFTER | INSTEAD OF } {[INSERT] [,] [UPDATE] [,] [DELETE]}
        [ WITH APPEND ]
        [ NOT FOR REPLICATION ]
        AS
        [ { IF UPDATE ( column )
            [ { AND | OR } UPDATE ( column ) ]
                [ ...n ]
        | IF ( COLUMNS_UPDATED ( ) { bitwise_operator } updated_bitmask )
                { comparison_operator } column_bitmask [ ...n ]
        } ]
        sql_statement [ ...n ]
    }
}
```

Where:
trigger_name: The name of the trigger

Table | view: The table or view on which the trigger is executed

WITH ENCRYPTION: Encrypts the text of the trigger in the SYSCOMMENTS table. It also prevents the trigger from being replicated.
AFTER: Specifies that the trigger is fired only when all operations specified have executed successfully. All referential cascade actions and constraint checks also must complete successfully. For example "AFTER UPDATE" fires after the entire update operation is completed.

- AFTER is the default, if FOR is the only keyword specified.
- AFTER triggers cannot be defined on views.

INSTEAD OF = Specifies that the trigger is executed *instead of* the triggering SQL statement.
Restrictions:

- Only one INSTEAD OF trigger can be defined per INSERT, UPDATE, or DELETE statement. However, you can define views in which each view has its own INSTEAD OF trigger.
- INSTEAD OF is not allowed on views declared with the WITH CHECK OPTION.
- The DELETE option is not allowed on tables that have a referential relationship specifying a cascade action ON DELETE.
- The UPDATE option is not allowed on tables that have a referential relationship specifying a cascade action ON UPDATE.

{ [DELETE] [,] [INSERT] [,] [UPDATE] } = The keywords that specify which DML statements fire the trigger. At least one must be specified.

WITH APPEND = Specifies that an additional trigger of an existing type should be added. This is used for backwards compatibility with SQL Server 6.5.
Restrictions:

- Cannot be used with INSTEAD OF triggers
- Cannot be used if the AFTER trigger is explicitly stated
- WITH APPEND can be used only when FOR is specified (without INSTEAD OF or AFTER) for backward compatibility reasons

NOT FOR REPLICATION = The trigger should not be fired when a replication process modifies the table involved in the trigger.

AS = Actions the trigger is to perform

sql_statement = The actual condition(s) and action(s) to be performed

NOTE

WITH APPEND and FOR (which is interpreted as AFTER) will not be supported in future releases.

Trigger Conditions

A trigger condition is a way to prevent a trigger from firing if the UPDATE, INSERT, or DELETE statement does not meet the trigger condition criteria. There are two intrinsic functions that you can test for UPDATE or INSERT triggers that allow you to identify which columns are being updated. These functions are UPDATE() and COLUMNS_UPDATED().

UPDATE(*column*), used in an IF statement, returns TRUE if the column specified in the *column* argument is being updated, or inserted into. The function will return TRUE if a null or default value is implicitly inserted as a result of an INSERT DML statement.

COLUMNS_UPDATED(), also used in an IF statement, returns a **varbinary** bit pattern that indicates which column or columns are being updated or inserted into. The bits are in order from left to right, with the least significant bit being the first bit on the left. This bit represents the first column in the table; the next bit to the right represents the second column, and so on. COLUMNS_UPDATED returns multiple bytes if the table contains more than eight columns, with the least significant byte being on the far left. Like the UPDATE() function, COLUMNS_UPDATED returns TRUE if a NULL or default value is implicitly inserted.

Because the return value is of type varbinary, you use bitwise operators to test the results. For example, consider the DRUG_COMPANY table in the PHARMACY database. The table contains three columns:

- DRUG_COMPANY_ID
- DRUG_COMPANY_NAME
- PHONE

If you want to test for changes made to the DRUG_COMPANY_NAME or PHONE columns, you test the binary value against the lowest value. To get the binary value, first determine the binary value of the maximum return value, then use the "&" BITWISE operator to invert the bits. So, in our example:

The return value = 011

Remember, the far-left bit is the least significant. However, in the binary system, the far-right bit is the least significant. So, to compare this numerically, you need to invert the bits:

011 becomes 00000110, or 6 in decimal (2^1 + 2^2)

To do this in code, you use the & operator:

COLUMNS_UPDATED() & 6

So, your condition looks like this:

IF (COLUMNS_UPDATED() & 6) > 0

This evaluates whether any of the tested columns are updated. To evaluate whether all of them were updated, use this statement:

IF (COLUMNS_UPDATED() & 6) = 6

Logical Tables

Two logical tables are used in CREATE TRIGGER statements: **DELETED** and **INSERTED**. In structure, they are identical to the table the trigger is created on. The logical table DELETED contains the original data or what the row looked like before the DML was executed, and contains only data on an update or delete action. The logical table INSERTED contains the new data or what the row looks like (or would look like for an INSTEAD OF trigger) and contains only data on UPDATE or INSERT.

You can use these tables to identify which action is being performed in a single trigger that fires on more than one action.

For example:

```
create trigger gAudit_Drug_Company on Drug_Company
for insert, update, delete
as
       --if update
       if exists (select * from deleted) and exists (select * from inserted)
       begin
              /* statements to execute on update */
       end

       --if delete
       if exists (select * from deleted) and not exists (select * from inserted)
       begin
              /* statements to execute on delete */
       end

       --if insert
       if not exists (select * from deleted) and exists (select * from inserted)
       begin
              /* statements to execute on insert */
       end

go
```

SQL Server 2000 does not allow references to TEXT, NTEXT, or IMAGE columns in the INSERTED and DELETED tables. That's not to say that you can't use these logical tables for base tables that contain these data types; you just can't reference these columns.

Restrictions on Triggers

Because the TRUNCATE TABLE statement does not take place inside the transaction log, it is not caught by a DELETE trigger.

These Transact-SQL statements are not allowed in a trigger:

- ALTER DATABASE
- CREATE DATABASE
- DISK INIT
- DISK RESIZE
- DROP DATABASE
- LOAD DATABASE
- LOAD LOG
- RECONFIGURE
- RESTORE DATABASE
- RESTORE LOG

Implementation of an Historical Model with SQL Server

The historical model is commonly used for applications in which auditing requirements mandate the tracking of modified or deleted rows. To implement this model in SQL Server 2000, you use triggers and leverage the logical table, DELETED.

Take, for example the DRUG_COMPANY table in your PHARMACY database. You create a history table with the same structure as the original table with the addition of a HISTORY_ID column for use as the primary key. Although a primary key isn't required, it makes querying the table later much faster.

```
create table Drug_Company_History (
        history_id uniqueidentifier not null primary key default newid(),
        drug_company_id  int,
        drug_company_name varchar(20),
        phone int
)
go
```

After the HISTORY table is established, you can create a trigger on the ENTITY table to insert the original row into the HISTORY table:

```
create trigger gAudit_Drug_Company
on Drug_Company
for update, delete
as
        insert into Drug_Company_History (drug_company_id, drug
_company_name, phone)
        select * from deleted;
go
```

Now, each time data is modified or deleted from the table, you'll have a history of what it used to look like.

Implementation of Application Actions Model with SQL Server

Sometimes, there is a need within an application to audit not only changes at the data level, but actions taken within the application itself. The example mentioned in Chapter 7 was a credit or debit to an invoice.

To implement this model, first create a set of invoice tables in the PHARMACY database that would be used to invoice doctor's offices and hospitals.

```
create table Invoice_Header (
        Invoice_Id      int not null primary key identity(1,1),
        Customer_Id     int not null,
        Invoice_Date    datetime,
        Due_Date        datetime
)
go
```

```
create table Invoice_LineItem (
        LineItem_Id     int not null primary key identity(1,1),
        Invoice_Id      int not null,
        Amount money
)
go
```

Now, build the AUDIT tables. Assume the application has a database user for each end-user:

```
create table App_Audit_Actions (
        Action_Id       int not null primary key identity(1,1),
        Action_Desc     varchar(255)
)
go

create table App_Data_Dictionary(
        Class_Id        int not null primary key identity(1,1),
        Class_Desc      varchar(128)
)
go

create table App_Audit_Trail (
        Audit_Trail_Id int not null primary key identity(1,1),
        Class_Id               int not null references App_Data_Dictionary(class_id),
        Action_Id              int not null references App_Audit_Actions(action_id),
        Object_Id              int not null,
        Reason                 varchar(255) not null,
        CTL_UPD_DTTM           datetime not null default getdate(),
        CTL_UPD_USER           varchar(128) not null default user
)
go
```

You need to populate the DATA DICTIONARY, ACTIONS, and INVOICE tables, so you create an action called CREDIT INVOICE.

```
insert into App_Audit_Actions (Action_Desc) values ('Credit Invoice')
insert into App_Data_Dictionary(Class_Desc) values ('Invoice')
go

insert into Invoice_Header (Customer_Id, Invoice_Date, Due_Date) values (1, getdate(),
dateadd(d, 30, getdate()))
go
insert into Invoice_LineItem (Invoice_Id, Amount) values (@@identity, 5000)
go
```

With the data set up, you next create the stored procedure your developers will use to credit an invoice:

```
create procedure pCredit_Invoice (@id int, @amt money, @reason varchar(255))
as
      insert into Invoice_LineItem (invoice_id, amount) values (@id, @amt);

      declare @action_id int, @class_id int;
      select @action_id = action_id
      from App_Audit_Actions
      where Action_Desc = 'Credit Invoice';

      select @class_id = class_id
      from App_Data_Dictionary
      where class_desc = 'Invoice';

      insert into App_Audit_Trail ( Action_Id, Class_Id, Object_Id, Reason)
      values ( @action_id, @class_id, @id, @reason );
go
```

Now to test:

```
exec pCredit_Invoice (1, -200, 'Volume Discount')

select * from App_Audit_Trail
```

You must have noticed that SQL Server and Oracle10*g* triggers are similar in concept as well as in implementation. Both provide you with the capability to implement auditing models. However, SQL Server does not go beyond triggers to offer you the fine-grained auditing on tables and columns that Oracle10*g* offers. The next section presents a step-by-step illustration of how to implement fine-grained auditing with Oracle10*g*.

Fine-grained Auditing (FGA) with Oracle

Oracle is a complete database system, especially when it comes to auditing. As stated previously, Oracle provides column-level auditing, known as fine-grained auditing, which keeps an audit trail of all DML activities. FGA is simply an internal mechanism that allows the database administrator to set up auditing policies on tables. This is similar to the security policies introduced in Chapter 6, "Virtual Private Databases." FGA allows administrators to generate an audit trail of DML activities to operating system files or database tables. FGA is capable of auditing columns or tables using the Oracle PL/SQL-supplied package called DBMS_FGA. The feature not only records an audit trail of activities, it can also alert administrators or perform an action when an auditable event occurs. For example, suppose the administrator has added an update policy on the column CREDIT_LIMIT of the CUSTOMERS table. This policy records when the value for this column changes, as well as alerts the administrator through e-mail or pager that the change occurred. This section demonstrates how to use this alert feature. Again, refer to Oracle10g documentation found under *otn.oracle.com* for more details.

The four DBMS_FGA procedures used to implement DML auditing are:

- ADD_POLICY—Add an auditing policy
- DISABLE_POLICY—Disable an existing policy
- DROP_POLICY—Drop an existing policy
- ENABLE_POLICY—Enable an existing policy

```
PROCEDURE ADD_POLICY
 Argument Name                  Type                    In/Out Default?
 ------------------------------ --------                ------ --------
 OBJECT_SCHEMA                  VARCHAR2                IN     DEFAULT
 OBJECT_NAME                    VARCHAR2                IN
 POLICY_NAME                    VARCHAR2                IN
 AUDIT_CONDITION                VARCHAR2                IN     DEFAULT
 AUDIT_COLUMN                   VARCHAR2                IN     DEFAULT
 HANDLER_SCHEMA                 VARCHAR2                IN     DEFAULT
 HANDLER_MODULE                 VARCHAR2                IN     DEFAULT
 ENABLE                         BOOLEAN                 IN     DEFAULT
 STATEMENT_TYPES                VARCHAR2                IN     DEFAULT

PROCEDURE DISABLE_POLICY
 Argument Name                  Type                    In/Out Default?
 ------------------------------ --------                ------ --------
 OBJECT_SCHEMA                  VARCHAR2                IN     DEFAULT
 OBJECT_NAME                    VARCHAR2                IN
 POLICY_NAME                    VARCHAR2                IN

PROCEDURE DROP_POLICY
 Argument Name                  Type                    In/Out Default?
 ------------------------------ --------                ------ --------
 OBJECT_SCHEMA                  VARCHAR2                IN     DEFAULT
 OBJECT_NAME                    VARCHAR2                IN
 POLICY_NAME                    VARCHAR2                IN

PROCEDURE ENABLE_POLICY
 Argument Name                  Type                    In/Out Default?
 ------------------------------ --------                ------ --------
 OBJECT_SCHEMA                  VARCHAR2                IN     DEFAULT
 OBJECT_NAME                    VARCHAR2                IN
 POLICY_NAME                    VARCHAR2                IN
 ENABLE                         BOOLEAN                 IN     DEFAULT
```

Suppose you have business requirements that specify the following:

- Generate an audit trail for all SELECT activities on the CUSTOMERS table.
- Generate an audit trail when CREDIT_LIMIT is set to a value above $10,000.00.
- Generate an audit trail when a CUSTOMER row is deleted.

For practice, implement these policies by performing the following steps:

1. Work through a user other than SYSTEM or SYS that has privileges to CREATE TABLE and EXECUTE on DBMS_FGA. Create a CUSTOMERS table. If this table exists, remove it or use another user for this demonstration.

```
SQL> CREATE TABLE CUSTOMERS
  2  (
  3    CUSTOMER_ID     NUMBER PRIMARY KEY,
  4    CUSTOMER_NAME   VARCHAR2(20),
  5    CREDIT_LIMIT    NUMBER
  6  );

Table created.
```

2. Populate the CUSTOMERS table with two rows.

```
SQL> INSERT INTO CUSTOMERS VALUES( 100, 'TOM JONES', 2000);

1 row created.

SQL> INSERT INTO CUSTOMERS VALUES( 101, 'Joan Collins', 1000);

1 row created.

SQL> COMMIT;

Commit complete.
```

3. Now you need to add the auditing policies as specified previously. In this step, you need to use the ADD_POLICY procedure found in DBMS_FGA. The procedure requires the following parameters:

- OBJECT_SCHEMA—Name of the user that the desired object is to audit
- OBJECT_NAME—Name of the object that you want to audit
- POLICY_NAME—Unique name of the audit policy
- AUDIT_CONDITION—Condition on the row used as audit criteria. NULL means no condition.
- AUDIT_COLUMN—Name of columns that you would like to audit. NULL means all columns.
- HANDLER_SCHEMA—Name of the user name that owns the procedure that is used as the event handler. NULL indicates current schema.

- HANDLER_MODULE—Name of the procedure that fires when the audit is exercised. This is called the EVENT HANDLER procedure. It can be used to page or e-mail or any desired action. NULL means no event handler.
- ENABLE—Indication of whether the audit policy is enabled or not. The default value is TRUE, which means enabled.
- STATEMENT_TYPES—Specification of what DML statements you would like to audit. Allowed values are INSERT, UPDATE, DELETE, or SELECT, and the default value is SELECT.

```
SQL> EXEC DBMS_FGA.ADD_POLICY( OBJECT_SCHEMA  => 'DBSEC', -
>                              OBJECT_NAME    => 'CUSTOMERS', -
>                              POLICY_NAME    => 'AUDIT_POLICY_1_SELECT', -
>                              AUDIT_CONDITION=> NULL, -
>                              AUDIT_COLUMN   => NULL, -
>                              HANDLER_SCHEMA => NULL, -
>                              HANDLER_MODULE => NULL, -
>                              ENABLE         => TRUE, -
>                              STATEMENT_TYPES=> 'SELECT')

PL/SQL procedure successfully completed.

SQL> EXEC DBMS_FGA.ADD_POLICY( OBJECT_SCHEMA  => 'DBSEC', -
>                              OBJECT_NAME    => 'CUSTOMERS', -
>                              POLICY_NAME    => 'Audit_Policy_2_CREDIT', -
>                              AUDIT_CONDITION=> 'CREDIT_LIMIT >= 10000', -
>                              AUDIT_COLUMN   => NULL, -
>                              HANDLER_SCHEMA => NULL, -
>                              HANDLER_MODULE => NULL, -
>                              ENABLE         => TRUE, -
>                              STATEMENT_TYPES=> 'INSERT,UPDATE')

PL/SQL procedure successfully completed.

SQL> EXEC DBMS_FGA.ADD_POLICY( OBJECT_SCHEMA  => 'DBSEC', -
>                              OBJECT_NAME    => 'CUSTOMERS', -
>                              POLICY_NAME    => 'Audit_Policy_3_DELETE', -
>                              AUDIT_CONDITION=> NULL, -
>                              AUDIT_COLUMN   => NULL, -
>                              HANDLER_SCHEMA => NULL, -
>                              HANDLER_MODULE => NULL, -
>                              ENABLE         => TRUE, -
>                              STATEMENT_TYPES=> 'DELETE')

PL/SQL procedure successfully completed.
```

4. You need to verify that the policies were created. There is no shortage of data dictionary views provided by Oracle. You can use USER_AUDIT_POLICIES to view all auditing policies that you own.

```
SQL> SELECT OBJECT_NAME,
  2         POLICY_NAME,
  3         POLICY_TEXT,
  4         SEL, INS, UPD, DEL
  5    FROM USER_AUDIT_POLICIES;

OBJECT_NAME     POLICY_NAME               POLICY_TEXT                   SEL INS UPD DEL
--------------- ------------------------- ----------------------------- --- --- --- ---
CUSTOMERS       AUDIT_POLICY_1_SELECT                                   YES NO  NO  NO
CUSTOMERS       AUDIT_POLICY_2_CREDIT     CREDIT_LIMIT >= 10000         YES YES YES NO
CUSTOMERS       AUDIT_POLICY_3_DELETE                                   NO  NO  NO  YES
```

5. As SYS, you must turn on "auditing" on the DML statements, using the AUDIT command.

```
SQL> CONN SYS@SEC AS SYSDBA
Enter password: ******
Connected.
SQL> AUDIT SELECT, INSERT, UPDATE, DELETE ON SYS.AUD$ BY ACCESS;

Audit succeeded.
```

6. Now you need to perform some DML actions on the CUSTOMERS table. Issue the following statements:

```
SQL> SELECT * FROM CUSTOMERS;

CUSTOMER_ID CUSTOMER_NAME        CREDIT_LIMIT
----------- -------------------- ------------
        100 Tom Jones                    2000
        101 Joan Collins                 1000

SQL> UPDATE CUSTOMERS SET
  2         CREDIT_LIMIT = 11000
  3   WHERE CUSTOMER_ID = 101
  4  /

1 row updated.

SQL> UPDATE CUSTOMERS SET
  2         CREDIT_LIMIT = 5000
  3   WHERE CUSTOMER_ID = 101
  4  /

1 row updated.
```

```
SQL> INSERT INTO CUSTOMERS VALUES(102, 'Sam Afyouni', '50000');

1 row created.

SQL> INSERT INTO CUSTOMERS VALUES(103, 'Linda Evans', '2000');

1 row created.

SQL> DELETE CUSTOMERS WHERE CUSTOMER_ID = 103;

1 row deleted.

SQL> COMMIT;

Commit complete.

SQL> SELECT * FROM CUSTOMERS;

CUSTOMER_ID CUSTOMER_NAME        CREDIT_LIMIT
----------- -------------------- ------------
        100 Tom Jones                    2000
        101 Joan Collins                 5000
        102 Sam Afyouni                 50000
```

What do you notice about all of the DML activities? The database allowed the setting of the CREDIT_LIMIT column to be a value above 10000. However, the update is recorded as part of the audit trail.

7. Now check the contents of DBA_FGA_AUDIT_TRAIL to view the audit trail of the DML activities performed on the CUSTOMERS table shown in the previous example. The result is displayed in Figure 8-5.

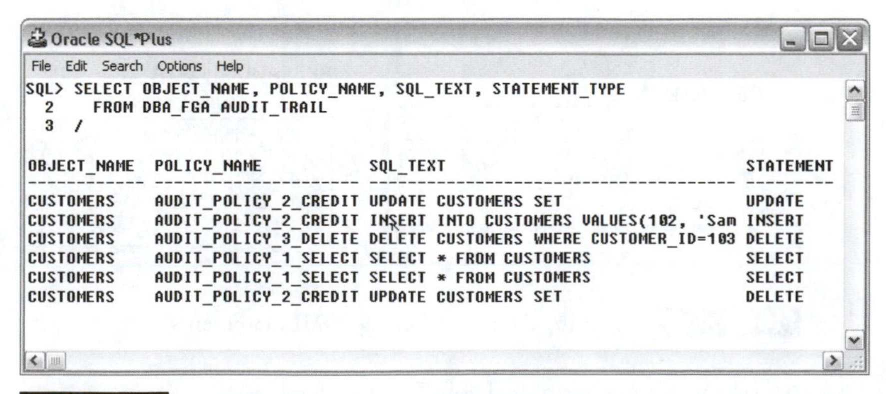

FIGURE 8-5 Content of the DBA_FGA_AUDIT_TRAIL data dictionary view

This view consists of many columns. For more information, refer to Oracle documentation found at *otn.oracle.com*.

DML Statement Audit Trail

Wissam is a database architect, who works for a mid-size fishery. He designed a terabyte transactional database application. A few days ago he attended a meeting about new projects. In the meeting the vice president of engineering asked Wissam to be the architect for a project that would track the type, time, and date of all DML activities. He immediately started to craft several designs that would capture the DML activities. After the design was completed, he presented it to the development and database team and asked for comments. One of the database operators instantly commented that the database system already had this auditing feature, as a built-in functionality. She asked, "Why reinvent the wheel?" Wissam replied quickly, "We need this module to be database independent. It must also have a front-end interface that will make it easy to administer."

Many companies frequently abandon built-in features of the database for home-grown solutions to perform the same tasks. It would not be unusual to find yourself in Wissam's situation of having to design a solution that enhances an existing application. To do this, you need to understand the different design implementations used for DML actions and auditing changes.

DML Action Auditing with Oracle

Oracle10*g* provides functionality to implement auditing schemes from basic to advanced. Oracle uses a combination of database objects, such as triggers, tables, and stored procedures. The purpose of auditing DML statements is to record the data changes occurring on the table, including the name of the person who made the data change, the date, and the time of the change. In this model the before or after value of the columns are not recorded. Figure 8-6 represents the data model for this. On the left of the diagram is the DEPARTMENTS table that will contain application data, whereas the table on the right is the auditing table APP_AUDIT_DATA that will contain an audit trail of all data change operations on the DEPARTMENTS table. Of course, APP_AUDIT_DATA can be used for other tables that need to be audited.

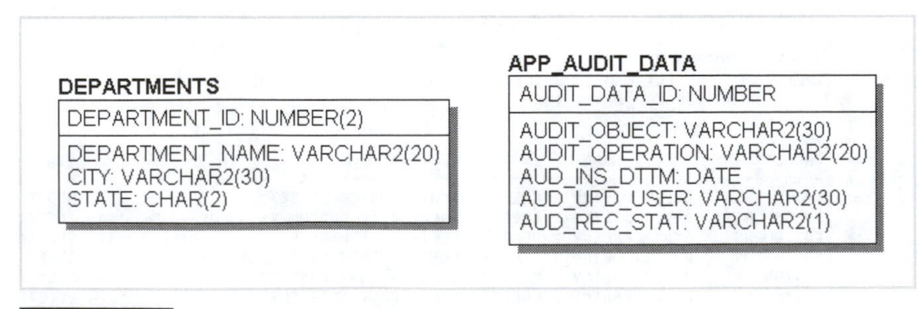

FIGURE 8-6 **Auditing data model for the DML statements**

To implement this model, follow these steps:

1. Using a user other than SYSTEM or SYS with privileges to create TABLES, SEQUENCES, and TRIGGERS, create the DEPARTMENTS table.

```
SQL> CREATE TABLE DEPARTMENTS
  2  (
  3    DEPARTMENT_ID    NUMBER(2) PRIMARY KEY,
  4    DEPARTMENT_NAME  VARCHAR2(20),
  5    CITY             VARCHAR2(30),
  6    STATE            CHAR(2)
  7  );

Table created.
```

2. Create the auditing table APP_AUDIT_DATA (see Table 8-1 for a description of the columns).

Table 8-1 APP_AUDIT_DATA columns description

Column Name	Column Description
AUDIT_DATA_ID	Unique identifier of the table; this column is generated automatically, using a SEQUENCE object
AUDIT_OBJECT	The name of the table that was audited
AUDIT_OPERATION	The operation that was performed on the object
AUD_INS_DTTM	Date and time when the operation occurred on the object
AUD_UPD_USER	Username of the account that performed the operation on the object
AUD_REC_STAT	Used for many purposes, such as security, or simply to flag the record to indicate whether it is active or not

```
SQL> CREATE TABLE APP_AUDIT_DATA
  2  (
  3    AUDIT_DATA_ID    NUMBER PRIMARY KEY,
  4    AUDIT_OBJECT     VARCHAR2(30),
  5    AUDIT_OPERATION  VARCHAR2(20),
  6    AUD_INS_DTTM     DATE,
  7    AUD_UPD_USER     VARCHAR2(30),
  8    AUD_REC_STAT     VARCHAR2(1)
  9  );

Table created.
```

3. Create a sequence object that will be used for the AUDIT_DATA_ID column of the APP_AUDIT_DATA table. The sequence will generate unique values.

```
SQL> CREATE SEQUENCE SEQ_APP_AUDIT_DATA
  2         INCREMENT BY 1
  3         START WITH 1
  4         MINVALUE 1
  5         NOCYCLE
  6         CACHE 20
  7         NOORDER;

Sequence created.
```

4. Create the trigger on the DEPARTMENTS table that will record the DML operations that occur on it. Note that the name of the trigger has a suffix AIUD, which is an acronym for "After Insert Update Delete" to help distinguish the trigger. This is a convention that you can use, or you may create your own convention as long as you are consistent in naming triggers and other database objects.

```
SQL> CREATE TRIGGER TRG_DEPARTMENT_AIUD
  2    AFTER INSERT OR UPDATE OR DELETE
  3    ON DEPARTMENTS
  4  DECLARE
  5
  6     V_OPR      VARCHAR2(20);
  7
  8  BEGIN
  9
 10     IF INSERTING THEN
 11        V_OPR := 'INSERT';
 12     ELSIF UPDATING THEN
 13        V_OPR := 'UPDATE';
 14     ELSE
 15        V_OPR := 'DELETE';
 16     END IF;
 17
 18     INSERT INTO APP_AUDIT_DATA(
 19                                     AUDIT_DATA_ID,
 20                     AUDIT_OBJECT,
 21                                     AUDIT_OPERATION,
 22                                     AUD_INS_DTTM,
 23           AUD_UPD_USER,
 24                                     AUD_REC_STAT
 25                                    ) VALUES
 26          (
 27                                     SEQ_APP_AUDIT_DATA.NEXTVAL,
 28                                     'DEPARTMENTS',
 29                                     V_OPR,
 30                                     SYSDATE,
```

```
31                                        USER,
32                                        'A'
33                                       );
34
35  EXCEPTION WHEN OTHERS THEN
36
37    -- TO SURPRESS THE ERROR IN CASE YOU DON'T
38    -- WANT ANYONE TO KNOW THAT TABLE IS BEING AUDITED
39    NULL;
40
41  END;
42  /

Trigger created.
```

5. Now you are ready to see what is recorded when you insert, modify, or delete rows on the DEPARTMENTS table. First you try INSERT. When executing these statements, make sure you pause for a minute after each statement and then commit your changes.

```
SQL> INSERT INTO DEPARTMENTS ( DEPARTMENT_ID, DEPARTMENT_NAME, CITY, STATE )
  2                    VALUES ( 10, 'Accounting', 'Boston', 'NV');

1 row created.

SQL> INSERT INTO DEPARTMENTS ( DEPARTMENT_ID, DEPARTMENT_NAME, CITY, STATE )
  2                    VALUES ( 11, 'Production', 'Redlands', 'CA');

1 row created.

SQL>
SQL> COMMIT;

Commit complete.
```

6. Now try an UPDATE on the DEPARTMENTS table.

```
SQL> UPDATE DEPARTMENTS SET
  2       CITY = 'Dallas',
  3       STATE = 'TX'
  4      WHERE DEPARTMENT_ID = 10;

1 row updated.

SQL>
SQL>  COMMIT;

Commit complete.
```

7. Try the DELETE statement again on the DEPARTMENTS table.

```
SQL> DELETE FROM DEPARTMENTS
  2   WHERE DEPARTMENT_ID = 11;

1 row deleted.

SQL>
SQL> COMMIT;

Commit complete.
```

8. This last step allows you to view the contents of the auditing table APP_AUDIT_DATA.

```
SQL> SELECT AUDIT_DATA_ID ID,
  2         AUDIT_OBJECT TABLE_NAME,
  3         AUDIT_OPERATION OPERATION,
  4         TO_CHAR(AUD_INS_DTTM, 'DD-MON-YYYY HH24:MI:SS') CREATE_DATE,
  5         AUD_UPD_USER USERNAME,
  6         AUD_REC_STAT ROW_STATUS
  7    FROM APP_AUDIT_DATA
  8  /

ID TABLE_NAME  OPERATION CREATE_DATE          USERNAME ROW_STATUS
-- ----------- --------- -------------------- -------- ----------
 1 DEPARTMENTS INSERT    04-JUL-2005 12:40:41 DBSEC    A
 2 DEPARTMENTS INSERT    04-JUL-2005 12:41:43 DBSEC    A
 3 DEPARTMENTS UPDATE    04-JUL-2005 12:43:54 DBSEC    A
 4 DEPARTMENTS DELETE    04-JUL-2005 12:45:19 DBSEC    A
```

The auditing table does not indicate what column name or column values are being modified. It serves merely to allow you to view DML activities on the DEPARTMENTS table.

Data Manipulation History

It is early morning, and you arrive at work to start a typical day as one of your company's database administrators. As you approach the door, you notice that the police are preventing employees from entering the premises. In addition, two men and a woman in dark blue suits are asking for the CEO. When the CEO arrives, they take her inside with them. You and the other employees wait with the police, until the CEO returns. She announces that everyone should go home for the day, except for the database management team, which includes you.

You, another administrator, and your manager enter the conference room, where the CEO explains what is happening."As you all know, we keep all transaction records and changes for our entire customer base. The day has come for recognizing how important this historical data is.The FBI is here because they have a court order to collect the records for one of our clients. We must provide this data immediately. Let me explain why I dismissed the rest of the employees. We have to be sure that no data changes occur today. The three of you will work from one computer to extract this data. You will be accompanied by the FBI representatives. I want to thank you in advance for your cooperation."

This scene is not as uncommon as you might think. Good record keeping, as well as thorough data auditing, has many important uses, including investigative support.

Almost every financial company implements some form of data manipulation history (also known as an audit trail). This is because almost all data in the financial world has monetary impact and is therefore sensitive. A data manipulation history provides a complete trail of all changes that are applied to data. The history contains either the before or the after value of the data, as well as a record of the person who made the change and date and time it occurred. The benefit of such an audit is to reconcile and verify current values. For example, suppose that the SALARY column in the EMPLOYEES table is being audited and you want to see if the current value is accurate. You can verify the value with the historical data of SALARY, particularly the documentation that authorizes promotions. Other uses of historical data include auditing employees, preventing fraud, and analyzing statistics.

History Auditing Model Implementation Using Oracle

Historical data auditing is simple to implement. Its main components are the TRIGGER and TABLE objects. Consider the following scenario. A small retail company decides to develop an historical audit solution to keep track of its customers' phone numbers, addresses, and the names of their sales representatives. Figure 8-7 is the customer table (named CUSTOMERS). Note that the last four columns are used as control columns, containing auditing information. A quick review of the description for these columns is represented in Table 8-2.

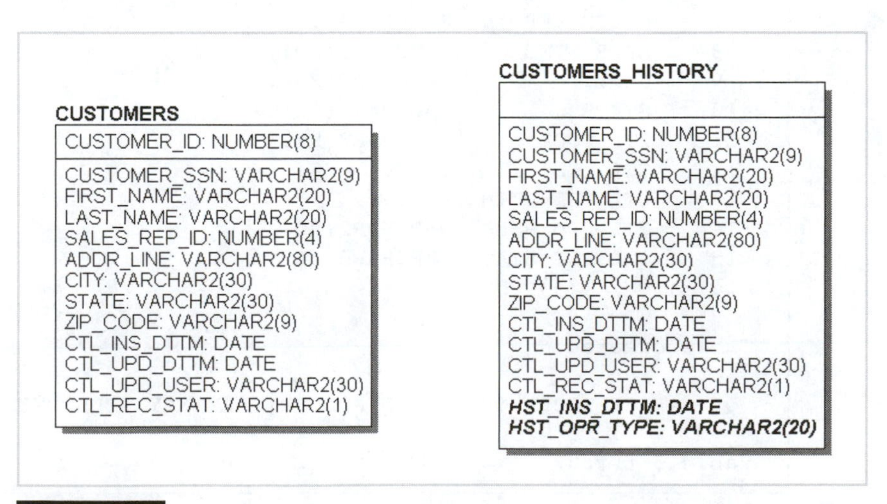

FIGURE 8-7 **Historical audit data model for the CUSTOMER table**

Table 8-2 Control columns description

Column Name	Description
CTL_INS_DTTM	Contains the date that the record (row) was created
CTL_UPD_DTTM	Contains the date that the record was modified
CTL_UPD_USER	Contains the user name that created or modified the record
CTL_REC_STAT	Can be used for many purposes such as security (refer to Chapters 5 and 6) or simply as an indicator to flag the record

Figure 8-7 shows a table that contains the historical data of all changes to the CUSTOMERS table. Notice that it is an identical copy of the CUSTOMERS table, except that two columns have been added: HST_INS_DTTM and HST_OPR_TYPE. These represent, respectively, the date and time the copy of the customer record was captured, and the type of operation applied to that record.

Now, follow these steps to implement historical auditing for the CUSTOMERS table shown in Figure 8-7.

1. To create the CUSTOMERS tables, use any database user account other than SYSTEM or SYS that has privileges to create tables and triggers. This table will hold customer data and will be audited for changes.

```
SQL> CREATE TABLE CUSTOMERS
  2  (
  3    CUSTOMER_ID    NUMBER(8) NOT NULL,
  4    CUSTOMER_SSN   VARCHAR2(9),
  5    FIRST_NAME     VARCHAR2(20),
  6    LAST_NAME      VARCHAR2(20),
  7    SALES_REP_ID   NUMBER(4),
  8    ADDR_LINE      VARCHAR2(80),
  9    CITY           VARCHAR2(30),
 10    STATE          VARCHAR2(30),
 11    ZIP_CODE       VARCHAR2(9),
 12    CTL_INS_DTTM   DATE,
 13    CTL_UPD_DTTM   DATE,
 14    CTL_UPD_USER   VARCHAR2(30),
 15    CTL_REC_STAT   VARCHAR2(1)
 16  );

Table created.

SQL> ALTER TABLE CUSTOMERS ADD PRIMARY KEY (CUSTOMER_ID);

Table altered.
```

2. You will be creating the history table for CUSTOMERS, as shown in the data model in the previous example. This table will contain all changes to the CUSTOMERS

table before an UPDATE, INSERT, or DELETE occurs. Note there is no primary key on this table because the same row in CUSTOMERS is modified several times.

```
SQL> CREATE TABLE CUSTOMERS_HISTORY
  2  (
  3    CUSTOMER_ID     NUMBER(8) NOT NULL,
  4    CUSTOMER_SSN    VARCHAR2(9),
  5    FIRST_NAME      VARCHAR2(20),
  6    LAST_NAME       VARCHAR2(20),
  7    SALES_REP_ID    NUMBER(4),
  8    ADDR_LINE       VARCHAR2(80),
  9    CITY            VARCHAR2(30),
 10    STATE           VARCHAR2(30),
 11    ZIP_CODE        VARCHAR2(9),
 12    CTL_INS_DTTM    DATE,
 13    CTL_UPD_DTTM    DATE,
 14    CTL_UPD_USER    VARCHAR2(30),
 15    CTL_REC_STAT    VARCHAR2(1),
 16    HST_INS_DTTM    DATE,
 17    HST_OPR_TYPE    VARCHAR2(20)
 18  );

Table created.
```

3. Create the trigger on the CUSTOMERS table to track changes and record all the values of the columns before they were changed. Note that all columns will be recorded, even if only one column is being changed.

```
SQL> CREATE OR REPLACE TRIGGER TRG_CUSTOMERS_BIUR
  2    BEFORE UPDATE OR INSERT OR DELETE ON CUSTOMERS
  3    FOR EACH ROW
  4  DECLARE
  5     V_CUSTOMER_ID      CUSTOMERS_HISTORY.CUSTOMER_ID%TYPE;
  6     V_CUSTOMER_SSN     CUSTOMERS_HISTORY.CUSTOMER_SSN%TYPE;
  7     V_FIRST_NAME       CUSTOMERS_HISTORY.FIRST_NAME%TYPE;
  8     V_LAST_NAME        CUSTOMERS_HISTORY.LAST_NAME%TYPE;
  9     V_SALES_REP_ID     CUSTOMERS_HISTORY.SALES_REP_ID%TYPE;
 10     V_ADDR_LINE        CUSTOMERS_HISTORY.ADDR_LINE%TYPE;
 11     V_CITY             CUSTOMERS_HISTORY.CITY%TYPE;
 12     V_STATE            CUSTOMERS_HISTORY.STATE%TYPE;
 13     V_ZIP_CODE         CUSTOMERS_HISTORY.ZIP_CODE%TYPE;
 14     V_CTL_INS_DTTM     CUSTOMERS_HISTORY.CTL_INS_DTTM%TYPE;
 15     V_CTL_UPD_DTTM     CUSTOMERS_HISTORY.CTL_UPD_DTTM%TYPE;
 16     V_CTL_UPD_USER     CUSTOMERS_HISTORY.CTL_UPD_USER%TYPE;
 17     V_CTL_REC_STAT     CUSTOMERS_HISTORY.CTL_REC_STAT%TYPE;
 18     V_HST_OPR_TYPE     CUSTOMERS_HISTORY.HST_OPR_TYPE%TYPE;
 19  BEGIN
 20     IF INSERTING THEN
```

```
21          :NEW.CTL_INS_DTTM := SYSDATE;
22          :NEW.CTL_UPD_DTTM := NULL;
23         :NEW.CTL_REC_STAT := 'N';
24          V_HST_OPR_TYPE     := 'INSERT';
25       ELSIF UPDATING THEN
26          :NEW.CTL_UPD_DTTM := SYSDATE;
27         V_CTL_UPD_DTTM     := :NEW.CTL_UPD_DTTM ;
28          V_HST_OPR_TYPE     := 'UPDATE';
29       ELSIF DELETING THEN
30          V_CUSTOMER_ID  := :OLD.CUSTOMER_ID;
31          V_CUSTOMER_SSN := :OLD.CUSTOMER_SSN;
32          V_FIRST_NAME   := :OLD.FIRST_NAME;
33          V_LAST_NAME    := :OLD.LAST_NAME;
34          V_SALES_REP_ID := :OLD.SALES_REP_ID;
35          V_ADDR_LINE    := :OLD.ADDR_LINE;
36          V_CITY         := :OLD.CITY;
37          V_STATE        := :OLD.STATE;
38          V_ZIP_CODE     := :OLD.ZIP_CODE;
39          V_CTL_INS_DTTM := :OLD.CTL_INS_DTTM;
40          V_CTL_UPD_DTTM := :OLD.CTL_UPD_DTTM;
41          V_CTL_UPD_USER := :OLD.CTL_UPD_USER;
42          V_CTL_REC_STAT := :OLD.CTL_REC_STAT;
43          V_HST_OPR_TYPE := 'DELETE';
44       END IF;
45       IF INSERTING OR UPDATING THEN
46          :NEW.CTL_UPD_USER := USER;
47          V_CUSTOMER_ID  := :NEW.CUSTOMER_ID;
48          V_CUSTOMER_SSN := :NEW.CUSTOMER_SSN;
49          V_FIRST_NAME   := :NEW.FIRST_NAME;
50          V_LAST_NAME    := :NEW.LAST_NAME;
51          V_SALES_REP_ID := :NEW.SALES_REP_ID;
52          V_ADDR_LINE    := :NEW.ADDR_LINE;
53          V_CITY         := :NEW.CITY;
54          V_STATE        := :NEW.STATE;
55          V_ZIP_CODE     := :NEW.ZIP_CODE;
56          V_CTL_INS_DTTM := :NEW.CTL_INS_DTTM;
57          V_CTL_UPD_DTTM := :NEW.CTL_UPD_DTTM;
58          V_CTL_UPD_USER := :NEW.CTL_UPD_USER;
59          V_CTL_REC_STAT := :NEW.CTL_REC_STAT;
60       END IF;
61       INSERT INTO CUSTOMERS_HISTORY
62       (
63        CUSTOMER_ID,
64        CUSTOMER_SSN,
65        FIRST_NAME,
66        LAST_NAME,
67        SALES_REP_ID,
68        ADDR_LINE,
69        CITY,
70        STATE,
71        ZIP_CODE,
```

```
 72       CTL_INS_DTTM,
 73       CTL_UPD_DTTM,
 74       CTL_UPD_USER,
 75       CTL_REC_STAT,
 76       HST_INS_DTTM,
 77      HST_OPR_TYPE
 78      )
 79      VALUES
 80      (
 81       V_CUSTOMER_ID,
 82       V_CUSTOMER_SSN,
 83       V_FIRST_NAME,
 84       V_LAST_NAME,
 85       V_SALES_REP_ID,
 86       V_ADDR_LINE,
 87       V_CITY,
 88       V_STATE,
 89       V_ZIP_CODE,
 90       V_CTL_INS_DTTM,
 91       V_CTL_UPD_DTTM,
 92       V_CTL_UPD_USER,
 93       V_CTL_REC_STAT,
 94      SYSDATE,
 95      V_HST_OPR_TYPE
 96      );
 97   EXCEPTION WHEN OTHERS THEN
 98      RAISE_APPLICATION_ERROR(-20000,SQLERRM);
 97   END;
 98   /

Trigger created.
```

4. Now you insert two rows into the CUSTOMERS table.

```
SQL> INSERT INTO CUSTOMERS (CUSTOMER_ID, CUSTOMER_SSN, FIRST_NAME, LAST_NAME, SALES_REP_ID,
  2                         ADDR_LINE, CITY, STATE, ZIP_CODE)
  3                 VALUES (201340,'969996970','Jeffrey','Antoine',6459,'9938 Moreno St.',
  4                         'Champagne','SD',' 43172');

1 row created.

SQL> INSERT INTO CUSTOMERS ( CUSTOMER_ID, CUSTOMER_SSN, FIRST_NAME, LAST_NAME, SALES_REP_ID,
  2                         ADDR_LINE, CITY, STATE, ZIP_CODE)
  3                 VALUES (801349,'716647546','Cordell','Ayres',2200,'37 Noyes Street',
  4                         'Narod','NC',' 15199');
```

```
1 row created.

SQL> COMMIT;

Commit complete.
```

5. Retrieve all rows from the CUSTOMERS and CUSTOMERS_HISTORY tables to verify that the rows were added to the CUSTOMERS table and that the trigger is populating the CUSTOMERS_HISTORY table, before the data is inserted into the CUSTOMERS table.

```
SQL> SELECT CUSTOMER_ID CUS_ID, FIRST_NAME, LAST_NAME,
  2         TO_CHAR(CTL_INS_DTTM, 'DD-MON-YYYY HH24:MI:SS') CTL_INS_DTTM,
  3         CTL_UPD_DTTM, CTL_UPD_USER, CTL_REC_STAT
  4    FROM CUSTOMERS;

CUS_ID FIRST_NAME   LAST_NAME    CTL_INS_DTTM         CTL_UPD_DTTM CTL_UPD_US CTL_REC_STAT
------------------- ------------ -------------------- ------------ ---------- ------------
201340 Jeffrey      Antoine      23-JUN-2005 12:55:06              SEC        N
801349 Cordell      Ayres        23-JUN-2005 12:55:13              SEC        N

SQL> SELECT CUSTOMER_ID CUS_ID, FIRST_NAME, LAST_NAME,
  2         TO_CHAR(CTL_INS_DTTM, 'DD-MON-YYYY HH24:MI:SS') CTL_INS_DTTM,
  3         CTL_UPD_DTTM, CTL_UPD_USER USER,
  4         TO_CHAR(HST_INS_DTTM, 'DD-MON-YYYY HH24:MI:SS') HST_INS_DTTM,
  5         HST_OPR_TYPE OPER
  6    FROM CUSTOMERS_HISTORY;

CUS_ID FIRST_NAME LAST_NAME CTL_INS_DTTM         CTL_UPD_DTTM USER HST_INS_DTTM         OPER
------ ---------- --------- -------------------- ------------ ---- -------------------- ------
201340 Jeffrey    Antoine   23-JUN-2005 12:55:06              SEC  23-JUN-2005 12:55:06 INSERT
801349 Cordell    Ayres     23-JUN-2005 12:55:13              SEC  23-JUN-2005 12:55:13 INSERT
```

6. Now update the street address for customer 201340.

```
SQL> UPDATE CUSTOMERS SET
  2         ADDR_LINE = '123 Hello street'
  3   WHERE CUSTOMER_ID = 201340
  4  /

1 row updated.

SQL> COMMIT;

Commit complete.
```

7. Examine the value of the STREET ADDRESS column for customer 201340 and its value before it was updated.

```
SQL> SELECT CUSTOMER_ID CUS_ID, ADDR_LINE,
  2         TO_CHAR(CTL_INS_DTTM, 'DD-MON-YYYY HH24:MI:SS') CTL_INS_DTTM,
  3         TO_CHAR(CTL_UPD_DTTM, 'DD-MON-YYYY HH24:MI:SS') CTL_UPD_DTTM,
  4         CTL_UPD_USER, CTL_REC_STAT
  5    FROM CUSTOMERS
  6   WHERE CUSTOMER_ID = 201340
  7 /

CUS_ID ADDR_LINE        CTL_INS_DTTM         CTL_UPD_DTTM         CTL_UPD_US CTL_INS_DTTM
------ ---------------- -------------------- -------------------- ---------- ------------
201340 123 Hello street 23-JUN-2005 12:55:06 23-JUN-2005 12:59:51 SEC        N

SQL> SELECT CUSTOMER_ID CUS_ID, ADDR_LINE,
  2         TO_CHAR(CTL_INS_DTTM, 'DD-MON-YYYY HH24:MI:SS') CTL_INS_DTTM,
  3         TO_CHAR(CTL_UPD_DTTM, 'DD-MON-YYYY HH24:MI:SS') CTL_UPD_DTTM,
  4         TO_CHAR(HST_INS_DTTM, 'DD-MON-YYYY HH24:MI:SS') HST_INS_DTTM,
  5         CTL_UPD_USER USERNAME, HST_OPR_TYPE OPER
  6    FROM CUSTOMERS_HISTORY
  7   WHERE CUSTOMER_ID = 201340
  8   ORDER BY HST_INS_DTTM
  9 /

CUS_ID ADDR_LINE        CTL_INS_DTTM         CTL_UPD_DTTM         USER HST_INS_DTTM         OPER
------ ---------------- -------------------- -------------------- ------------------------- ------
201340 9938 Moreno St.  23-JUN-2005 12:55:06                      SEC  23-JUN-2005 12:55:06 INSERT
201340 123 Hello street 23-JUN-2005 12:55:06 23-JUN-2005 12:59:51 SEC  23-JUN-2005 12:59:51 UPDATE
```

8. Now see what happens when you delete customer 201340 and then verify it was deleted. Is that row recorded in the history table?

```
SQL> DELETE CUSTOMERS WHERE CUSTOMER_ID = 201340;

1 row deleted.

SQL> COMMIT;

Commit complete.

SQL> SELECT * FROM CUSTOMERS
  2   WHERE CUSTOMER_ID = 201340;

no rows selected

SQL> SELECT CUSTOMER_ID, ADDR_LINE,
  2         TO_CHAR(CTL_INS_DTTM, 'DD-MON-YYYY HH24:MI:SS') CTL_INS_DTTM,
  3         TO_CHAR(CTL_UPD_DTTM, 'DD-MON-YYYY HH24:MI:SS') CTL_UPD_DTTM,
  4         CTL_UPD_USER USERNAME,
  5         TO_CHAR(HST_INS_DTTM, 'DD-MON-YYYY HH24:MI:SS') HST_INS_DTTM,
  6         HST_OPR_TYPE
```

```
  7     FROM CUSTOMERS_HISTORY
  8    WHERE CUSTOMER_ID = 201340
  9    ORDER BY HST_INS_DTTM
 10   /

CUS_ID ADDR_LINE        CTL_INS_DTTM         CTL_UPD_DTTM         USER HST_INS_DTTM         OPER
------ ---------------- -------------------- -------------------- ---- -------------------- ------
201340 9938 Moreno St.  23-JUN-2005 12:55:06                      SEC  23-JUN-2005 12:55:06 INSERT
201340 123 Hello street 23-JUN-2005 12:55:06 23-JUN-2005 12:59:51 SEC  23-JUN-2005 13:01:43 UPDATE
201340 123 Hello street 23-JUN-2005 12:55:06 23-JUN-2005 12:59:51 SEC  23-JUN-2005 13:09:12 DELETE
```

Simple isn't it? It is more difficult to review the history and implement a purging scheme to keep only what is required. The more the activity on the table, the larger the history table grows. If all historical data must be kept based on business requirements or government mandate, then you should think of implementing table partitioning to enhance the data retrieval performance for the history table.

DML Auditing Using Repository with Oracle (Simple 1)

The model in this section was presented in Chapter 7 as Simple Auditing Model 1. The main purpose of this model is to flag users, tables, or columns for auditing. In other words, this model serves as a mechanism to audit by changing the registry entry for the user in the repository, without changing code. For example, suppose TOM was suspected of suspicious activities. All you need to do is add an entry in the repository for that user and automatically any DML activities by this user are recorded. You might ask how this differs from the Oracle auditing mechanism of DML statements. The difference is in the administrative skill level needed to conduct an audit. Database administration skills are required to use the Oracle auditing mechanism. In this DML model, an application administrator can manage the audit because a user interface is built on top of the repository. Figure 8-8 illustrates the data model of this DML auditing structure.

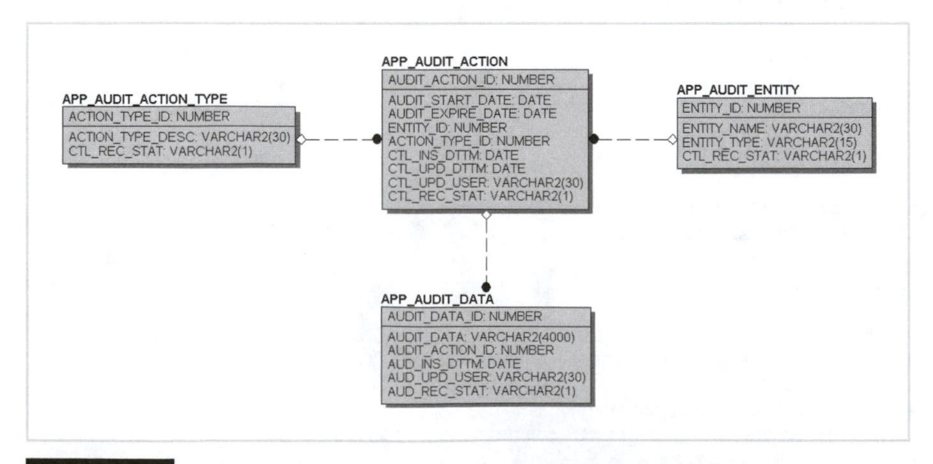

FIGURE 8-8 DML auditing using repository (Simple 1) data model

Before you start developing DML auditing with a repository, you need to know the rules for the flagging hierarchy. For example, suppose TOM is registered in the repository and the DEPARTMENTS table is also registered. What happens if SCOTT modifies DEPARTMENTS? And what happens if TOM updates DEPARTMENTS and EMPLOYEES? The answer to both questions depends on the flagging hierarchy that is in the procedure used by the trigger. Audit flagging rules are flexible. Therefore, in this implementation you use the rules in Figure 8-9. The rules state that if the user is registered, it does not matter whether the table is registered or not. Actions by the user are audited. However, if the user is not registered, actions performed on the table are audited. Note that this model does not record before or after column values. It only registers the type of DML operation occurring on the table.

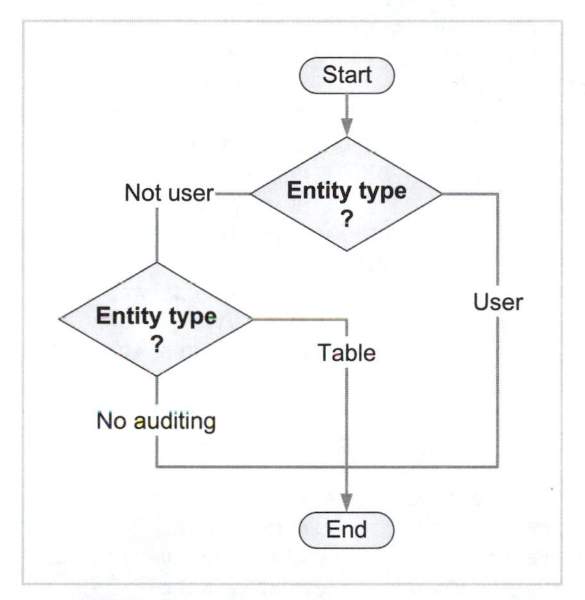

FIGURE 8-9 Flowchart of auditing flagging hierarchy

1. Select a user that is not SYSTEM or SYS and create the trigger for the procedures and privileges table; in this way you build the repository for this auditing mechanism.

```
SQL> CREATE TABLE APP_AUDIT_ENTITY (
  2          ENTITY_ID                NUMBER NULL,
  3          ENTITY_NAME              VARCHAR2(30) NULL,
  4          ENTITY_TYPE              VARCHAR2(15) NULL,
  5          CTL_REC_STAT             VARCHAR2(1) NULL,
  6          PRIMARY KEY (ENTITY_ID)
  7  );

Table created.
```

```
SQL> CREATE TABLE APP_AUDIT_ACTION_TYPE (
  2         ACTION_TYPE_ID          NUMBER NULL,
  3         ACTION_TYPE_DESC        VARCHAR2(30) NULL,
  4         CTL_REC_STAT            VARCHAR2(1) NULL,
  5         PRIMARY KEY (ACTION_TYPE_ID)
  6  );

Table created.

SQL> CREATE TABLE APP_AUDIT_ACTION (
  2         AUDIT_ACTION_ID         NUMBER NULL,
  3         AUDIT_START_DATE        DATE NULL,
  4         AUDIT_EXPIRE_DATE       DATE NULL,
  5         CTL_INS_DTTM            DATE NULL,
  6         CTL_UPD_DTTM            DATE NULL,
  7         CTL_UPD_USER            VARCHAR2(30) NULL,
  8         CTL_REC_STAT            VARCHAR2(1) NULL,
  9         ACTION_TYPE_ID          NUMBER NULL,
 10         ENTITY_ID               NUMBER NULL,
 11         PRIMARY KEY (AUDIT_ACTION_ID),
 12         FOREIGN KEY (ENTITY_ID)
 13                                   REFERENCES APP_AUDIT_ENTITY,
 14         FOREIGN KEY (ACTION_TYPE_ID)
 15                                   REFERENCES APP_AUDIT_ACTION_TYPE
 16  );

Table created.

SQL> CREATE TABLE APP_AUDIT_DATA (
  2         AUDIT_DATA_ID           NUMBER NULL,
  3         AUDIT_DATA              VARCHAR2(4000) NULL,
  4         AUDIT_ACTION_ID         NUMBER NULL,
  5         AUD_INS_DTTM            DATE NULL,
  6         AUD_UPD_USER            VARCHAR2(30) NULL,
  7         AUD_REC_STAT            VARCHAR2(1) NULL,
  8         PRIMARY KEY (AUDIT_DATA_ID),
  9         FOREIGN KEY (AUDIT_ACTION_ID)
 10                                   REFERENCES APP_AUDIT_ACTION
 11  );

Table created.
```

2. Now you need to create a sequence object for the AUDIT_DATA_ID column. If this sequence already exists, you can skip this step. Otherwise execute the following statement to create the sequence.

```
SQL> CREATE SEQUENCE SEQ_APP_AUDIT_DATA
  2         INCREMENT BY 1
  3         START WITH 1
  4         MINVALUE 1
  5         NOCYCLE
  6         CACHE 20
  7         NOORDER;

Sequence created.
```

3. Next, build two tables to use for applications.

```
SQL> CREATE TABLE APP_TBL1(
  2      CODE        NUMBER PRIMARY KEY,
  3      DESCRIPTION VARCHAR2(20)
  4    );

Table created.

SQL>
SQL>  CREATE TABLE APP_TBL2(
  2      ID          NUMBER PRIMARY KEY,
  3      NAME        VARCHAR2(35),
  4      PHONE       VARCHAR2(14)
  5   );

Table created.
```

4. Populate the application tables with data.

```
SQL> INSERT INTO APP_TBL1 ( CODE, DESCRIPTION ) VALUES ( 1, 'Description #1');

1 row created.

SQL> INSERT INTO APP_TBL1 ( CODE, DESCRIPTION ) VALUES ( 2, 'Description #2');

1 row created.

SQL> INSERT INTO APP_TBL2 ( ID, NAME, PHONE )
  2                VALUES ( 100, 'Tom Jones', '123-123-1234');

1 row created.

SQL> INSERT INTO APP_TBL2 ( ID, NAME, PHONE )
  2                VALUES ( 101, 'Linda Evans', '234-234-2345');

1 row created.
```

```
SQL> INSERT INTO APP_TBL2 ( ID, NAME, PHONE )
  2                VALUES ( 102, 'Joan Collins', '345-345-3456');

1 row created.

SQL> COMMIT;

Commit complete.
```

5. Populate the auditing repository with metadata.

```
SQL> INSERT INTO APP_AUDIT_ACTION_TYPE ( ACTION_TYPE_ID, ACTION_TYPE_DESC, CTL_REC_STAT )
  2                        VALUES ( 1, 'UPDATE', 'A');

1 row created.

SQL> INSERT INTO APP_AUDIT_ACTION_TYPE ( ACTION_TYPE_ID, ACTION_TYPE_DESC, CTL_REC_STAT )
  2                        VALUES ( 2, 'DELETE', 'A');

1 row created.

SQL> INSERT INTO APP_AUDIT_ACTION_TYPE ( ACTION_TYPE_ID, ACTION_TYPE_DESC, CTL_REC_STAT )
  2                        VALUES ( 3, 'INSERT', 'A');

1 row created.

SQL>
SQL> INSERT INTO APP_AUDIT_ENTITY ( ENTITY_ID, ENTITY_NAME, ENTITY_TYPE, CTL_REC_STAT )
  2                   VALUES ( 1, 'DBSEC', 'USER', 'A');

1 row created.

SQL> INSERT INTO APP_AUDIT_ENTITY ( ENTITY_ID, ENTITY_NAME, ENTITY_TYPE, CTL_REC_STAT )
  2                   VALUES ( 2, 'APP_TBL2', 'TABLE', 'A');

1 row created.

SQL> INSERT INTO APP_AUDIT_ENTITY ( ENTITY_ID, ENTITY_NAME, ENTITY_TYPE, CTL_REC_STAT )
  2                   VALUES ( 3, 'APP_TBL1', 'TABLE', 'A');

1 row created.

SQL>
SQL> INSERT INTO APP_AUDIT_ACTION ( AUDIT_ACTION_ID, AUDIT_START_DATE, AUDIT_EXPIRE_DATE,
  2                                 CTL_INS_DTTM, CTL_UPD_DTTM, CTL_UPD_USER, CTL_REC_STAT,
  3                                 ACTION_TYPE_ID, ENTITY_ID )
  4                    VALUES ( 1, TO_Date('06/28/2005 12:00:00 AM','MM/DD/YYYY HH:MI:SS AM'),
  5                               TO_Date('07/28/2005 12:00:00 AM','MM/DD/YYYY HH:MI:SS AM'),
  6                               TO_Date('06/28/2005 12:00:00 AM','MM/DD/YYYY HH:MI:SS AM'),
  7                               TO_Date('06/28/2005 12:00:00 AM','MM/DD/YYYY HH:MI:SS AM'),
  8                            'DBSEC', 'A', 1, 1);

1 row created.
```

```
SQL> INSERT INTO APP_AUDIT_ACTION ( AUDIT_ACTION_ID, AUDIT_START_DATE, AUDIT_EXPIRE_DATE,
  2                                 CTL_INS_DTTM, CTL_UPD_DTTM, CTL_UPD_USER, CTL_REC_STAT,
  3                                 ACTION_TYPE_ID, ENTITY_ID )
  4                    VALUES ( 2, TO_Date('06/28/2005 12:00:00 AM','MM/DD/YYYY HH:MI:SS AM'),
  5                                TO_Date('06/28/2005 12:00:00 AM','MM/DD/YYYY HH:MI:SS AM'),
  6                                TO_Date('06/28/2005 12:00:00 AM','MM/DD/YYYY HH:MI:SS AM'),
  7                                TO_Date('06/28/2005 12:00:00 AM','MM/DD/YYYY HH:MI:SS AM'),
  8                             'DBSEC', 'A', 2, 3);

1 row created.

SQL> commit;

Commit complete.
```

6. Now you need to create the stored package to be used with the trigger.

```
SQL> CREATE OR REPLACE PACKAGE Pkg_App_Audit IS
  2
  3     PROCEDURE INSERT_DATA(
  4                            P_TABLE_NAME VARCHAR2,
  5                            P_OPERATION  VARCHAR2
  6         );
  7
  8     FUNCTION AUDIT_CHECK (
  9                            P_USER_NAME  VARCHAR2,
 10                            P_TABLE_NAME VARCHAR2,
 11         P_OPERATION  VARCHAR2
 12        )RETURN NUMBER;
 13
 14 END;
 15 /

Package created.

SQL> CREATE OR REPLACE PACKAGE BODY Pkg_App_Audit IS
  2
  3     --~~~~~~~~~~~~~~~~~~~~~~~~~~~~~~~~~~~~~~~~~~
  4     -- CHECKS IF TABLE IS AUDITIED.
  5     --~~~~~~~~~~~~~~~~~~~~~~~~~~~~~~~~~~~~~~~~~~
  6     FUNCTION AUDIT_CHECK(
  7                            P_USER_NAME   VARCHAR2,
  8                            P_TABLE_NAME  VARCHAR2,
  9         P_OPERATION   VARCHAR2
 10          ) RETURN NUMBER IS
 11
 12        V_ID   NUMBER := 0;
 13
 14     BEGIN
```

```
15
16        BEGIN
17           SELECT A.AUDIT_ACTION_ID
18             INTO V_ID
19             FROM APP_AUDIT_ENTITY E,
20                  APP_AUDIT_ACTION A,
21                  APP_AUDIT_ACTION_TYPE T
22            WHERE A.ACTION_TYPE_ID = T.ACTION_TYPE_ID
23              AND A.ENTITY_ID = E.ENTITY_ID
24              AND E.ENTITY_NAME = P_USER_NAME
25              AND E.ENTITY_TYPE = 'USER'
26           AND T.ACTION_TYPE_DESC = UPPER(P_OPERATION)
27              AND SYSDATE BETWEEN AUDIT_START_DATE AND AUDIT_EXPIRE_DATE;
28     EXCEPTION WHEN OTHERS THEN
29        RETURN NULL;
30        END;
31
32     IF V_ID IS NULL THEN
33
34        BEGIN
35              SELECT A.AUDIT_ACTION_ID
36                INTO V_ID
37                FROM APP_AUDIT_ENTITY E,
38                     APP_AUDIT_ACTION A,
39                     APP_AUDIT_ACTION_TYPE T
40               WHERE A.ACTION_TYPE_ID = T.ACTION_TYPE_ID
41                 AND A.ENTITY_ID = E.ENTITY_ID
42           AND E.ENTITY_NAME = P_TABLE_NAME
43           AND E.ENTITY_TYPE = 'TABLE'
44                 AND T.ACTION_TYPE_DESC = UPPER(P_OPERATION)
45           AND SYSDATE BETWEEN AUDIT_START_DATE AND AUDIT_EXPIRE_DATE;
46     EXCEPTION WHEN OTHERS THEN
47        RETURN NULL;
48     END;
49     END IF;
50
51     RETURN V_ID;
52    END;
53     --~~~~~~~~~~~~~~~~~~~~~~~~~~~~~~~~~~~~~~~~~~
54
55     --~~~~~~~~~~~~~~~~~~~~~~~~~~~~~~~~~~~~~~~~~~
56     -- INSERT OLD VALUES INTO AUDIT TABLE
57     --~~~~~~~~~~~~~~~~~~~~~~~~~~~~~~~~~~~~~~~~~~
58     PROCEDURE INSERT_DATA(
59                           P_TABLE_NAME VARCHAR2,
60                           P_OPERATION  VARCHAR2
61                          ) IS
62        V_ID NUMBER;
63
```

```
 64     BEGIN
 65
 66        V_ID := AUDIT_CHECK(USER, P_TABLE_NAME, P_OPERATION);
 67
 68     IF V_ID IS NULL THEN
 69        RETURN;
 70     END IF;
 71
 72        INSERT INTO APP_AUDIT_DATA(AUDIT_DATA_ID, AUDIT_DATA, AUDIT_ACTION_ID,
 73                                   AUD_INS_DTTM, AUD_UPD_USER, AUD_REC_STAT)
 74                            VALUES(SEQ_APP_AUDIT_DATA.NEXTVAL, P_OPERATION, V_ID,
 75                                   SYSDATE, USER, 'A');
 76     END;
 77     --~~~~~~~~~~~~~~~~~~~~~~~~~~~~~~~~~~~~~~~~
 78
 79  END;
 80  /

Package body created.
```

7. Now create the trigger for APP_TBL1. Use the trigger template provided in the following code.

```
CREATE OR REPLACE TRIGGER TRG_TableName_BUDIR
  BEFORE UPDATE OR DELETE OR INSERT ON TableName
  FOR EACH ROW
DECLARE

   V_OPERATION  VARCHAR2(20);

BEGIN

   IF INSERTING THEN
      V_OPERATION := 'INSERT';
   ELSIF UPDATING THEN
      V_OPERATION := 'UPDATE';
   ELSE
      V_OPERATION := 'DELETE';
   END IF;

   PKG_APP_AUDIT.INSERT_DATA('TableName', V_OPERATION);

END;
/
```

8. At this point, you need only replace the bold and italics with the table name APP_TBL1, and another trigger is created for APP_TBL2.

```
SQL> CREATE OR REPLACE TRIGGER TRG_APP_TBL1_BUDIR
  2    BEFORE UPDATE OR DELETE OR INSERT ON APP_TBL1
  3    FOR EACH ROW
  4  DECLARE
  5
  6     V_OPERATION  VARCHAR2(20);
  7
  8  BEGIN
  9
 10     IF INSERTING THEN
 11        V_OPERATION := 'INSERT';
 12     ELSIF UPDATING THEN
 13        V_OPERATION := 'UPDATE';
 14     ELSE
 15        V_OPERATION := 'DELETE';
 16     END IF;
 17
 18     PKG_APP_AUDIT.INSERT_DATA('APP_TBL1', V_OPERATION);
 19
 20  END;
 21  /

Trigger created.

SQL> CREATE OR REPLACE TRIGGER TRG_APP_TBL2_BUDIR
  2    BEFORE UPDATE OR DELETE OR INSERT ON APP_TBL2
  3    FOR EACH ROW
  4  DECLARE
  5
  6     V_OPERATION  VARCHAR2(20);
  7
  8  BEGIN
  9
 10     IF INSERTING THEN
 11        V_OPERATION := 'INSERT';
 12     ELSIF UPDATING THEN
 13        V_OPERATION := 'UPDATE';
 14     ELSE
 15        V_OPERATION := 'DELETE';
 16     END IF;
 17
 18     Pkg_App_Audit.INSERT_DATA('APP_TBL2', V_OPERATION);
 19
 20  END;
 21  /

Trigger created.
```

9. Now you are ready to test. You will be updating and inserting into APP_TBL1. Also you will be updating and deleting from APP_TBL2. The result of these operations is based on the metadata in the APP_AUDIT_ACTION table. It should generate two rows only: one for the UPDATE on APP_TBL1 and one for the DELETE on APP_TBL2.

```
SQL> UPDATE APP_TBL1 SET
  2          DESCRIPTION = 'NEW DESCRIPTION #1'
  3   WHERE CODE = 1
  4  /

1 row updated.

SQL> INSERT INTO APP_TBL1
  2          VALUES(3, 'DESCRIPTION #3')
  3  /

1 row created.

SQL> UPDATE APP_TBL2 SET
  2          PHONE = '567-567-5678'
  3   WHERE ID = 100;

1 row updated.

SQL> DELETE APP_TBL2
  2   WHERE ID = 102
  3  /

1 row deleted.

SQL> COMMIT;

Commit complete.
```

10. Now check the content of APP_AUDIT_DATA.

```
SQL> SELECT * FROM APP_AUDIT_DATA
  2  /

AUDIT_DATA_ID AUDIT_DATA AUDIT_ACTION_ID AUD_INS_D AUD_UPD_USER AUD_REC_STAT
------------- ---------- --------------- --------- ------------ ------------
           26 UPDATE                   1 05-JUL-05 DBSEC        A
           27 UPDATE                   1 05-JUL-05 DBSEC        A
```

At this point you should feel a sense of accomplishment. The next section moves on to the Simple 2 Model.

DML Auditing Using Repository with Oracle (Simple 2)

This model was presented in Chapter 7 as the Simple Auditing Model 2. The word "simple" does not indicate that the development of this model is easy. In fact it is far from simple. Implementation requires a high level of expertise in PL/SQL. The main objective of this model is to use the auditing repository to store two types of data:

- **Audit data**—This data represents the values before or after a DML statement is applied. The audit data is mainly stored in one column called AUDIT_DATA, which uses the CLOB data type to store more data. This data is stored in a table called APP_AUDIT_DATA.
- **Audit table**—This data contains the names of the tables to be audited. Please note that values are not audited if either of these conditions exists: the table that is being updated does not exist in this table; or, the date when the table is being modified does not fall between the START_DATE and END_DATE column values. Note that the AUDIT_FLAG column value is used to turn auditing on or off for any table, providing the table to be audited is registered and has an auditing trigger. All metadata for this model is stored in a table called APP_AUDIT_TABLE.

Figure 8-10 represents these two tables.

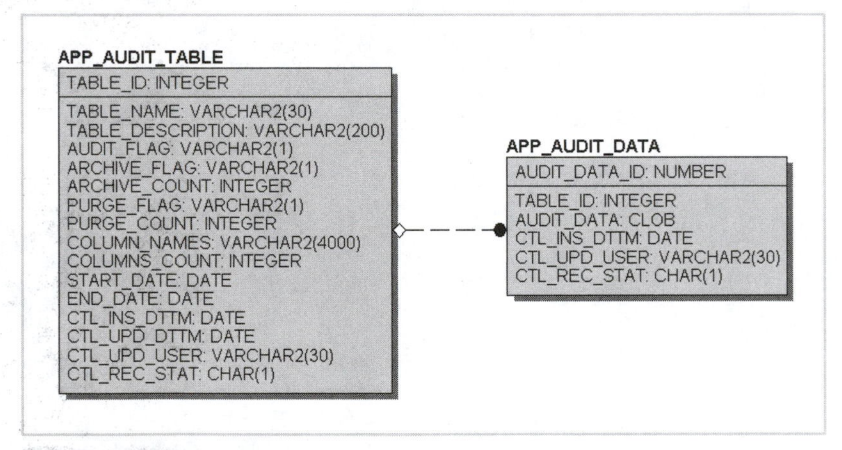

FIGURE 8-10 **Data model for the DML auditing repository**

To start implementing this model, follow these steps:

1. Select a user other than SYSTEM or SYS that has CREATE TABLE and TRIGGER privileges. Create the auditing repository represented in Figure 8-10. If any of these tables exist from previous demonstrations, you can either select a different user or drop the tables before proceeding.

```
SQL> CREATE TABLE APP_AUDIT_DATA
  2  (
  3     AUDIT_DATA_ID           NUMBER PRIMARY KEY,
  4     TABLE_ID                NUMBER,
  5     AUDIT_DATA              CLOB,
  6     CTL_INS_DTTM            DATE,
  7     CTL_UPD_USER            VARCHAR2(30),
  8     CTL_REC_STAT            CHAR(1)
  9  );

Table created.

SQL> CREATE TABLE APP_AUDIT_TABLE
  2  (
  3     TABLE_ID                NUMBER PRIMARY KEY,
  4     TABLE_NAME              VARCHAR2(30) NOT NULL,
  5     TABLE_DESCRIPTION       VARCHAR2(200),
  6     AUDIT_FLAG              VARCHAR2(1),
  7     ARCHIVE_FLAG            VARCHAR2(1),
  8     ARCHIVE_COUNT           INTEGER,
  9     PURGE_FLAG              VARCHAR2(1),
 10     PURGE_COUNT             INTEGER,
 11     COLUMN_NAMES            VARCHAR2(4000),
 12     COLUMNS_COUNT           INTEGER,
 13     START_DATE              DATE,
 14     END_DATE                DATE,
 15     CTL_INS_DTTM            DATE,
 16     CTL_UPD_DTTM            DATE,
 17     CTL_UPD_USER            VARCHAR2(30),
 18     CTL_REC_STAT            CHAR(1)
 19  );

Table created.
```

2. Establish a foreign key in APP_AUDIT_DATA, referencing APP_AUDIT_TABLE.

```
SQL> ALTER TABLE APP_AUDIT_DATA
  2        ADD  ( FOREIGN KEY (TABLE_ID)
  3               REFERENCES APP_AUDIT_TABLE) ;

Table altered.
```

3. Now you need to create a sequence object for the AUDIT_DATA_ID column. If this sequence already exists, you can skip this step. Otherwise, execute the following statement to create it.

```
SQL> CREATE SEQUENCE SEQ_APP_AUDIT_DATA
  2          INCREMENT BY 1
  3          START WITH 1
  4          MINVALUE 1
  5          NOCYCLE
  6          CACHE 20
  7          NOORDER;

Sequence created.
```

4. Create the application schema as shown in Figure 8-11, which consists of two tables, EMPLOYEES and DEPARTMENTS.

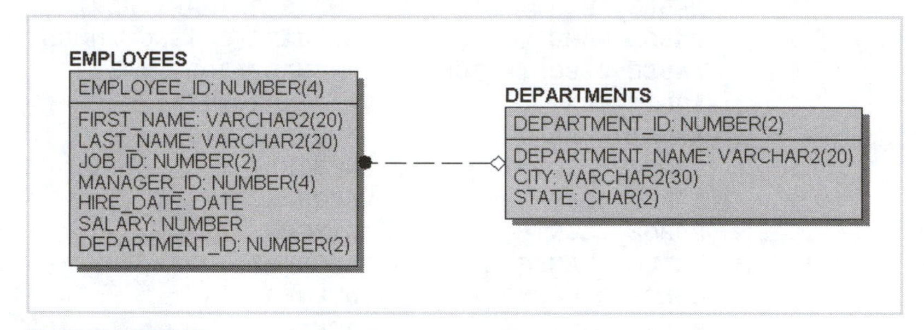

FIGURE 8-11 Data model for EMPLOYEES application schema

```
SQL> CREATE TABLE EMPLOYEES
  2  (
  3    EMPLOYEE_ID    NUMBER(4) PRIMARY KEY,
  4    FIRST_NAME     VARCHAR2(20),
  5    LAST_NAME      VARCHAR2(20),
  6    JOB_ID         NUMBER(2),
  7    MANAGER_ID     NUMBER(4),
  8    HIRE_DATE      DATE,
  9    SALARY         NUMBER,
 10    DEPARTMENT_ID  NUMBER(2)
 11  );

Table created.

SQL>
SQL> CREATE TABLE DEPARTMENTS
  2  (
  3    DEPARTMENT_ID    NUMBER(2) PRIMARY KEY,
  4    DEPARTMENT_NAME  VARCHAR2(20),
  5    CITY             VARCHAR2(30),
  6    STATE            CHAR(2)
  7  );

Table created.
```

```
SQL>
SQL> ALTER TABLE EMPLOYEES
  2         ADD  ( FOREIGN KEY (DEPARTMENT_ID)
  3                REFERENCES DEPARTMENTS) ;

Table altered.
```

5. Add data to the DEPARTMENTS table.

```
SQL> INSERT INTO DEPARTMENTS ( DEPARTMENT_ID, DEPARTMENT_NAME, CITY, STATE )
  2                   VALUES ( 10, 'Accounting', 'Pisgah', 'NV');

1 row created.

SQL> INSERT INTO DEPARTMENTS ( DEPARTMENT_ID, DEPARTMENT_NAME, CITY, STATE )
  2                   VALUES ( 11, 'Production', 'Redlands', 'CA');

1 row created.

SQL> INSERT INTO DEPARTMENTS ( DEPARTMENT_ID, DEPARTMENT_NAME, CITY, STATE )
  2                   VALUES ( 12, 'Engineering', 'Dallas', 'TX');

1 row created.

SQL> COMMIT;

Commit complete.
```

6. Add data to the EMPLOYEES table.

```
SQL> INSERT INTO EMPLOYEES ( EMPLOYEE_ID, FIRST_NAME, LAST_NAME, JOB_ID,
  2                          MANAGER_ID, HIRE_DATE, SALARY, DEPARTMENT_ID )
  3                   VALUES ( 8377, 'Delphine', 'Wenger', 54, 26,
  4                          TO_Date( '01/09/1998 01:11:12 PM', 'MM/DD/YYYY HH:MI:SS AM'),
  5                          102412.44, 10);

1 row created.

SQL> INSERT INTO EMPLOYEES ( EMPLOYEE_ID, FIRST_NAME, LAST_NAME, JOB_ID,
  2                          MANAGER_ID, HIRE_DATE, SALARY, DEPARTMENT_ID )
  3                   VALUES ( 2723, 'Pasty', 'Gaudet', 96, 19,
  4                          TO_Date( '04/05/1998 01:11:12 PM', 'MM/DD/YYYY HH:MI:SS AM'),
  5                          87662.97, 10);

1 row created.
```

```
SQL> INSERT INTO EMPLOYEES ( EMPLOYEE_ID, FIRST_NAME, LAST_NAME, JOB_ID,
  2                          MANAGER_ID, HIRE_DATE, SALARY, DEPARTMENT_ID )
  3                   VALUES ( 4034, 'Mario', 'Atwell', 2, 9,
  4                          TO_Date( '12/08/1997 01:11:12 PM', 'MM/DD/YYYY HH:MI:SS AM'),
  5                          109001.61, 11);

1 row created.

SQL> COMMIT;

Commit complete.
```

7. Now you need a stored PL/SQL package that will be used for auditing within the triggers. Note that a full explanation of the code in this step is presented in Table 8-3.

```
SQL> CREATE OR REPLACE PACKAGE Pkg_App_Audit IS
  2
  3     PROCEDURE INSERT_DATA  (P_TABLE_NAME VARCHAR2, P_COLUMN_VAL CLOB);
  4     FUNCTION GET_OLD_CLAUSE(P_TABLE_NAME VARCHAR2) RETURN VARCHAR2;
  5     FUNCTION GET_COL_CLAUSE(P_TABLE_NAME VARCHAR2) RETURN VARCHAR2;
  6     FUNCTION AUDIT_CHECK   (P_TABLE_NAME VARCHAR2) RETURN BOOLEAN;
  7     FUNCTION PARSE_VALUES  (P_STR VARCHAR2)        RETURN VARCHAR2;
  8
  9  END;
 10  /

Package created.


SQL> CREATE OR REPLACE PACKAGE BODY Pkg_App_Audit IS
  2
  3     G_TABLE_ID     NUMBER := NULL;
  4     G_COLUMN_SPEC  VARCHAR2(4000) := NULL;
  5
  6     --~~~~~~~~~~~~~~~~~~~~~~~~~~~~~~~~~~~~~~~~
  7     -- this function returns the column_clause
  8     -- of the SELECT statement for EXECUTE IMMDIATE
  9     --~~~~~~~~~~~~~~~~~~~~~~~~~~~~~~~~~~~~~~~~
 10     FUNCTION GET_COL_CLAUSE(P_TABLE_NAME VARCHAR2) RETURN VARCHAR2 IS
 11
 12        V_TMP     VARCHAR2(4000) := NULL;
 13        V_QUOTES  VARCHAR2(3);
 14        V_CNT     NUMBER := 0;
 15
```

```
16      BEGIN
17
18      SELECT COUNT(*)
19        INTO V_CNT
20           FROM USER_TAB_COLUMNS
21         WHERE TABLE_NAME = UPPER(P_TABLE_NAME)
22       ORDER BY COLUMN_ID;
23
24         FOR I IN 1..V_CNT LOOP
25
26            V_TMP    := V_TMP || V_QUOTES || ':' || I || V_QUOTES || ' || ''~'' || ';
27
28         END LOOP;
29         V_TMP := SUBSTR(V_TMP, 1, LENGTH(V_TMP) –length('||''~''||');
30
31         RETURN V_TMP;
32
33      END;
34      --~~~~~~~~~~~~~~~~~~~~~~~~~~~~~~~~~~~~~~~~~~
35
36      --~~~~~~~~~~~~~~~~~~~~~~~~~~~~~~~~~~~~~~~~~~
37      -- this function returns the USING clause
38      -- of the EXECVTE IMMDIATE statement
39      --~~~~~~~~~~~~~~~~~~~~~~~~~~~~~~~~~~~~~~~~~~
40      FUNCTION GET_OLD_CLAUSE(P_TABLE_NAME VARCHAR2) RETURN VARCHAR2 IS
41
42         V_REC      APP_AUDIT_TABLE%ROWTYPE;
43         V_TMP      VARCHAR2(4000) := NULL;
44         V_START    NUMBER := 1;
45         V_END      NUMBER;
46
47      BEGIN
48
49         FOR I IN (SELECT COLUMN_NAME
50                     FROM USER_TAB_COLUMNS
51                    WHERE TABLE_NAME = UPPER(P_TABLE_NAME)
52                    ORDER BY COLUMN_ID) LOOP
53
54            V_TMP    := V_TMP || ':OLD.' || I.COLUMN_NAME || ',';
55
56         END LOOP;
57         V_TMP := SUBSTR(V_TMP, 1, LENGTH(V_TMP) -1);
58
59         RETURN V_TMP;
60
61      END;
```

```
 62     --~~~~~~~~~~~~~~~~~~~~~~~~~~~~~~~~~~~~~~~~~~~
 63
 64     --~~~~~~~~~~~~~~~~~~~~~~~~~~~~~~~~~~~~~~~~~~~
 65     -- CHECKS IF TABLE IS AUDITIED.
 66     --~~~~~~~~~~~~~~~~~~~~~~~~~~~~~~~~~~~~~~~~~~~
 67     FUNCTION AUDIT_CHECK( P_TABLE_NAME  VARCHAR2 ) RETURN BOOLEAN IS
 68
 69        V_REC     APP_AUDIT_TABLE%ROWTYPE;
 70
 71     BEGIN
 72
 73        BEGIN
 74           SELECT *
 75             INTO V_REC
 76             FROM APP_AUDIT_TABLE
 77            WHERE TABLE_NAME= UPPER(P_TABLE_NAME);
 78        EXCEPTION WHEN OTHERS THEN
 79           RETURN FALSE;
 80        END;
 81
 82        IF (SYSDATE <= V_REC.START_DATE OR
 83         SYSDATE >= V_REC.END_DATE) OR
 84            V_REC.AUDIT_FLAG <> 'Y' THEN
 85           RETURN FALSE;
 86        END IF;
 87        G_TABLE_ID    := V_REC.TABLE_ID;
 88     G_COLUMN_SPEC := V_REC.COLUMN_NAMES;
 89     RETURN TRUE;
 90    END;
 91     --~~~~~~~~~~~~~~~~~~~~~~~~~~~~~~~~~~~~~~~~~~~
 92
 93     --~~~~~~~~~~~~~~~~~~~~~~~~~~~~~~~~~~~~~~~~~~~
 94     -- GETS ONLY REQUIRED VALUES.
 95     --~~~~~~~~~~~~~~~~~~~~~~~~~~~~~~~~~~~~~~~~~~~
 96     FUNCTION PARSE_VALUES( P_STR  VARCHAR2 ) RETURN VARCHAR2 IS
 97
 98        V_VAL   VARCHAR2(4000) := NULL;
 99     V_STR   VARCHAR2(4000);
100     V_POS1  NUMBER;
101     V_POS2  NUMBER;
102     V_CNT   NUMBER := 0;
103
104     BEGIN
105
106        IF G_COLUMN_SPEC IS NULL OR
107        UPPER(G_COLUMN_SPEC) = 'ALL' THEN
108        RETURN P_TMP;
109     END IF;
110
```

```
111        V_POS1 := 1;
112
113        LOOP
114        V_POS2 := INSTR(P_TMP, '~', V_POS1);
115     EXIT WHEN V_POS2 = 0;
116     V_STR := SUBSTR(P_TMP, V_POS1, V_POS2 - V_POS1);
117     DBMS_OUTPUT.PUT_LINE(V_STR);
118            V_POS1 := V_POS2 + 1;
119     V_CNT  := V_CNT + 1;
120
121     IF INSTR(G_COLUMN_SPEC, '['||V_CNT||']') > 0 THEN
122        V_VAL := V_VAL || V_STR || '~';
123     END IF;
124        END LOOP;
125
126        V_STR := SUBSTR(P_TMP, V_POS1);
127
128     RETURN SUBSTR(V_VAL, 1, LENGTH(V_VAL) -1);
129    END;
130     --~~~~~~~~~~~~~~~~~~~~~~~~~~~~~~~~~~~~~~~~
131
132     --~~~~~~~~~~~~~~~~~~~~~~~~~~~~~~~~~~~~~~~~
133     -- INSERT OLD VALUES INTO AUDIT TABLE
134     --~~~~~~~~~~~~~~~~~~~~~~~~~~~~~~~~~~~~~~~~
135     PROCEDURE INSERT_DATA(
136                           P_TABLE_NAME  VARCHAR2,
137                           P_COLUMN_VAL  CLOB
138                          ) IS
139
140        V_REC     APP_AUDIT_TABLE%ROWTYPE;
141
142     BEGIN
143
144        INSERT INTO APP_AUDIT_DATA(AUDIT_DATA_ID, TABLE_ID, AUDIT_DATA,
145                                   CTL_INS_DTTM, CTL_UPD_USER, CTL_REC_STAT)
146                            VALUES(SEQ_APP_AUDIT_DATA.NEXTVAL, G_TABLE_ID,
147                P_COLUMN_VAL, SYSDATE, USER, 'A');
148     END;
149     --~~~~~~~~~~~~~~~~~~~~~~~~~~~~~~~~~~~~~~~~
150
151  END;
152  /

Package body created.
```

This package requires some explanation. Start with Table 8-3 and read the brief description of each function in this package.

Table 8-3 Package specification for PKG_APP_AUDIT

<table>
<tr><th>Function/Procedure Name</th><th>Description</th></tr>
<tr><td>FUNCTION AUDIT_CHECK
returns BOOLEAN</td><td>Based on all three of the following criteria, this function is used to check whether a table is audited or not:
1. The table name exists in the auditing repository; meaning an entry of the table name is registered in APP_AUDIT_TABLE.
2. The AUDIT_FLAG must be set to value Y.
3. The date of the modification is within the range of the START_DATE and END_DATE column values.
This function returns TRUE, if it meets all criteria, otherwise it returns FALSE. This function must be used with the trigger.</td></tr>
<tr><td>FUNCTION GET_COL_CLAUSE
returns VARCHAR2</td><td>This function is optional. You don't need to use it within the template trigger. However, you might want to use it, because it makes writing the trigger easier. This function returns a string containing a number for each column position in the table. For example, if a table contains three columns (id, name, phone), this function returns the following string: :1 ||'~'|| :2 ||'~'|| :3</td></tr>
<tr><td>FUNCTION GET_OLD_CLAUSE
returns VARCHAR2</td><td>This function is not used within the template trigger. However, it is recommended when constructing the trigger. This functions returns all column names prefixed with "OLD." and delimited with a comma. For example, if you execute the function passing TEST as a value for the P_TABLE_NAME parameter, it returns the following string:
:OLD.ID,:OLD.NAME,:OLD.PHONE</td></tr>
<tr><td>FUNCTION PARSE_VALUES
returns VARCHAR2</td><td>This function is used as part of the trigger template. It returns only the column values that are registered in APP_AUDIT_TABLE. For example, if you are interested in capturing all column values for a table, you register in COLUMN_NAMES the value ALL. You can also keep the value as NULL. If you are interested in capturing only two column values, you must register the position of these columns as they are created in the database. Each column is enclosed within brackets []. For example, [2][3] captures only the column values for the second and third column in the registered table.</td></tr>
<tr><td>PROCEDURE INSERT_DATA</td><td>This procedure is used to record the before value in the audited table.</td></tr>
</table>

Now look at the prototype for the PKG_APP_AUDIT package.

```
SQL> DESC PKG_APP_AUDIT

FUNCTION AUDIT_CHECK RETURNS BOOLEAN
 Argument Name                  Type                    In/Out Default?
 ------------------------------ ----------------------- ------ --------
 P_TABLE_NAME                   VARCHAR2                IN

FUNCTION GET_COL_CLAUSE RETURNS VARCHAR2
 Argument Name                  Type                    In/Out Default?
 ------------------------------ ----------------------- ------ --------
 P_TABLE_NAME                   VARCHAR2                IN

FUNCTION GET_OLD_CLAUSE RETURNS VARCHAR2
 Argument Name                  Type                    In/Out Default?
 ------------------------------ ----------------------- ------ --------
 P_TABLE_NAME                   VARCHAR2                IN

PROCEDURE INSERT_DATA
 Argument Name                  Type                    In/Out Default?
 ------------------------------ ----------------------- ------ --------
 P_TABLE_NAME                   VARCHAR2                IN
 P_COLUMN_VAL                   CLOB                    IN

FUNCTION PARSE_VALUES RETURNS VARCHAR2
 Argument Name                  Type                    In/Out Default?
 ------------------------------ ----------------------- ------ --------
 P_STR                          VARCHAR2                IN
```

8. In this step you use the trigger template to create one trigger for the DEPARTMENTS table and one for the EMPLOYEES table. The following code is a trigger template. All keywords that are bold and italic should be replaced with values. You need to provide the values according to the table name being audited.

```
CREATE OR REPLACE TRIGGER TRG_TableName_BUDR
  BEFORE UPDATE OR DELETE ON TableName
  FOR EACH ROW
DECLARE

   V_TMP  CLOB;
   V_FLAG BOOLEAN;
   V_COL  VARCHAR2(4000);
```

```
BEGIN

   V_FLAG := PKG_APP_AUDIT.AUDIT_CHECK( 'TableName' );
   IF V_FLAG = TRUE THEN
      V_COL := PKG_APP_AUDIT.GET_COL_CLAUSE( 'TableName' );

      EXECUTE IMMEDIATE 'SELECT ' || V_COL || ' FROM DUAL'
        INTO V_TMP
       USING :OLD.ColumnName1, :OLD.ColumnName2, … , :OLD.ColumnNameN;

      V_TMP := PKG_APP_AUDIT.PARSE_VALUES(V_TMP);
      PKG_APP_AUDIT.INSERT_DATA('TableName', V_TMP);
   END IF;

END;
/
```

The trigger for the DEPARTMENTS table is as follows:

```
SQL> CREATE OR REPLACE TRIGGER TRG_DEPARTMENTS_BUDR
  2    BEFORE UPDATE OR DELETE ON DEPARTMENTS
  3    FOR EACH ROW
  4  DECLARE
  5
  6     V_TMP  CLOB;
  7     V_FLAG BOOLEAN;
  8     V_COL  VARCHAR2(4000);
  9
 10  BEGIN
 11
 12     V_FLAG := PKG_APP_AUDIT.AUDIT_CHECK( 'DEPARTMENTS' );
 13     IF V_FLAG = TRUE THEN
 14        V_COL := PKG_APP_AUDIT.GET_COL_CLAUSE( 'DEPARTMENTS' );
 15
 16        EXECUTE IMMEDIATE 'SELECT ' || V_COL || ' FROM DUAL'
 17          INTO V_TMP
 18      USING :OLD.DEPARTMENT_ID,:OLD.DEPARTMENT_NAME,:OLD.CITY,:OLD.STATE;
 19
 20     V_TMP := PKG_APP_AUDIT.PARSE_VALUES(V_TMP);
 21        PKG_APP_AUDIT.INSERT_DATA('DEPARTMENTS', V_TMP);
 22     END IF;
 23
 24  END;
 25  /

Trigger created.
```

The following is the trigger for the EMPLOYEES table:

```
SQL> CREATE OR REPLACE TRIGGER TRG_EMPLOYEES_BUDR
  2    BEFORE UPDATE OR DELETE ON EMPLOYEES
  3    FOR EACH ROW
  4  DECLARE
  5
  6     V_TMP  CLOB;
  7     V_FLAG BOOLEAN;
  8     V_COL  VARCHAR2(4000);
  9
 10  BEGIN
 11
 12     V_FLAG := PKG_APP_AUDIT.AUDIT_CHECK( 'EMPLOYEES' );
 13     IF V_FLAG = TRUE THEN
 14        V_COL := PKG_APP_AUDIT.GET_COL_CLAUSE( 'EMPLOYEES' );
 15
 16        EXECUTE IMMEDIATE 'SELECT ' || V_COL || ' FROM DUAL'
 17          INTO V_TMP
 18      USING :OLD.EMPLOYEE_ID,:OLD.FIRST_NAME,:OLD.LAST_NAME,:OLD.JOB_ID,
 19            :OLD.MANAGER_ID,:OLD.HIRE_DATE,:OLD.SALARY,:OLD.DEPARTMENT_ID;
 20
 21     V_TMP := PKG_APP_AUDIT.PARSE_VALUES(V_TMP);
 22        PKG_APP_AUDIT.INSERT_DATA('EMPLOYEES', V_TMP);
 23     END IF;
 24
 25  END;
 26  /

Trigger created.
```

9. You're almost finished. In this step you add the auditing metadata. Note that the bolded text in the code represents the columns that you want to audit. The first INSERT statement tells you that the columns LAST_NAME, HIRE_DATE, and SALARY in the EMPLOYEES table are to be audited, whereas the second INSERT statement indicates that all columns in the DEPARTMENTS table will be audited.

```
SQL> INSERT INTO APP_AUDIT_TABLE ( TABLE_ID, TABLE_NAME, TABLE_DESCRIPTION, AUDIT_FLAG,
  2                                ARCHIVE_FLAG, ARCHIVE_COUNT, PURGE_FLAG, PURGE_COUNT,
  3                                COLUMN_NAMES, COLUMNS_COUNT, START_DATE, END_DATE,
  4                                CTL_INS_DTTM, CTL_UPD_DTTM, CTL_UPD_USER, CTL_REC_STAT )
  5                       VALUES ( 1, 'EMPLOYEES', 'EMPLOYEE TABLE HOLDING STAFF DATA',
  6                                'Y', 'Y', 5, 'Y', 3, '[3][6][7]', 2,
  7                                TRUNC(SYSDATE), TRUNC(SYSDATE+5),
  8                                SYSDATE, SYSDATE, 'DBSEC', 'A');

1 row created.
```

```
SQL> INSERT INTO APP_AUDIT_TABLE ( TABLE_ID, TABLE_NAME, TABLE_DESCRIPTION, AUDIT_FLAG,
  2                                ARCHIVE_FLAG, ARCHIVE_COUNT, PURGE_FLAG, PURGE_COUNT,
  3                                COLUMN_NAMES, COLUMNS_COUNT, START_DATE, END_DATE,
  4                                CTL_INS_DTTM, CTL_UPD_DTTM, CTL_UPD_USER, CTL_REC_STAT )
  5                       VALUES ( 2, 'DEPARTMENTS', 'DEPARTMENT TABLE', 'Y', 'N', 0, 'N',
  6                                0, 'ALL', 3,  TRUNC(SYSDATE), TRUNC(SYSDATE+5),
  7                                SYSDATE, SYSDATE, 'DBSEC', 'A');

1 row created.

SQL> COMMIT;

Commit complete.
```

10. In this step you modify one row in the EMPLOYEES table and delete one row from the DEPARTMENTS table.

```
SQL> UPDATE EMPLOYEES SET
  2         SALARY = 1000
  3   WHERE EMPLOYEE_ID = 8377
  4  /

1 row updated.

SQL> DELETE FROM DEPARTMENTS
  2   WHERE DEPARTMENT_ID = 12
  3  /

1 row deleted.
```

11. This is the last step. Review the content of APP_AUDIT_DATA.

```
SQL> SELECT T.TABLE_NAME,
  2         D.AUDIT_DATA,
  3         TO_CHAR(D.CTL_INS_DTTM, 'DD-MON-YYYY HH24:MI:SS') DATE_CREATED,
  4         D.CTL_UPD_USER USERNAME
  5    FROM APP_AUDIT_TABLE T,
  6         APP_AUDIT_DATA D
  7   WHERE T.TABLE_ID = D.TABLE_ID
  8  /

TABLE_NAME  AUDIT_DATA                 DATE_CREATED         USERNAME
----------- -------------------------- -------------------- --------
EMPLOYEES   Wenger~09-JAN-98~102412.44 04-JUL-2005 20:58:11 DBSEC
DEPARTMENTS 12~Engineering~Dallas~TX   04-JUL-2005 20:59:37 DBSEC
```

There are other ways to implement this model. It all depends on your creativity and expertise in PL/SQL.

Auditing Application Errors with Oracle

Suppose you are part of a team developing a reservation system, and one of the requirements dictates that any error that occurs within the application, regardless of the type, must be recorded in a store for further analysis. For example, if the booking module of the system generates an error while saving, a record of the error is stored in a table. All record errors are then reported to an application administrator and analyzed to produce an enhancement list, bug list, or action list to prevent these errors from happening again. Business requirements often mandate that you keep an audit trail of all application errors caused by data manipulation. To do this you need to create a repository consisting of one table and a methodology for your application. To capture DML errors, use the following steps:

1. Select a user other than SYS or SYSTEM that has CREATE TABLE and PROCEDURE privileges. Create the CUSTOMERS table. If this table already exists, you might want to change the user.

```
SQL> CREATE TABLE CUSTOMERS
  2  (
  3     ID           NUMBER,
  4     NAME         VARCHAR2(10),
  5     CREDIT_LIMIT NUMBER
  6  );

Table created.
```

2. Populate the CUSTOMERS table with one row.

```
SQL> INSERT INTO CUSTOMERS ( ID, NAME, CREDIT_LIMIT )
  2                 VALUES ( 1, 'Tom Jones', 1000);

1 row created.

SQL> COMMIT;

Commit complete.
```

3. Create the ERROR table used to store errors caused by DML statements.

```
SQL> CREATE TABLE APP_AUDIT_ERRORS
  2  (
  3    TABLE_NAME      VARCHAR2(30)                    NOT NULL,
  4    ERROR_CODE      NUMBER                          NOT NULL,
  5    ERROR_MSG       VARCHAR2(2000)                  NOT NULL,
  6    ROW_VALUES      VARCHAR2(4000)                  NOT NULL,
  7    CTL_INS_DTTM    DATE,
  8    CTL_INS_USER    VARCHAR2(30),
  9    CTL_OPS_USER    VARCHAR2(30),
 10    CTL_IP_ADDR     VARCHAR2(255)
 11  );

Table created.
```

4. Create a stored package to perform the UPDATE statement.

```
SQL> CREATE OR REPLACE PACKAGE APP_AUDIT_DML IS
  2
  3     PROCEDURE CUSTOMERS_UPDATE(
  4                               p_ID             NUMBER,
  5             P_NAME          VARCHAR2,
  6             P_CREDIT_LIMIT NUMBER,
  7          P_COMMIT       BOOLEAN DEFAULT TRUE
  8          );
  9
 10  END;
 11  /

Package created.

SQL> CREATE OR REPLACE PACKAGE BODY APP_AUDIT_DML IS
  2
  3     --~~~~~~~~~~~~~~~~~~~~~~~~~~~~~~~~~~~~~~~~~~~~~~~~~~~~~~~~~
  4     -- INSERT ERROR CAUSED BY DML INTO APP_AUDIT_ERRORS
  5     --~~~~~~~~~~~~~~~~~~~~~~~~~~~~~~~~~~~~~~~~~~~~~~~~~~~~~~~~~
  6     PROCEDURE INSERT_ERROR(
  7                               P_TABLE VARCHAR2,
  8          P_CODE   NUMBER,
  9          P_MSG    VARCHAR2,
 10          P_VALS   VARCHAR2
 11          ) IS
 12
 13        PRAGMA AUTONOMOUS_TRANSACTION;
 14
```

```
15    BEGIN
16
17       INSERT INTO APP_AUDIT_ERRORS(
18                                  TABLE_NAME, ERROR_CODE,
19                                  ERROR_MSG, ROW_VALUES,
20            CTL_INS_DTTM, CTL_INS_USER,
21            CTL_OPS_USER,
22            CTL_IP_ADDR
23           )
24          VALUES (
25                  P_TABLE, P_CODE,
26                  P_MSG, P_VALS,
27            SYSDATE, USER,
28            (SELECT SYS_CONTEXT('USERENV', 'OS_USER') FROM DUAL),
29            (SELECT SYS_CONTEXT('USERENV', 'IP_ADDRESS') FROM DUAL)
30           );
31       COMMIT;
32
33    END;
34    --~~~~~~~~~~~~~~~~~~~~~~~~~~~~~~~~~~~~~~~~~~~~~~~~
35
36
37    --~~~~~~~~~~~~~~~~~~~~~~~~~~~~~~~~~~~~~~~~~~~~~~~~
38    -- PERFORMS UPDATE OPERATION.
39    --~~~~~~~~~~~~~~~~~~~~~~~~~~~~~~~~~~~~~~~~~~~~~~~~
40    PROCEDURE CUSTOMERS_UPDATE(
41                               P_ID           NUMBER,
42            P_NAME         VARCHAR2,
43            P_CREDIT_LIMIT NUMBER,
44         P_COMMIT       BOOLEAN DEFAULT TRUE
45        ) IS
46
47     V_STMT     VARCHAR2(4000);
48     E_ID_NULL  EXCEPTION;
49
50     PRAGMA EXCEPTION_INIT(E_ID_NULL, -200001);
51
52  BEGIN
53
54        IF P_ID IS NULL THEN
55        RAISE E_ID_NULL;
56     END IF;
57
58        UPDATE CUSTOMERS SET
59               ID           = NVL(P_ID          , ID          ),
60               NAME         = NVL(P_NAME        , NAME        ),
61               CREDIT_LIMIT = NVL(P_CREDIT_LIMIT, CREDIT_LIMIT)
```

```
62      WHERE ID = P_ID;
63
64     IF P_COMMIT THEN
65          COMMIT;
66     END IF;
67
68  EXCEPTION
69     WHEN E_ID_NULL THEN
70        V_STMT := P_ID            || '|' ||
71                  P_NAME          || '|' ||
72            P_CREDIT_LIMIT;
73
74          INSERT_ERROR('CUSTOMERS', SQLCODE, SQLERRM, V_STMT);
75          RAISE_APPLICATION_ERROR(-20001, 'Error: ' || SQLERRM);
76     WHEN OTHERS THEN
77        V_STMT := P_ID            || '|' ||
78                  P_NAME          || '|' ||
79            P_CREDIT_LIMIT;
80
81          INSERT_ERROR('CUSTOMERS', SQLCODE, SQLERRM, V_STMT);
82          RAISE_APPLICATION_ERROR(-20002, 'Error: ' || SQLERRM);
83     END;
84    --------------------------------------------------
85
86 END;
87 /

Package body created.
```

5. Perform an update using the CREATE package. To generate an error, you create a value for a name that exceeds the length of the NAME column, and then create another UPDATE that does not cause an error.

```
SQL> EXEC APP_AUDIT_DML.CUSTOMERS_UPDATE( 1, 'Tom Jones Jr.', null)
BEGIN APP_AUDIT_DML.CUSTOMERS_UPDATE( 1, 'Tom Jones Jr.', null); END;

*
ERROR at line 1:
ORA-20002: Error: ORA-01401: inserted value too large for column
ORA-06512: at "SEC.APP_AUDIT_DML", line 82
ORA-06512: at line 1

SQL> EXEC APP_AUDIT_DML.CUSTOMERS_UPDATE( 1, NULL, 10000)

PL/SQL procedure successfully completed.
```

6. Check the content of APP_AUDIT_ERRORS. You see an entry for the error caused by the UPDATE statement.

```
SQL> SELECT TABLE_NAME,
  2         ERORR_MSG,
  3         ROW_VALUES,
  4         TO_CHAR(CTL_INS_DTTM, 'DD-MON-YYYY HH24:MI:SS') DATE_CREATED,
  5         CTL_IP_ADDR
  6    FROM APP_AUDIT_ERRORS
  7  /

TABLE_NAME   ERROR_MSG                 ROW_VALUES        DATE_CREATED         CTL_IP_ADDR
------------ ------------------------- ----------------- -------------------- -----------
CUSTOMERS    ORA-01401: inserted value 1|Tom Jones Jr.|  06-JUL-2005 13:30:39 127.0.0.1
             column too large for
```

As you can see, the error is captured in a table, which is a very simple repository. There are sophisticated designs for error auditing. These designs consist of many tables with complex code. You should always consider your business requirements before deciding which designs are best for your environment.

Oracle PL/SQL Procedure Authorization

Oracle PL/SQL stored procedures are the mainstay of implementing business rules. This is because PL/SQL procedures are used in a wide variety of applications. The following code snippet represents the syntax for creating these procedures. It is important to notice that the bolded code, AUTHID, requires a value of either DEFINER or CURRENT_USER.

```
CREATE OR REPLACE PROCEDURE procedure_name[(parameter[, parameter]...)]
                            [AUTHID {DEFINER | CURRENT_USER}] {IS | AS}
```

These values are used to indicate what security mode is being used. There are two modes:

- **Invoker rights**—Indicates that the procedure is executed using the security credentials of the caller, not the owner of the procedure. Suppose Tom creates a procedure that references one of his tables and suppose that Tom grants EXECUTE privileges to Scott. If Scott tries to execute the procedure in this mode (invoker rights) and he does not have SELECT privileges on the table referenced in the procedure, Scott receives an error message.
- **Definer rights**—Indicates that the procedure is executed using the security credentials of the user that owns the procedure. Tom creates a procedure using this mode (definer rights), which references a table that he owns and grants Scott EXECUTE

privileges on the procedure, but not the table. When Scott executes the procedure, he does not get an error because the procedure is using Tom's security credentials.

This can be confusing. For clarification follow these steps:

1. Create a new user called CH8USER.

```
SQL> CREATE USER CH8USER IDENTIFIED BY CH8USER
  2  /

User created.

SQL> GRANT CREATE SESSION TO CH8USER
  2  /

Grant succeeded.
```

2. Select a user with CREATE TABLE and PROCEDURE privileges (other than SYS or SYSTEM). Create the CUSTOMERS table. If this table already exists, you may want to change the user.

```
SQL> CREATE TABLE CUSTOMERS
  2  (
  3      ID           NUMBER,
  4      NAME         VARCHAR2(10),
  5      CREDIT_LIMIT NUMBER
  6  );

Table created.
```

3. Populate the CUSTOMERS table with one row.

```
SQL> INSERT INTO CUSTOMERS ( ID, NAME, CREDIT_LIMIT )
  2                   VALUES ( 1, 'Tom Jones', 1000);

1 row created.

SQL> COMMIT;

Commit complete.
```

4. Create a stored procedure to select every row in the CUSTOMERS table using DEFINER rights.

```
SQL> CREATE OR REPLACE PROCEDURE DISPLAY_CUSTOMERS AUTHID DEFINER IS
  2
  3  BEGIN
  4
  5     FOR C IN (SELECT * FROM CUSTOMERS) LOOP
  6
  7     DBMS_OUTPUT.PUT_LINE(C.ID || ', ' || C.NAME || ',  ' || C.CREDIT_LIMIT);
  8
  9     END LOOP;
 10
 11  END;
 12  /

Procedure created.
```

5. Grant EXECUTE privileges on the new procedure you created in Step 4.

```
SQL> GRANT EXECUTE ON DISPLAY_CUSTOMERS TO CH8USER
  2  /

Grant succeeded.
```

6. Log on as CH8USER and query the CUSTOMERS table. As you might expect, you are not able to query this table.

```
SQL> CONN CH8USER/CH8USER@SEC
Connected.
SQL> SELECT * FROM DBSEC.CUSTOMERS
  2  /
SELECT * FROM DBSEC.CUSTOMERS
                    *
ERROR at line 1:
ORA-00942: table or view does not exist
```

7. Now execute the procedure created by DBSEC.DISPLAY_CUSTOMERS.

```
SQL> SET SERVEROUTPUT ON SIZE 1000000
SQL> EXEC DBSEC.DISPLAY_CUSTOMERS
1, Tom Jones,  10000

PL/SQL procedure successfully completed.
```

You reached the end of this chapter after lengthy and detailed step-by-step implementation of several auditing models that were presented in Chapter 7 to audit DML operations. You have mastered not only the basic concept but also the practical aspects of auditing applications activities. Chapter 9 covers audit database activities such as database startup, shutdown, logon, and logoff.

Chapter Summary

- This chapter presents two approaches to auditing DML statements. The first approach is to set up an audit trail for DML activities that occur on tables. The second approach is to register all column values either before or after the DML statement is applied to the table.
- Recording of the before value of the modified column(s) is referred to as column-level auditing.
- In Oracle, column-level auditing is known as fine-grained auditing (FGA).
- Auditing of DML changes is widely used wherever highly sensitive data exists.
- A trigger is a stored PL/SQL procedure that is executed automatically whenever a DML operation occurs or a specific database event occurs.
- Oracle has six Data Manipulation Language (DML) events, also known as trigger timings for INSERT, UPDATE, and DELETE.
- The FOR EACH ROW clause indicates that a trigger fires for each row affected by a DML statement.
- Omitting the FOR EACH ROW clause indicates that the trigger fires only once, regardless of the number of rows affected by the statement. This is known as the STATEMENT LEVEL trigger.
- You can view all triggers created on a table by using the USER_TRIGGERS data dictionary view.
- The body of a trigger is contained in the TRIGGER_BODY column of USER_TRIGGERS view.
- To create a trigger in Microsoft SQL Server 2000, you use the CREATE TRIGGER DDL statement.
- There are two intrinsic functions that you can test for UPDATE or INSERT triggers that allow you to identify which columns are being updated. These functions are UPDATE() and COLUMNS_UPDATED().

- The TRUNCATE TABLE statement does not take place inside the transaction log. It is not caught by a DELETE trigger.
- FGA allows administrators to generate an audit trail of DML activities to operating system files or database tables.
- FGA is capable of auditing columns or tables using the Oracle PL/SQL-supplied package called DBMS_FGA.
- Invoker rights indicates that a procedure is executed using the security credentials of the caller, not the credentials of the procedure owner.
- Definer rights indicates that a procedure is executed using the security credentials of the user that owns the procedure.

Review Questions

1. List the six DML triggers that Oracle offers.
2. List all DML triggers available in Microsoft SQL Server.
3. Explain the difference between row-level and statement-level triggers.
4. List five statements that are not allowed in Microsoft SQL Server triggers.
5. Briefly state how to use each procedure of the DBMS_FGA package.
6. What are the restrictions of the INSTEAD OF clause in the Microsoft SQL Server triggers?
7. Identify four drawbacks of the DML audit model, Simple 2.
8. Give an example (not a full implementation) of how to use a trigger condition in Microsoft SQL Server trigger.
9. Provide a step-by-step demonstration of a procedure using invoker rights.
10. What is the use of logical tables in Microsoft SQL Server triggers?
11. Log in as SYSTEM and list all the triggers in your database.
12. Demonstrate in detail the difference between statement-level and row-level triggers.

Hands-on Projects

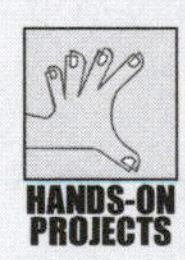

Hands-on Project 8-1

Use Oracle10*g* to modify the DML audit model (Simple 2) to include the INSERT statement.

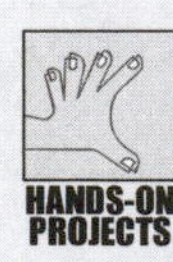

Hands-on Project 8-2

Produce an audit trail for DML statement activities on one table using the Oracle10*g* database feature. Use any table in any schema other than SYS or SYSTEM.

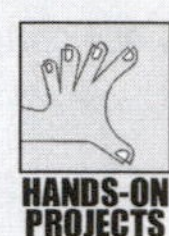

Hands-on Project 8-3

Use Oracle10*g* to modify the DML audit model (Simple 2) to register a type of operation applied on the row.

Hands-on Project 8-4

Use Oracle10*g* to develop a data history model based on the AFTER INSERT, UPDATE, or DELETE trigger.

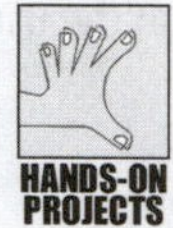

Hands-on Project 8-5

Using the Auditing Simple 1 model, modify the PL/SQL code presented for this model in the chapter to include auditing columns.

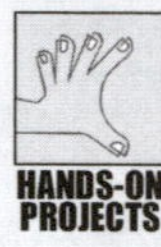

Hands-on Project 8-6

As a newly hired Oracle10*g* database administrator, your first assignment is to implement the Oracle10*g* fine-grained auditing feature on application tables. Use any table in any schema other than SYS or SYSTEM. List each step of the implementation.

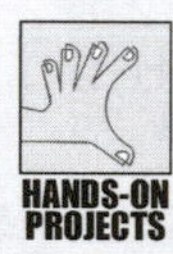

Hands-on Project 8-7

Using Oracle10*g* and any of the audit designs presented in this chapter or Chapter 7, produce an audit trail of any DML statement activities on any table belonging to a schema other than SYS or SYSTEM.

Hands-on Project 8-8

Create a design that uses Oracle FGA to capture all DML statements issued against tables. The tables must be in a repository other than Oracle.

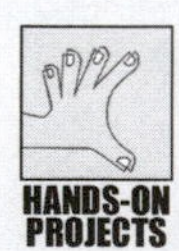

Hands-on Project 8-9

Your manager read this book and was impressed with the benefits of auditing application errors. He asked you to modify the PL/SQL package presented in this chapter to add INSERT and UPDATE DML operations, using definer rights. List all the steps you would take to implement your manager's request.

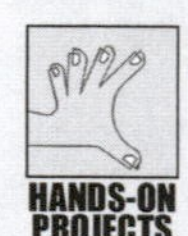

Hands-on Project 8-10

Using Microsoft SQL Server, implement the history data auditing model for DELETE actions on any table in the NORTHWIND database.

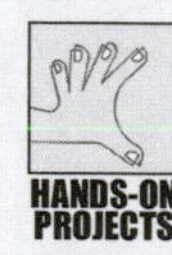

Hands-on Project 8-11

Using Microsoft SQL Server, implement application action auditing model on any table in the NORTHWIND database.

Case Projects

Case 8-1 History Auditing Model with Oracle

After reading Chapters 7 and 8, you've decided to push the envelope and implement the auditing model presented in Figure 8-12. Provide a step-by-step summary for implementing this model.

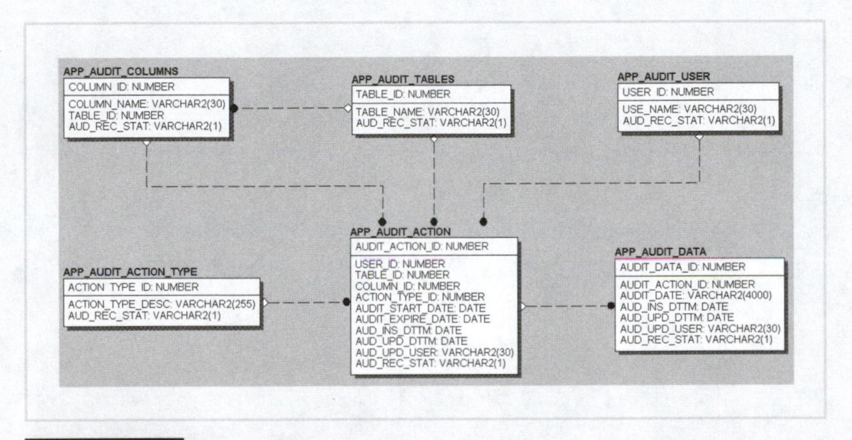

FIGURE 8-12 Advanced auditing data model

Case 8-2 History Auditing Model with SQL Server Case

Now that you have read Chapters 7 and 8, implement the history auditing model for all DML activities on one table in the PHARMACY database to generate a step-by-step template for any application requiring history auditing.

9

Auditing Database Activities

LEARNING OBJECTIVES

Upon completion of this material, you should be able to:

- Use Oracle database activities
- Learn how to create DLL triggers with Oracle
- Audit database activities using Oracle
- Audit server activities with Microsoft SQL Server 2000
- Audit database activities using Microsoft SQL Profiler
- Use SQL Server for security auditing

Introduction

A database manager for a small biomedical company implemented stringent security policies. These policies were so strict that they made it cumbersome for the development team to perform its tasks. One day the development team was working on the migration of a database application to new hardware and software. The migration process was stalled when one of the developers could not execute her script because she did not have all the necessary privileges. When her manager inquired why this happened, he learned from the database manager that no one was granted read or write privileges. The development manager asked him if these policies could be changed during the migration. The answer was "No!"

A day later the database and development teams held a meeting to find a solution for this problem that would not compromise the security policies. The meeting was tense, and no solutions surfaced. Just when the meeting was about to adjourn, one of the database administrators suggested the use of triggers. "Triggers would track data changes, grant privileges, and create database objects," she explained. The team immediately appreciated what this suggestion meant. Triggers would allow developers to have the privileges they needed, and at the same time the database team could trace any suspicious activities.

Chapter 8 presented a detailed implementation of DML statement auditing. This chapter augments your knowledge and background with the second half of auditing, which is tracking of DDL statements and database events. Chapter 9 is the final part of the auditing section of this book. In this chapter you learn how to selectively audit various database activities, as well as create and maintain their audit trails.

The main purpose of this chapter is to equip you with all the necessary background you need to understand and implement database auditing and to track security violations.

Using Oracle Database Activities

In Chapter 8 you learned how to audit DML statements, but in fact, auditing DML statements is only one of several types of activities that can be applied to the database. There are three types of operations involved with database activities:

1. *Application activities*—Encompass SQL statements issued against application tables.
2. *Administration activities*—Encompass commands issued by the database administrators or operators for maintenance and administration purposes. Some of the commands and statements in this category are actually SQL statements.
3. *Database events*—Events that occur when a specific activity occurs; for example, when a user logs on or logs off, when the database is started or is shut down, or when an error is generated by a command or statement.

Understanding these activities will help you design an auditing scheme based on activities that are dictated by business requirements. By this point in the book, you have learned how to audit DML activities. In this chapter you learn how to audit DDL, DCL, and database event activities.

Creating DLL Triggers with Oracle

There are many good reasons for auditing database activities. A good audit program provides an audit trail for all activities, especially those that challenge the system's security. It also creates the opportunity for using process controls, such as alerts that signal the database administrator when certain activities occur. Because the alert is in place, an opportunity is created to lessen the controls on other activities, such as password verification. The introductory scenario is a good example of this.

It is possible to track not only data changes (DML statements) through Oracle triggers, but also main database activities (DDL statements). In the last chapter you were shown how to work with Oracle DML triggers. In this section, you are shown how DDL triggers work. You are also shown how to create a complete audit trail of these activities. First, here is a summary of database activities in addition to DML statements:

- ***Data Definition Language (DDL) statements***—Including CREATE, ALTER, and DROP commands
- ***Data Control Language (DCL) statements***—Including GRANT and REVOKE commands
- ***Database events***—Including such events as AFTER LOGON and BEFORE LOGON
- ***SQL statements audit trail***—Including the audit trail—a history of all statements issued by a specific user on any table

The following is the Oracle10*g* CREATE TRIGGER syntax for DDL statements and database events:

```
CREATE [ OR REPLACE ] TRIGGER [ schema. ]trigger
   { BEFORE | AFTER | INSTEAD OF }
   | { ddl_event [ OR ddl_event ]...
     | database_event [ OR database_event ]...
     }
ON { [ schema. ]SCHEMA
       | DATABASE
       }
   }
   [ WHEN (condition) ]
   { pl/sql_block | call_procedure_statement } ;
```

WHERE:

ddl_event is ALTER, ANALYZE, ASSOCIATE STATISITICS, AUDIT, COMMENT, CREATE, DISASSOCIATE STATISTICS, DROP, GRANT, NOAUDIT, RENAME, REVOKE, TRUNCATE, or DDL.

database_event is one of the following events:

- SERVERERROR is any database server error except for these errors: ORA-01403, ORA-01422, ORA-01423, ORA-01034, or ORA-04030.
- LOGON event fires the associated trigger just after a user connection is established.
- LOGOFF event fires the associated trigger just before a user disconnects.
- STARTUP event fires the associated trigger just after the database server is started.
- SHUTDOWN event fires the associated trigger just before the database server is shut down.
- SUSPEND fires the associated trigger when the server suspends a transaction.

NOTE

The information in this table is derived from the online documentation that Oracle provides at the Oracle Technology Network site: *www.otn.oracle.com*.

Start by using these definitions in the two examples provided. The first is the LOGON and LOGOFF example, which records the user name, logon time, and IP address when a user logs on to a table (APP_AUDIT_LOGINS). In addition, the login record is updated using database events LOGON and LOGOFF, when the user logs off.

Example of LOGON and LOGOFF Database Events

As indicated earlier, there are many instances in which business requirements dictate that a database administrator capture all LOGON and LOGOFF activities to analyze database connectivity. In such instances, you need to follow the steps that follow:

1. Log on as SYSTEM and create the APP_AUDIT_LOGINS table and a sequence used to generate a unique ID number for each login record.

```
SQL> CREATE TABLE APP_AUDIT_LOGINS
  2  (
  3      LOGINS_ID          NUMBER,
  4      SESSION_ID         NUMBER,
  5      USERNAME           VARCHAR2(30),
  6      LOGON_TIME         DATE,
  7      LOGOFF_TIME        DATE,
  8      IP_ADDRESS         VARCHAR2(255),
  9      AUD_INS_DTTM       DATE,
 10      AUD_UPD_DTTM       DATE
 11  )
 12  /

Table created.

SQL> CREATE SEQUENCE SEQ_LOGIN_ID
  2  /

Sequence created.
```

2. Create two triggers, one that fires after the logon event and one that fires before the logoff event.

```
SQL> CREATE OR REPLACE TRIGGER TRG_AFTER_LOGON
  2     AFTER LOGON ON DATABASE
  3  BEGIN
  4     INSERT INTO APP_AUDIT_LOGINS VALUES
  5        (SEQ_LOGIN_ID.NEXTVAL,
  6         SYS_CONTEXT('USERENV', 'SESSIONID'),
  7         USER,
  8         SYSDATE,
  9         NULL,
 10         SYS_CONTEXT('USERENV', 'IP_ADDRESS'),
 11         SYSDATE,
 12         NULL
 13        );
 14  END;
 15  /

Trigger created.

SQL> CREATE OR REPLACE TRIGGER TRG_BEFORE_LOGOFF
  2     BEFORE LOGOFF ON DATABASE
  3  BEGIN
  4     UPDATE APP_AUDIT_LOGINS SET
  5            LOGOFF_TIME = SYSDATE,
  6            AUD_UPD_DTTM= SYSDATE
  7      WHERE SESSION_ID  = SYS_CONTEXT('USERENV', 'SESSIONID')
  8        AND USERNAME    = USER
  9        AND LOGOFF_TIME IS NULL;
 10  END;
 11  /

Trigger created.
```

3. Log on as DBSEC and then disconnect after a few minutes.

```
SQL> CONN DBSEC
Enter password: *****
Connected.
SQL> DISCONNECT
```

4. Log on as SYSTEM and view the contents of the APP_AUDIT_LOGINS table.

```
SQL> SELECT * FROM APP_AUDIT_LOGINS
  2  /

LOGINS_ID SESSION_ID USERNAME   LOGON_TIM LOGOFF_TI IP_ADDRESS         AUD_INS_D AUD_UPD_D
--------- ---------- ---------- --------- --------- ------------------ --------- ----------
        1        585 DBSEC      06-AUG-04 06-AUG-04 127.0.0.1          06-AUG-04 06-AUG-04
        2        586 SYSTEM     06-AUG-04           127.0.0.1          06-AUG-04
```

DDL Event Example

The second example is a DDL event. Create a trigger that prevents DBSEC from altering any of its tables.

1. Log on as SYSTEM and create a trigger that fires before an ALTER statement is completed.

```
SQL> CREATE OR REPLACE TRIGGER TRG_BEFORE_ALTER
 2     BEFORE ALTER ON DATABASE
 3 BEGIN
 4
 5  IF USER = 'DBSEC' THEN
 6    RAISE_APPLICATION_ERROR(-20000, 'YOU MAY NOT MODIFY STRUCTURE OF ANY TABLE');
 7     END IF;
 8  END;
 9  /

Trigger created.
```

2. Log on as DBSEC and alter the CUSTOMERS table. If the table does not exist, create one.

```
SQL> ALTER TABLE CUSTOMERS
  2     MODIFY NAME VARCHAR2(60)
  3  /
ALTER TABLE CUSTOMERS
*
ERROR at line 1:
ORA-00604: error occurred at recursive SQL level 1
ORA-20000: YOU MAY NOT MODIFY STRUCTURE OF ANY TABLE
ORA-06512: at line 4
```

You remember in previous chapters that Oracle10*g* provides two pseudocolumns NEW and OLD that capture the new and old column values, respectively. Similarly,

Oracle10*g* provides pseudocolumns for DDL activities that capture important values. The following list represents these pseudocolumns:

- ora_dict_obj_name—Contains the object name being operated on with a DDL statement
- ora_dict_obj_owner—Contains the owner of the object being operated on with a DDL statement
- ora_sysevent—Contains the name of the DDL operation performed on an object

Auditing Code with Oracle

In this section you are shown how to track all statements issued against a table and to store all auditing records in an auditing table.

1. Log on as DBSEC and create an auditing table in which all auditing records are stored.

```
SQL> CREATE TABLE APP_AUDIT_SQLS
  2  (
  3    TABLE_NAME     VARCHAR2(30)                            NOT NULL,
  4    SQL_STATEMENT  VARCHAR2(4000)                          NOT NULL,
  5    SQL_TYPE       VARCHAR2(10)                            NOT NULL,
  6    CTL_INS_DTTM   DATE,
  7    CTL_INS_USER   VARCHAR2(30),
  8    CTL_OPS_USER   VARCHAR2(30),
  9    CTL_IP_ADDR    VARCHAR2(255)
 10  );

Table created.
```

2. Create a CUSTOMERS table and populate two records.

```
SQL> CREATE TABLE CUSTOMERS
  2  (
  3    ID            NUMBER,
  4    NAME          VARCHAR2(10),
  5    CREDIT_LIMIT  NUMBER
  6  );

Table created.

SQL> INSERT INTO CUSTOMERS ( ID, NAME, CREDIT_LIMIT ) VALUES (
  2  1, 'Tom Jones', 2);

1 row created.

SQL> INSERT INTO CUSTOMERS ( ID, NAME, CREDIT_LIMIT ) VALUES (
  2  10, 'BLA AFENDI', 500);

1 row created.
```

```
SQL> COMMIT;

Commit complete.
```

3. Create a trigger to track code.

```
SQL> CREATE OR REPLACE TRIGGER TRG_CUSTOMER_BDIUR
  2     BEFORE UPDATE OR INSERT OR DELETE
  3     ON CUSTOMERS
  4     FOR EACH ROW
  5  DECLARE
  6     V_STMT VARCHAR2(4000);
  7     V_OPER VARCHAR2(10);
  8  BEGIN
  9
 10     IF INSERTING THEN
 11        V_OPER := 'INSERT';
 12     ELSIF UPDATING THEN
 13        V_OPER := 'UPDATE';
 14     ELSE
 15        V_OPER := 'DELETE';
 16     END IF;
 17
 18     SELECT Q.SQL_TEXT
 19       INTO V_STMT
 20       FROM V$SQL Q, V$SESSION S
 21      WHERE S.audsid = SYS_CONTEXT('USERENV','SESSIONID')
 22        AND Q.PARSING_USER_ID = SYS_CONTEXT('USERENV', 'CURRENT_USERID')
 23        AND Q.LAST_LOAD_TIME = (SELECT MAX(LAST_LOAD_TIME)
 24                                  FROM V$SQL
 25                                 WHERE PARSING_USER_ID = Q.PARSING_USER_ID
 26                                   AND UPPER(SQL_TEXT) LIKE '%' || V_OPER ||'%'
 27                                   AND UPPER(SQL_TEXT) NOT LIKE 'SELECT'||'%')
 28        AND UPPER(SQL_TEXT) NOT LIKE 'SELECT'||'%'
 29        AND UPPER(SQL_TEXT) LIKE '%' || V_OPER ||'%';
 30
 31     INSERT INTO APP_AUDIT_SQLS (TABLE_NAME, SQL_STATEMENT, SQL_TYPE,
 32                                 CTL_INS_DTTM, CTL_INS_USER,
 33                                 CTL_OPS_USER,
 34                                 CTL_IP_ADDR)
 35                         VALUES ('CUSTOMERS', V_STMT, V_OPER,
 36                                 SYSDATE, USER,
 37                                 SYS_CONTEXT('USERENV', 'OS_USER'),
 38                                 SYS_CONTEXT('USERENV', 'IP_ADDRESS'));
 39  END;
 40  /

Trigger created.
```

Note that SYS must provide SELECT privilege on V_$SQL and V_$SYSTEM to DBSEC.

```
GRANT SELECT ON V_$SQL TO DBSEC
/
GRANT SELECT ON V_$SESSION TO DBSEC
/
```

4. Now update the CUSTOMERS table.

```
SQL> UPDATE CUSTOMERS SET
  2          CREDIT_LIMIT = 1000
  3  /

2 rows updated.

SQL> COMMIT;

Commit complete.
```

5. Look at the contents of the APP_AUDIT_SQLS table.

```
SQL> SELECT TABLE_NAME, SQL_STATEMENT
  2    FROM APP_AUDIT_SQLS
  3  /

TABLE_NAME    SQL_STATEMENT
------------- --------------------------- -------------------
CUSTOMERS     UPDATE CUSTOMERS SET        CREDIT_LIMIT = 1000
CUSTOMERS     UPDATE CUSTOMERS SET        CREDIT_LIMIT = 1000
```

Auditing Database Activities with Oracle

During an economic downturn, Tom, a database administrator, was laid off and spent two years searching for a job. He finally got a job as a database administrator in a pharmaceutical company. During his first month, Tom was given time to become familiar with the huge database and its hundreds of tables. Soon, he was asked to work on database changes. Tom was very worried about being laid off again, and to make himself indispensable he decided to make intentional changes to objects in the QA and development databases. He would drop an index or a NOT NULL constraint. When developers got application errors, Tom would quickly resolve the issues and become a hero.

Soon a senior DBA realized that they never had these problems before Tom joined them. He quickly enabled auditing and monitored all of Tom's activities. Tom changed a column to a table a few days later. He was immediately summoned to his manager's office, and when he was unable to explain the changes, he was terminated without notice.

Oracle provides the mechanism for auditing everything: from tracking who is creating or modifying the structure to who is granting which privileges to whom. In this section you will learn to audit database activities. The activities are divided into two types based on the type of SQL command statement used: activities defined by DDL (Data Definition Language) and activities defined by DCL (Data Control Language).

Auditing DDL Activities

Oracle uses a SQL-based AUDIT command. Suppose you wanted to know when a specific user issued an ALTER statement or when a specific object was altered. If AUDIT is enabled, you can obtain this type of information easily. Before this is demonstrated, examine the full syntax of the AUDIT command. Figure 9-1 is adopted from Oracle10*g* documentation found at *www.otn.oracle.com*.

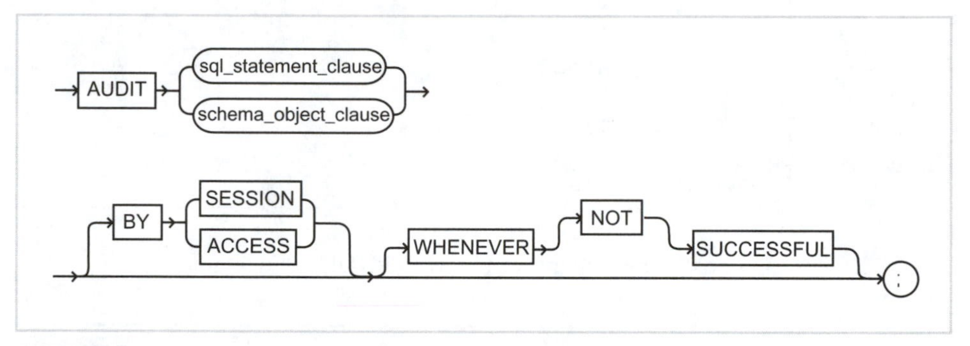

FIGURE 9-1 **Audit command syntax diagram**

The following is a text representation of the AUDIT command syntax presented in Figure 9-1.

```
AUDIT
  {
    { { statement_option | ALL }
        [, { statement_option | ALL } ]...
      | { system_privilege | ALL PRIVILEGES }
        [, { system_privilege | ALL PRIVILEGES } ]...
    }
      [ BY { proxy [, proxy ]...
           | user [, user ]...
           }
      ]
  |
    { object_option [, object_option ]... | ALL }
      ON { [ schema. ]object
           | DIRECTORY directory_name
           | DEFAULT
         }
```

```
}
[ BY { SESSION | ACCESS } ]
[ WHENEVER [ NOT ] SUCCESSFUL ] ;
```

Where:

statement_option—Tells Oracle to audit the specified DDL or DCL statement:
DDL statements—CREATE, DROP, ALTER, SET, and TRUNCATE
DCL statements—GRANT or REVOKE; note that to audit any GRANT or REVOKE statement, you use SYSTEM GRANT, for example: AUDIT SYSTEM GRANT BY ACCESS; this command audits both GRANT and REVOKE statements

system_privilege—Tells Oracle to audit the specified system privilege, such as SELECT, CREATE ANY, or ALTER ANY

object_option—Specifies the type of privilege for the specified object to be audited

BY SESSION—Tells Oracle to record audit data once per session even if the audited statement is issued multiple times in a session

BY ACCESS—Tells Oracle to record audit data every time the audited statement is issued

WHENEVER SUCCESSFUL—Tells Oracle to capture audit data only when the audited command is successful

WHENEVER NOT SUCCESSFUL—Tells Oracle to capture audit data only when the audited command fails

NOTE

The information in this table is derived from the online documentation that Oracle provides at the Oracle Technology Network site: *www.otn.oracle.com*.

Before you proceed with the demonstration, you must make sure that auditing is turned on. You can verify this by checking the AUDIT_TRAIL parameter. This parameter uses the following values:

- DB—Indicates that the audit trail will be stored in the database in a table called SYS.AUD$
- DB_EXTENDED—Indicates that the audit trail will be stored in the database, in the table, SYS.AUD$; this value enables the storing of bind variables
- OS—Indicates that the audit trail will be stored in a file
- NONE—Indicates that no auditing records will be stored

The following DDL and DCL activities demonstrate the AUDIT statement.

DDL Activities Example 1

Suppose you want to audit a table named CUSTOMER every time it is altered or every time a record from the table is deleted. The following steps show you how to do this. In order to perform this demonstration you need to drop or disable any triggers created from previous exercises or demonstrations.

1. Use any user other than SYS or SYSTEM to create the CUSTOMER table. If the CUSTOMER table already exists, you may want to drop it or use another schema that does not have this table. In this demonstration DBSEC user will be used.

```
SQL> CREATE TABLE CUSTOMER
  2  (
  3    ID          NUMBER,
  4    NAME        VARCHAR2(20),
  5    CR_LIMIT    NUMBER
  6  )
  7  /

Table created.
```

2. Add three rows into the CUSTOMER table and commit changes.

```
SQL> INSERT INTO CUSTOMER VALUES(1, 'TOM', 200);

1 row created.

SQL> INSERT INTO CUSTOMER VALUES(2, 'SUSAN', 130);

1 row created.

SQL> INSERT INTO CUSTOMER VALUES(3, 'LINDA', 230);

1 row created.

SQL> COMMIT;

Commit complete.
```

3. Now, log on as SYSTEM or SYS to enable auditing, as specified in this example. The first statement is for ALTER, the next for DELETE.

```
SQL> CONNECT SYSTEM@SEC
Enter password: ******
Connected.
SQL> AUDIT ALTER ON DBSEC.CUSTOMER BY ACCESS WHENEVER SUCCESSFUL
  2  /

Audit succeeded.

SQL> AUDIT DELETE ON DBSEC.CUSTOMER BY ACCESS WHENEVER SUCCESSFUL
  2  /

Audit succeeded.
```

4. Log in as the owner of the CUSTOMERS table, DBSEC, delete a row and modify the structure of the table, as specified in the following code:

```
SQL> CONN DBSEC@SEC
Enter password: *****
Connected.
SQL> DELETE FROM CUSTOMER WHERE ID = 2
  2  /

1 row deleted.

SQL> ALTER TABLE CUSTOMER MODIFY NAME VARCHAR2(30)
  2  /

Table altered.
```

5. In this step you will see the audit records stored in the auditing tables caused by the DELETE and ALTER statements issued in Step 4. Note that all auditing records are stored in the SYS.AUD$ table and can be viewed from DBA_AUDIT_TRAIL. You may truncate the AUD$ table whenever necessary; however, you must log in as SYS to do that. Log in as SYSTEM and view the DBA_AUDIT_TRAIL. You will find two records in Figure 9-2.

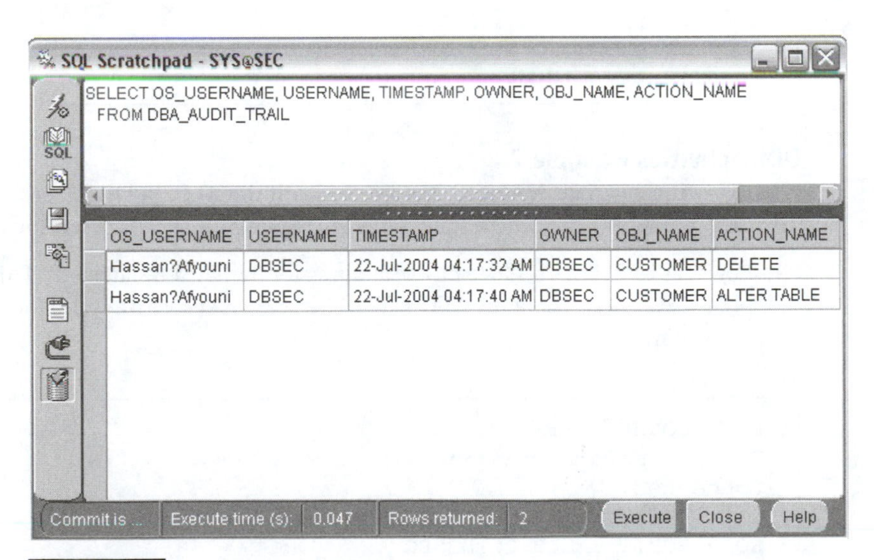

FIGURE 9-2 Contents of DBA_AUDIT_TRAIL

When you finish with auditing a specific object or command, you may turn it off by using the NOAUDIT statement. The following step turns off auditing on the two audit statements issued in Step 3. However, before issuing the NOAUDIT statement, you should look at the contents of DBA_AUDIT_OBJECT to see the auditing metadata. Refer to Figure 9-3.

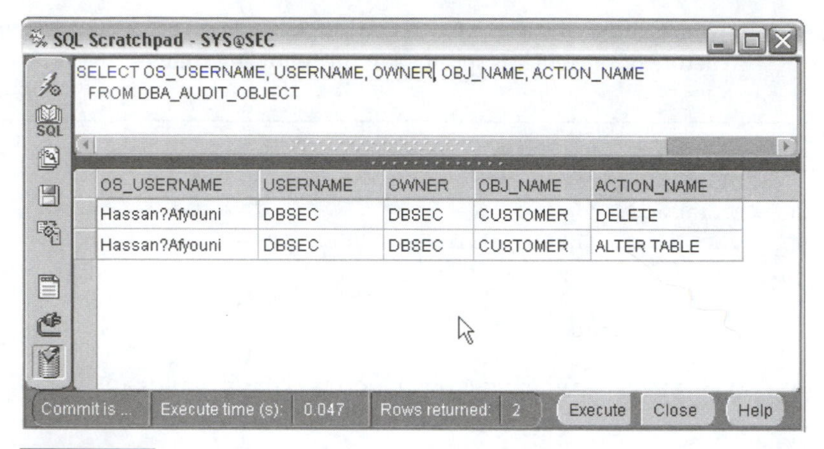

FIGURE 9-3 Contents of DBA_AUDIT_OBJECT

```
SQL> NOAUDIT ALTER ON DBSEC.CUSTOMER
  2  /

Noaudit succeeded.

SQL> NOAUDIT DELETE ON DBSEC.CUSTOMER
  2  /

Noaudit succeeded.
```

DDL Activities Example 2

The following example demonstrates how to audit the ALTER, DROP, or CREATE TABLE statements, when these statements are issued by a specific user, such as DBSEC:

1. Log in as SYSTEM or SYS to enable auditing for the TABLE statement. TABLE is a comprehensive audit trail of the ALTER TABLE, the CREATE TABLE, and the DROP TABLE statements.

```
SQL> CONNECT SYSTEM@SEC
Enter password: ******
Connected.

SQL> AUDIT TABLE BY DBSEC
  2  /

Audit succeeded.
```

2. Log on as DBSEC and create a table called TEMP, then drop it.

```
SQL> CONN DBSEC@SEC
Enter password: *****
Connected.
SQL> CREATE TABLE TEMP( NUM NUMBER )
  2  /

Table created.

SQL> INSERT INTO TEMP VALUES (1000)
  2  /

1 row created.

SQL> SELECT * FROM TEMP
  2  /

       NUM
----------
      1000

SQL> DROP TABLE TEMP
  2  /

Table dropped.
```

3. Log on as SYSTEM and view the content of DBA_AUDIT_TRAIL, which contains the auditing records of the TABLE statements.

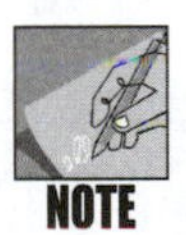

The rows shown are merely excerpts. You should delete entries from DBA_AUDIT_TRAIL after each set of steps in the chapter.

```
SQL> SELECT OS_USERNAME, USERNAME, TIMESTAMP, OWNER, OBJ_NAME, ACTION_NAME
  2     FROM DBA_AUDIT_TRAIL
  3  /

OS_USERNAME    USERNAME   TIMESTAMP OWNER      OBJ_NAME        ACTION_NAME
-------------- ---------- --------- ---------- --------------- ------------
Hassan?Afyouni DBSEC      05-AUG-05 DBSEC      TEMP            CREATE TABLE
Hassan?Afyouni DBSEC      05-AUG-05 DBSEC      TEMP            DROP TABLE
```

4. Now, turn off auditing for the TABLE statement.

```
SQL> NOAUDIT TABLE BY DBSEC
  2  /

Noaudit succeeded.
```

DCL Activities Example

This example shows you how to audit DCL statements. In this case, you are auditing the GRANT privilege issued on a TEMP table owned by DBSEC rr. The same process can be used for other privilege statements, such as REVOKE. In this demonstration, you need to recreate the same TEMP table as in the previous example.

1. Log on as SYSTEM or SYS and issue an AUDIT statement as follows:

```
SQL> CONN SYSTEM
Enter password: ******
Connected.
SQL> DELETE SYS.AUD$
  2  /

1 rows deleted.

SQL> COMMIT
  2  /

Commit complete.

SQL> AUDIT GRANT ON DBSEC.TEMP
/

Audit succeeded.
```

2. Log on as DBSEC and grant SELECT and UPDATE to SYSTEM.

```
SQL> CONN DBSEC
Enter password: *****
Connected.
SQL> GRANT SELECT ON TEMP TO SYSTEM
  2  /

Grant succeeded.

SQL> GRANT UPDATE ON TEMP TO SYSTEM
  2  /

Grant succeeded.
```

3. Log on as SYSTEM and display the contents of DBA_AUDIT_TRAIL.

```
SQL> SELECT USERNAME, TIMESTAMP, OWNER, OBJ_NAME
  2    FROM DBA_AUDIT_TRAIL
  /

USERNAME   TIMESTAMP OWNER      OBJ_NAME
---------- --------- ---------- --------
DBSEC      06-AUG-05 DBSEC      TEMP
DBSEC      06-AUG-05 DBSEC      TEMP
```

The complete details of the GRANT statement are not shown. This example shows only that a record of a GRANT statement was issued.

Before moving on to the next example, you should review the data dictionary views listed in Table 9-1.

Table 9-1 Audit data dictionary views

Table name	Description of contents
DBA_AUDIT_TRAIL	All audit trail records when the AUDIT_TRAIL initialization parameter is set to DB or DB_EXTENDED
DBA_AUDIT_OBJECT	All audit trail records for database objects
DBA_AUDIT_SESSION	All audit trail records for session connections and disconnections
DBA_AUDIT_STATEMENT	All audit trail records for GRANT, REVOKE, AUDIT, NOAUDIT, and ALTER SYSTEM

Note: The information in this table is derived from the online documentation that Oracle provides at the Oracle Technology Network site: *www.otn.oracle.com*. To find the relevant information search on DBA_AUDIT.

Example of Auditing User Activities

Follow these steps to audit all activities performed by the DBSEC user:

1. Log on as SYSTEM or SYS and issue the following audit statement:

```
SQL> AUDIT ALL BY DBSEC
  2  /

Audit succeeded.

SQL> DELETE SYS.AUD$
/

3 rows deleted.
```

Note that number of rows deleted may vary.

2. Now log on as DBSEC and create a temporary table.

```
SQL> CONN DBSEC
Enter password: *****
Connected.
SQL> CREATE TABLE TEMP2(NUM NUMBER)
  2  /

Table created.
```

3. Go back to SYSTEM to view the contents of DBA_AUDIT_TRAIL. Note that there are two records: one for logging on and one for creating a table.

```
OS_USERNAME    USERNAME   TIMESTAMP OWNER      OBJ_NAME       ACTION_NAME
-------------- ---------- --------- ---------- -------------- ------------
Hassan?Afyouni DBSEC      09-AUG-05                           LOGON
Hassan?Afyouni DBSEC      09-AUG-05 DBSEC      TEMP2          CREATE TABLE
```

Audit Trail File Destination

Do you remember that Oracle allows you to set the destination of the audit trail to an operating system file? The following steps show you how to send an audit trail to a file:

1. Modify the initialization parameter file, INIT.ORA. Then set the parameter AUDIT_TRAIL to the value OS.

```
AUDIT_TRAIL=OS
```

2. Create a folder/directory in which the auditing file will be created and stored. The location of the auditing files is under C:\ORACLE\ADMIN\SEC\ADUMP, where ADUMP stands for auditing dump.
3. You need to set another initialization parameter, AUDIT_FILE_DEST, to the directory set up in Step 2. If AUDIT_FILE_DEST is not specified, the audit file location by default becomes %ORACLE_HOME%\rdbms\audit.

```
AUDIT_FILE_DEST="c:\oracle\admin\sec\adump"
```

4. You must shut down and restart the database for the AUDIT_TRAIL parameter to take effect. For the AUDIT_FILE_DEST parameter, you may make modifications while the database is running by issuing an ALTER SYSTEM statement.
5. Connect as DBSEC. Because DBSEC is being audited in the previous demonstration, you will see a record in the auditing file.

Oracle Alert Log

A small financial company has hired a DBA, Bill, who is having some trouble getting organized. One day Bill is required to increase the size of a tablespace. Unable to focus, he creates one of the tablespace files in the wrong directory. After several incidents like this, Bill is placed on warning. Unfortunately, Bill is on call during a major holiday and one of the databases crashes. The next day the database manager summons Bill to his office to find out what happened. Bill says that the crash was caused by a hard drive failure. The manager asks another DBA to look into the matter. The other DBA inspects the operating system trace files and the Alert log. He notices something unusual. There were two date entries with a space between them. Typically, the Alert log contains activity entries in which each entry starts with date and a time stamp. The DBA requests the restoration of a backup of the Alert log, which confirms that Bill was lying about the hard drive failure. Bill is terminated on the spot for altering the Alert log.

In addition to the audit trail destination file, Oracle provides the Alert log for auditing database activities. In the Alert log you can find the following database activities:[1]

- *Errors*—All errors related to physical structure are recorded in the Alert log. For example, an error is recorded when a background process fails, when a table runs out of space, or when a deadlock occurs. There are many more examples, of course. Because errors have a systemwide (instance and database) effect, you need to monitor the Alert log, preferably every five minutes, or ten minutes at the most. Monitoring can be done through a UNIX or Windows script. You can also employ a third-party tool to monitor the file. Some errors are not registered in the file, such as a SQL statement that fails because of a syntactical error. Figure 9-4 shows an error entry. The format of the entry is the date and time line, followed by the trace file name and location, if applicable, and the ORA-error with an error description.

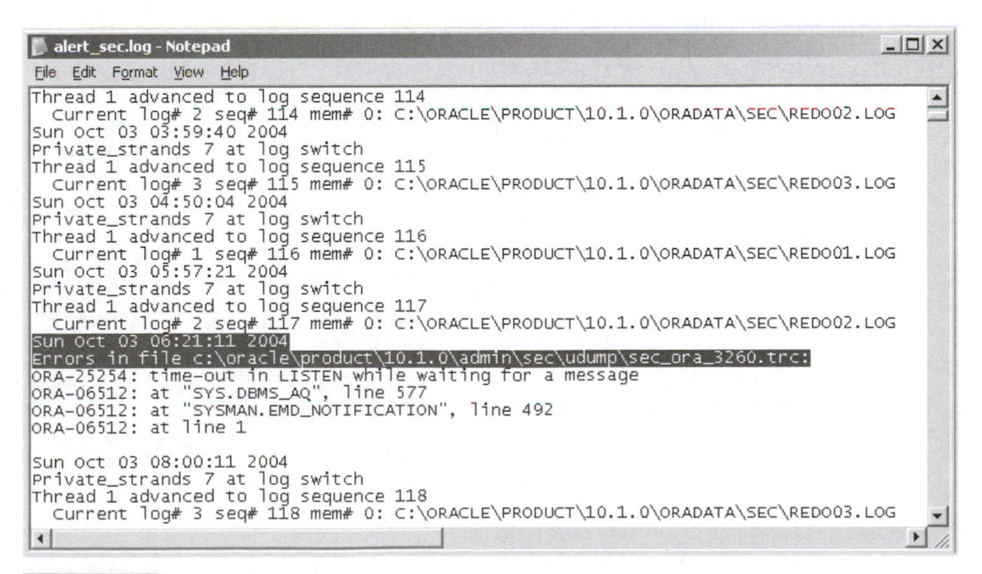

alert_sec.log - Notepad

File Edit Format View Help

```
Thread 1 advanced to log sequence 114
  Current log# 2 seq# 114 mem# 0: C:\ORACLE\PRODUCT\10.1.0\ORADATA\SEC\REDO02.LOG
Sun Oct 03 03:59:40 2004
Private_strands 7 at log switch
Thread 1 advanced to log sequence 115
  Current log# 3 seq# 115 mem# 0: C:\ORACLE\PRODUCT\10.1.0\ORADATA\SEC\REDO03.LOG
Sun Oct 03 04:50:04 2004
Private_strands 7 at log switch
Thread 1 advanced to log sequence 116
  Current log# 1 seq# 116 mem# 0: C:\ORACLE\PRODUCT\10.1.0\ORADATA\SEC\REDO01.LOG
Sun Oct 03 05:57:21 2004
Private_strands 7 at log switch
Thread 1 advanced to log sequence 117
  Current log# 2 seq# 117 mem# 0: C:\ORACLE\PRODUCT\10.1.0\ORADATA\SEC\REDO02.LOG
Sun Oct 03 06:21:11 2004
Errors in file c:\oracle\product\10.1.0\admin\sec\udump\sec_ora_3260.trc:
ORA-25254: time-out in LISTEN while waiting for a message
ORA-06512: at "SYS.DBMS_AQ", line 577
ORA-06512: at "SYSMAN.EMD_NOTIFICATION", line 492
ORA-06512: at line 1

Sun Oct 03 08:00:11 2004
Private_strands 7 at log switch
Thread 1 advanced to log sequence 118
  Current log# 3 seq# 118 mem# 0: C:\ORACLE\PRODUCT\10.1.0\ORADATA\SEC\REDO03.LOG
```

FIGURE 9-4 Sample contents of the Alert log file

- *Startup and shutdown*—The Alert log records the date and time of each occurrence of the database being signaled to shut down or start up.
- *Modified initialization parameters*—Each time a database is started, Oracle records all modified initialization parameters in the Alert log.
- *Checkpoints*—You can configure Oracle to record checkpoint time in the Alert log by setting the initialization parameter as follows: LOG_CHECKPOINT_TO_ALERT = TRUE. You can also determine when a checkpoint is completed.
- *Archiving*—You can view the timing for all redo log sequences, as well as archiving times (when an archive log is started and completed).
- *Physical database changes*—As stated previously, any change to the physical structure of the database (not objects) is recorded in the Alert log. The following actions are registered in the Alert log: creating a new tablespace, dropping a tablespace, resizing a data file, and adding a redo log file. Figure 9-5 shows entries for a tablespace that was dropped.

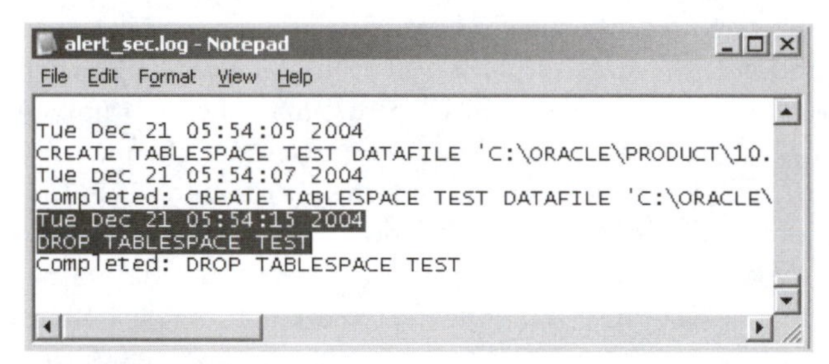

FIGURE 9-5 Alert log example showing an entry for tablespace changes

Auditing Server Activity with Microsoft SQL Server 2000

Microsoft SQL Server 2000 provides auditing as a way to track and log activity for each SQL Server occurrence. You must be a member of the sysadmin fixed server role to enable or modify auditing. Every modification of an audit is an auditable event.

In SQL Server 2000, there are two types of auditing for server events: auditing and C2 auditing. For more information on C2 Auditing, refer to Chapter 7.

Auditing can have a significant impact on performance. The impact varies depending on how many counters you have enabled and how many objects you are auditing. You must evaluate how many events need to be audited against their impact on performance. The audit trail analysis can also be costly in terms of system resources. It is recommended that SQL Profiler be run on a server separate from the production server.

Implementing SQL Profiler

One of the tools that accompanies SQL Server 2000 is SQL Profiler. This tool provides the user interface for auditing events. You can audit several types of events using SQL Profiler, including those listed in Table 9-2.

Table 9-2 SQL Server event descriptions

Event	Description
End user events	All SQL commands, LOGOUT/LOGIN, enabling of application roles
DBA events	DDL (other than security events), configuration (DB or server)
Security events	GRANT/REVOKE/DENY, LOGIN USER/ROLE ADD/REMOVE/CONFIGURE
Utility events	BACKUP/RESTORE/BULK INSERT/BCP/DBCC commands
Server events	SHUTDOWN, PAUSE, START
Audit events	ADD AUDIT, MODIFY AUDIT, STOP AUDIT

For each event, you can audit:

- Date and time of the event
- User who caused the event to occur
- Type of event
- Success or failure of the event
- Origin of the request
- Name of the object accessed
- Text of the SQL statement (passwords replaced with ****)

Later in this chapter, there is more information about using SQL Profiler to monitor database activities.

Security Auditing with SQL Server

Before you can audit security events in SQL Server 2000, you need to enable it. This is done by setting the security auditing level under the SQL Server properties in Enterprise Manager. Security events can be audited on success, failure, or both. Follow these steps:

1. Open Enterprise Manager.
2. Expand the appropriate SQL Server group.
3. Right-click on the desired server.
4. Click **Properties**.
5. On the security tab, select the desired security level, as shown in Figure 9-6.

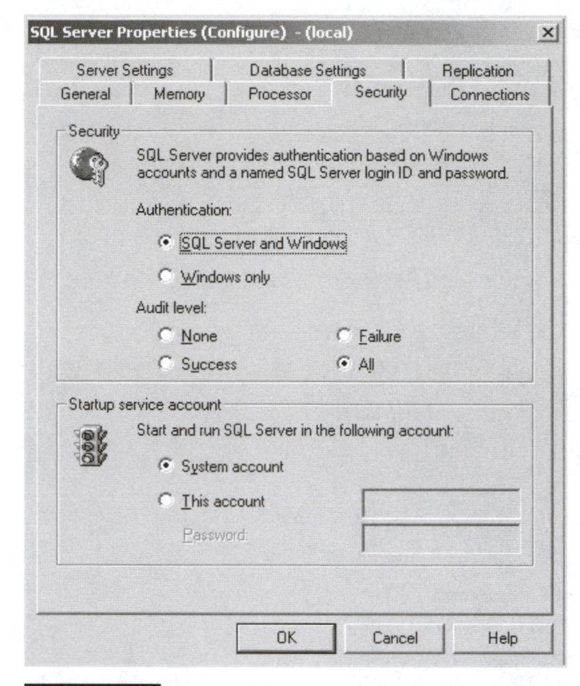

FIGURE 9-6 SQL Server configuration

After the audit level is set, you can then use SQL Profiler to monitor security events. The following events can be audited:

- ADD DB USER
- ADD LOGIN TO SERVER ROLE
- ADD MEMBER TO DB ROLE
- ADD ROLE
- APP ROLE CHANGE PASSWORD
- BACKUP/RESTORE
- CHANGE AUDIT
- DBCC
- LOGIN
- LOGOUT
- LOGIN CHANGE PASSWORD
- LOGIN CHANGE PROPERTY
- LOGIN FAILED
- Login GDR (GRANT, DENY, REVOKE)
- Object Derived Permissions
- Object GDR
- Object Permissions
- Server Start and Stop
- Statement GDR
- Statement Permission

You can start SQL Profiler by selecting it from the program group on the Start menu or from the Tools menu in Enterprise Manager. To start a new Audit Trace from the file menu, click New, then Trace. See Figure 9-7.

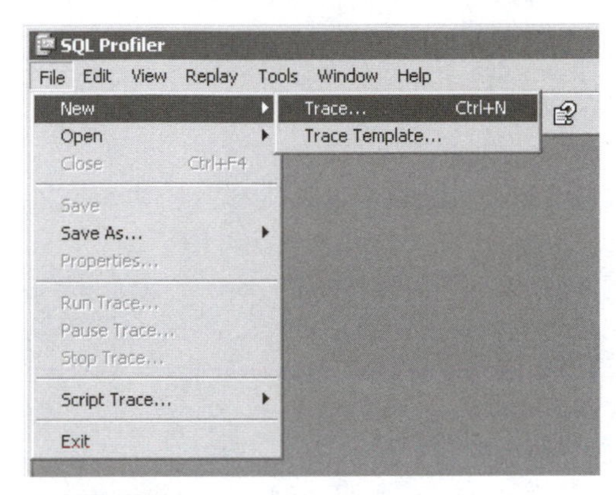

FIGURE 9-7 SQL Profiler main menu

The new trace dialog box appears, as shown in Figure 9-8. On the General tab, you provide:

- A name for the trace
- The server you want to audit
- The base template to start with
- Where to save the audit data, either to a file or to a database table
- A stop time, if you don't want the trace to run indefinitely

Now it should be clear why Microsoft recommends a separate server to actually perform the monitoring and recording of the auditing data. The constant write operations could cause severe disk contention, but that's better left for a performance and tuning course.

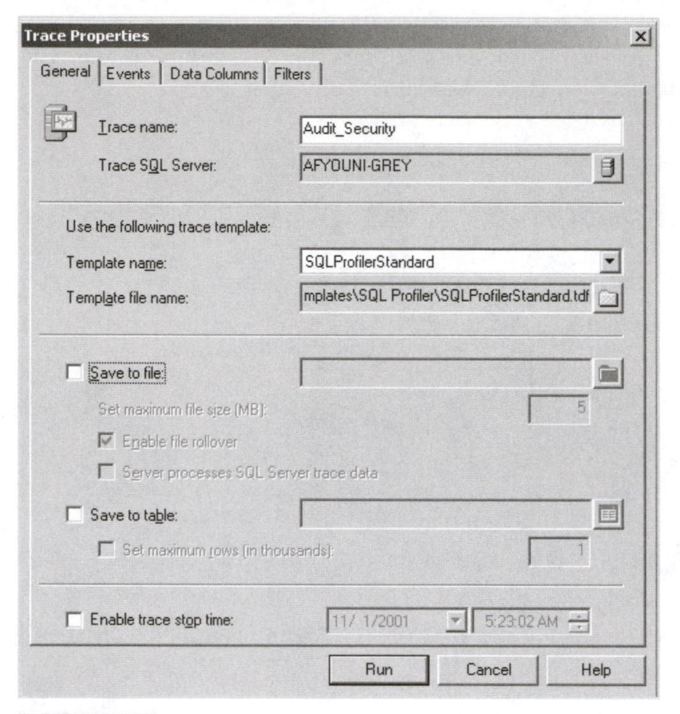

FIGURE 9-8 SQL Server Trace Properties dialog box

On the Events tab, you specify events to be audited and in which category they belong. As you can see in Figure 9-9, the standard template provides defaults in the Selected event classes list.

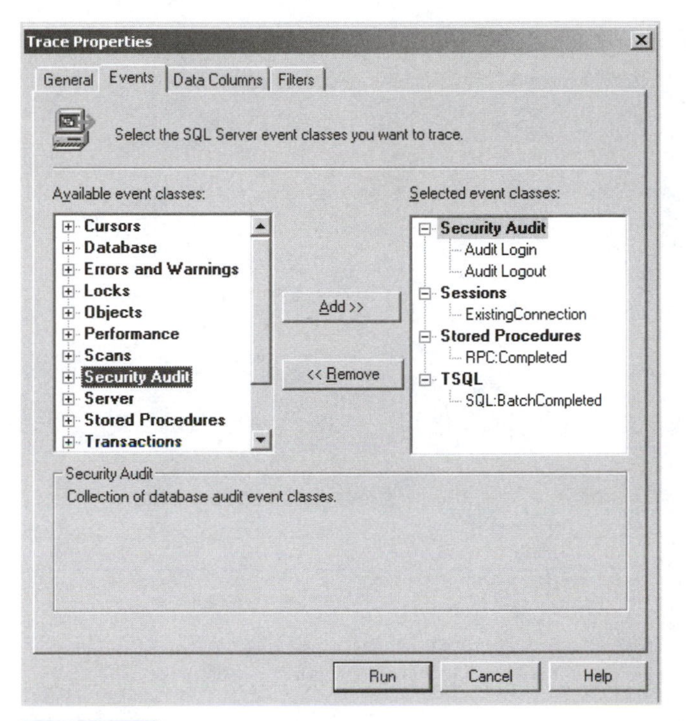

FIGURE 9-9 SQL Server trace configuration screen showing security audit selection

Add the Login Change Password security event to the trace by performing the following steps:

1. Expand the **Security Audit** node under Available event classes.
2. Click **Audit Login Change Password Event.**
3. Click the **Add** button.

Audit Login Change Password Event should now appear under Security Audit in Selected event classes, as shown in Figure 9-10.

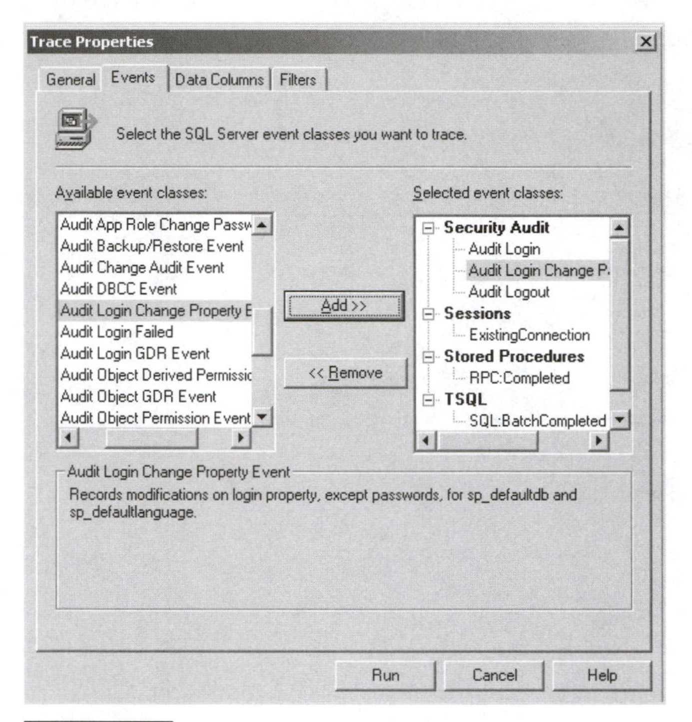

FIGURE 9-10 SQL Server trace configuration screen showing all available auditing options for security audit selection

Data Definition Auditing

To audit DDL statements, on the Events tab of your trace, you select Object:Created and Object:Deleted under the Objects category. These two events audit all CREATE and DROP statements. See Figure 9-11.

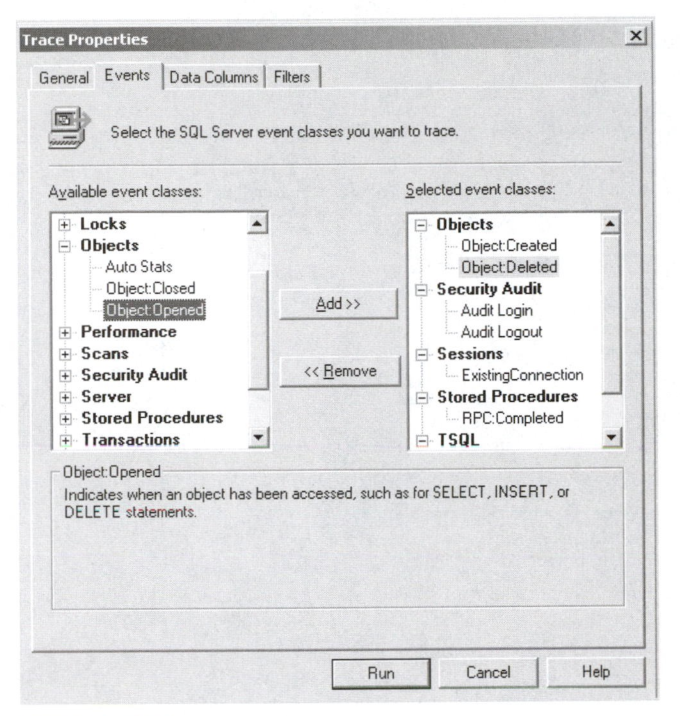

FIGURE 9-11 SQL Server trace configuration screen showing Object: Created by audit selection

Database Auditing with SQL Server

To audit operations to the database files, select events under the Database category, as shown in Figure 9-12.

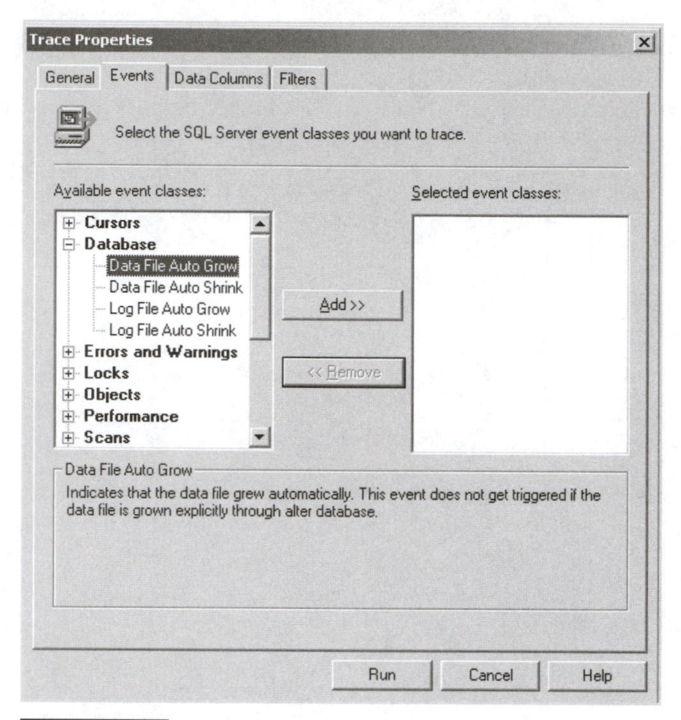

FIGURE 9-12 SQL Server trace configuration screen showing database audit selection

Database Errors Auditing with SQL Server

To audit errors that occur within the database, select the events under the Errors and Warnings category on the Events tab of your trace. See Figure 9-13.

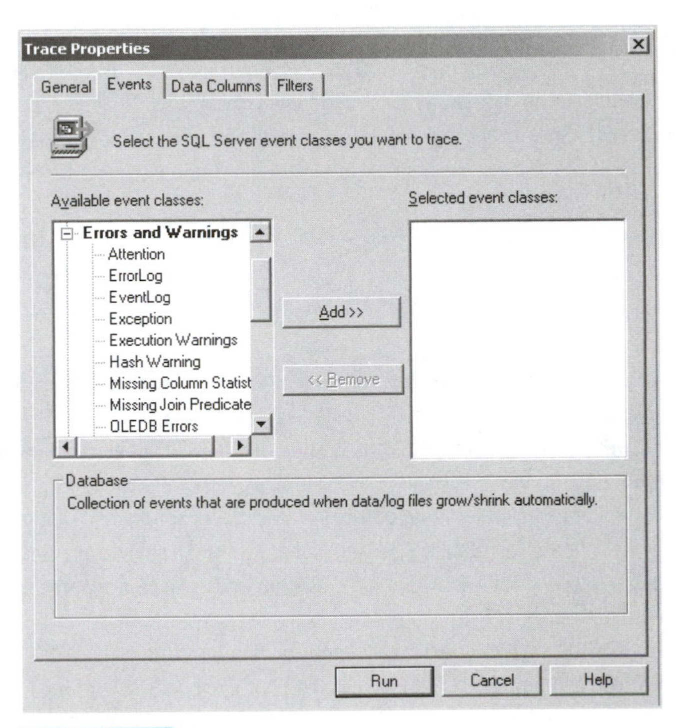

FIGURE 9-13 SQL Server trace configuration for adding events to be traced

Chapter Summary

- Database activities are classified into three types: application activities, administration activities, and database events.
- Oracle triggers provide a way to create an audit trail for DDL changes and database activities.
- Oracle database activities that can be tracked include user logon and logoff, as well as database startup and shutdown.
- Oracle provides a way to audit activities, such as who is creating or modifying the structure of a specific object and who granted which privileges to whom.
- Oracle provides the AUDIT command, which is a SQL command for auditing.
- The Oracle initialization parameter, AUDIT_TRAIL, specifies the destination of the audit trail file.
- The Oracle NOAUDIT statement is used to stop auditing.
- The Oracle DBA_AUDIT_TRAIL data dictionary view contains the audit trail data.
- The Oracle Alert log is another method of auditing database activities.
- Oracle Alert log contains information related to errors, such as database startup and shutdown events, modified initialization parameters, checkpoints, archiving, and physical database changes.

- Microsoft SQL Server 2000 provides auditing as a way to track and log each SQL Server activity.
- In SQL Server 2000, you must be a member of the sysadmin fixed server role to enable or modify auditing. Every modification of an audit is an auditable event.
- SQL Profiler is the SQL Server tool that provides the user an interface for auditing events.
- SQL Server allows you to audit errors that occur within the database using the SQL Profiler trace facility.

Review Questions

1. Oracle provides specific trigger events that enable you to produce an audit trail of database application errors. True or false?
2. Oracle provides specific triggers that enable auditing database activities, such as user logons and logoffs. True or false?
3. Oracle provides only one way of auditing database startup and shutdown. True or false?
4. SQL Server provides an interface called SQL Trace for auditing database activities. True or false?
5. After the Oracle audit trail database activities are stored in the AUD$ table, the audit trail records cannot be deleted. True or false?
6. The DBA_AUDIT_TRAIL data dictionary view cannot be deleted. True or false?
7. Using SQL Server you can audit DDL by selecting the TRACE tab in SQL Profiler. True or false?
8. The Oracle AUD$ table is owned by SYSTEM. True or false?
9. SQL Server allows you to audit the ADD_DB_USER event. True or false?
10. Oracle provides an interface called Oracle Enterprise Manager that allows you to audit database activities. True or false?
11. Using Oracle, shut down and start up the database and then locate the related audit records in the alert log.
12. Using an Oracle or SQL Server, select an existing user and audit all its activities.
13. For both Oracle and SQL Server, list the steps for determining the location of the audit trail records.
14. What is the purpose of the SQL Profiler tool?
15. Briefly explain the purpose of the Oracle Alert file.

Hands-on Projects

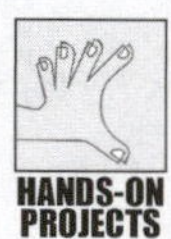

Hands-on Project 9-1

Your manager asked you to create a document outlining steps to audit programs that are being used to connect to the database. She wants to know the name of the program that connects to the database along with the date and time for all logon and logoff activities. If the program is not Oracle SQL*Plus your code should raise an exception.

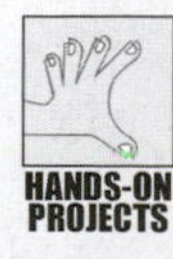

Hands-on Project 9-2

Your security policy requires you to track every table creation statement to a table, but only when the creation is not successful. List all steps necessary to audit every unsuccessful instance of a user attempting to create table.

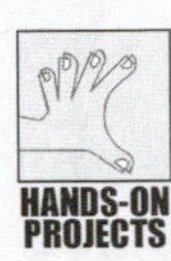

Hands-on Project 9-3

Your manager suspects that a user is tampering with the database. Provide a step-by-step solution to track the user activities using Oracle10g. You may use any schema except SYS and SYSTEM.

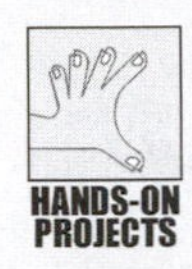

Hands-on Project 9-4

Outline the steps that are necessary to audit in Oracle the REVOKE statement activities.

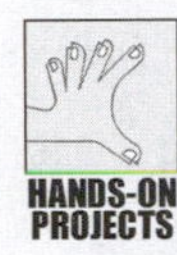

Hands-on Project 9-5

Within an Oracle environment, outline the steps necessary to prevent owners from modifying the structure of their tables. Record the date, time, and the object that was modified in an audit table.

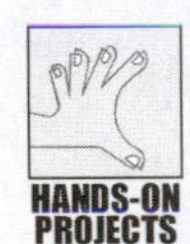

Hands-on Project 9-6

Using Oracle, shut down and start up the database and then locate the related startup and shutdown auditing information in the Alert log.

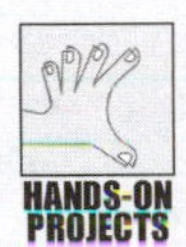

Hands-on Project 9-7

You were asked by a colleague to create a trigger that will track object creation activities in the database view. Using Oracle, outline all necessary steps to accomplish this task.

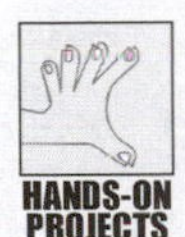

Hands-on Project 9-8

Using Oracle, provide the steps necessary to track all database server errors.

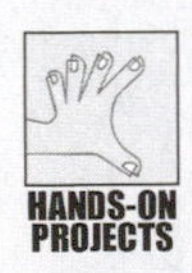

Hands-on Project 9-9

Using Oracle, create a trigger that enables an existing database role when a user logs on.

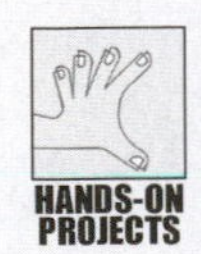

Hands-on Project 9-10

After an intensive investigation, you found out that some users were abusing their privileges. You decide to implement a solution in which users are audited when they create any database object. Using Oracle, outline the steps you will take to accomplish this task.

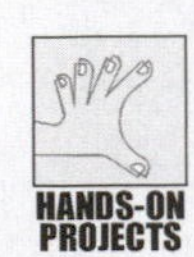

Hands-on Project 9-11

Your manager has asked you to produce a document that contains instructions for other DBAs to audit database errors. Using SQL Server, outline the steps to audit database errors.

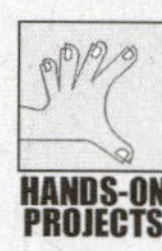

Hands-on Project 9-12

At last your wish comes true! Your manager has asked you to audit the DROP statement activities in the SQL Server production database. List a step-by-step procedure for accomplishing this task.

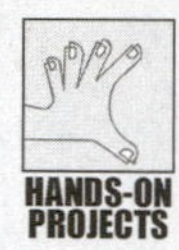

Hands-on Project 9-13

Using the SQL Server SQL Profiler tool, provide a step-by-step outline to audit the LOGIN events.

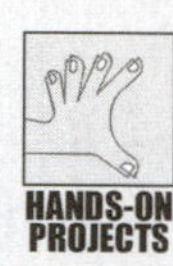

Hands-on Project 9-14

Using Oracle, check the status of the audit as well as the location of the audit trail. If auditing is turned off, turn it on and include a file destination.

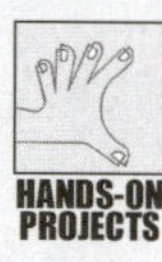

Hands-on Project 9-15

Using Oracle or SQL Server, select an existing user and audit all its activities.

Case Project

Case 9-1 The Auditing Game

A database specialist was hired by a company that manufactures video games. The new hire was directed to create and develop a marketing data warehouse. Five of the auditing requirements on his task list were:

1. Create the mechanism that will track all user SQL statements activity.
2. Record the times of logon and logoff, the program used to connect, and all available operating system information.
3. Disallow users from logging onto the database before 8:00 A.M. and after 5:00 P.M.
4. Disable any user account that has not been used within the last 17 working days.
5. Track all DDL activities to a user-specified table.

You may implement the solution for this case project in either Oracle 10*g* or SQL Server.

Endnotes

1 Alert log documentation is adopted from *Oracle9i Performance Tuning: Optimizing Database Productivity* by Hassan A. Afyouni.

PART THREE

Security and Auditing Project Cases

10

LEARNING OBJECTIVES

Upon completion of this material, you should be able to:

- Design and implement security and auditing solutions for many common business situations

Introduction

A database developer is assigned to a new database application project and is asked to develop an auditing scheme to comply with the industry standards. Because auditing is a new arena for the developer, he looks up titles on auditing in a nearby bookstore. After browsing through all the relevant titles looking for practical auditing solutions, he has learned nothing. So he visits the local university library to examine auditing materials that he can use, but finds that auditing theory requires hours to understand and practical suggestions are hard to find.

Developers often face this problem. Database administrators are often asked to provide an effective data security and auditing design. Database developers should be able to create and implement a sound security model and integrate auditing design into their applications. This chapter is designed to help database developers and administrators effectively respond to these types of demands. Instead of introducing new concepts on either database security or database auditing, this chapter offers you practical case studies that you can use in different circumstances during software development.

This chapter gives you the opportunity to teach yourself to create effective security models and audit systems when information is scarce. It consolidates your skills by helping you apply all the database security and audit concepts presented in this book. The five cases that follow require you to use these concepts, methods, and techniques to solve data accessibility, integrity, and confidentiality problems that are frequently encountered in the workplace. These cases can be implemented in either Oracle or SQL Server, however solutions are provided for Oracle only, not for SQL Server. Please note that each case takes two to six hours to complete, and you might want to make minor modifications to these projects depending on the feasibility of working through the problems on your database.

Case 1: Developing an Online Database

A new dot-com company has decided to launch an affiliated Web site, specifically for individuals interested in database issues. The main mission of the Web site is to provide a forum for database technical tips, issues, and scripts. The CIO and his technical team held a meeting to draft the requirements for the new Web site and decided that it would include the following:

- Technical documents
- A forum where members can exchange ideas and share experiences
- Online access so that members can query or try the site's technical examples and scripts
- A tips section
- Technical support for error messages

Immediately after the meeting, the newly appointed project manager asks you to implement security for the site. She mentions that the security of a public database is so important that the CIO himself has outlined the security requirements, as follows:

- The online database will have 10 public host database accounts that allow multiple sessions.
- The password of a public host account must be reset to its original setting whenever disconnects or logoffs occur.

- The maximum duration for a session is 45 minutes.
- Allocations will be set on memory and CPU usage to make sure the database is not overloaded due to the excessive burden caused by badly written queries.
- Storage for each public host account must be limited to 1 MB.
- The public host accounts will have privileges to create the most common database objects.
- All newly created database objects must be removed before logoff.
- The database must have the default human resources (HR) user account enabled. All other accounts will be removed. All public host accounts must be able to view the data owned by HR, but must not be allowed to modify the data or the structure of the database objects.
- When a member logs onto the database, all session information, such as IP address, terminal, and user session information, must be recorded for future analysis.

You may add other security or auditing features, as long as you do not overlook any of the requirements in this list.

Case 2: Taking Care of Payroll

Acme Payroll Systems is a small payroll services company that has been in business for two years and has had only one major customer. Suddenly, it lands a contract with another large corporation. The development director of the services company has hired you as a database consultant to design and implement a virtual private database for the existing payroll application. Figure 10-1 represents the existing application data model, and Table 10-1 describes the tables and columns presented in the data model. The main objective of the virtual private database feature is to allow each client to administer his own payroll data without violating the privacy of other clients.

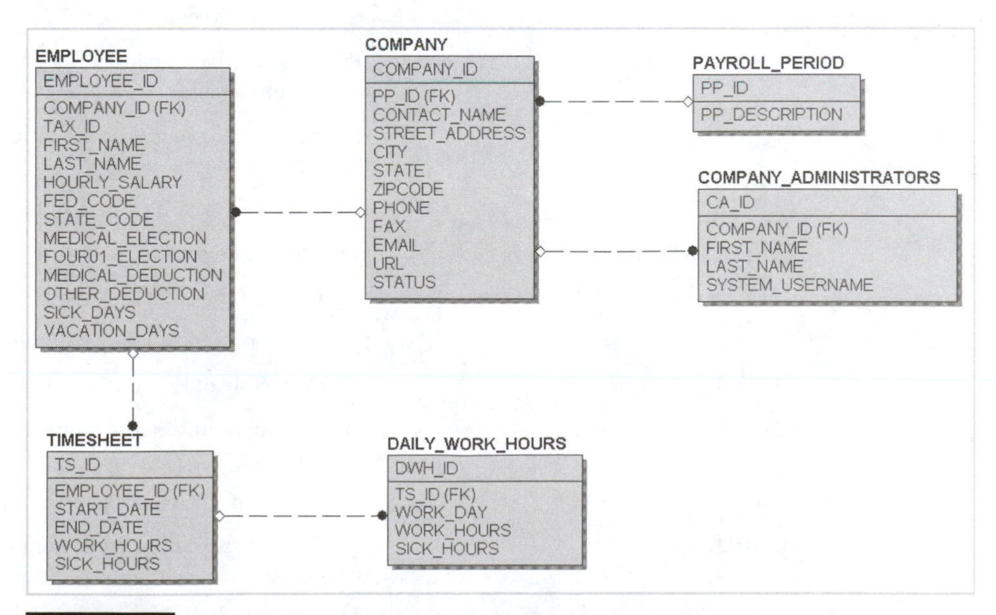

FIGURE 10-1 Payroll application data model for Case 2

Table 10-1 Payroll application model tables and columns

Table/Column	Description
EMPLOYEE table	Holds records for all employees of a specific client company
EMPLOYEE_ID	(Primary key) Unique identifier of EMPLOYEE table generated automatically by the application
COMPANY_ID	(Foreign key) Contains identification number of the company the employee works for
TAX_ID	Employee tax identification number
FIRST_NAME	First name of employee
LAST_NAME	Last name of employee
HOURLY_SALARY	Hourly salary rate of employee
FED_CODE	Federal income tax deduction code
STATE_CODE	State income tax deduction code
MEDICAL_ELECTION	Medical plan option
FOUR01_ELECTION	Retirement election value
MEDICAL_DEDUCTION	Medical tax deductible amount
OTHER_DEDUCTION	Other payroll deduction
SICK_DAYS	Number of eligible sick days
VACATION_DAYS	Number of vacation days
COMPANY table	Holds data for all companies that are clients of Acme Payroll Systems
COMPANY_ID	(Primary key) Unique identifier of COMPANY table automatically assigned by the application
PP_ID	(Foreign key) Contains identification number of the company payroll period
CONTACT_NAME	Full name of the contact person for the company
STREET_ADDRESS	Full street address of the company mailing address
CITY	City name of the company mailing address
STATE	State name of the company mailing address
ZIPCODE	Postal zip code of the company mailing address
PHONE	Phone number of the contact person of the company
FAX	Fax number of the contact person of the company
EMAIL	Electronic mail address of the contact person of the company
URL	Home page link of the company

Table 10-1 Payroll application model tables and columns (continued)

Table/Column	Description
STATUS	Status of the company (active or inactive)
PAYROLL_PERIOD table	Holds data for company payroll period
PP_ID	(Primary key) Unique identifier of PAYROLL_PERIOD table
PP_DESCRIPTION	Description of payroll period such as daily, weekly, biweekly, or monthly
COMPANY_ADMINISTRATORS table	Holds user names and accounts of administrators that are allowed to maintain their own employee and company data
CA_ID	(Primary key) Unique identification number of company administration
COMPANY_ID	(Foreign key) Company identification number
FIRST_NAME	Company administrator first name
LAST_NAME	Company administrator last name
SYSTEM_USERNAME	Company administrator user name used to log in to the application
TIMESHEET table	Holds data for employee timesheet
TS_ID	(Primary key) Unique identification number of timesheet number
EMPLOYEE_ID	(Foreign key) Employee identification number
START_DATE	Timesheet start date and time
END_DATE	Timesheet end date and time
WORK_HOURS	Total number of worked hours
SICK_HOURS	Total number of sick hours used
DAILY_WORK_HOURS table	Holds data for employee work hours
DWH_ID	(Primary key) Unique identification number of the daily work hours
TS_ID	(Foreign key) Timesheet identification number
WORK_DAY	Day of the year
WORK_HOURS	Number of total work hours
SICK_HOURS	Number of sick hours

As the database consultant for Acme Payroll Systems, you are to develop a virtual private database framework that will be adopted by your client's technical staff. The database framework is key to the ability of the client's technical staff to achieve their strategic goals. You may modify the structure of the tables, as necessary. You may use Oracle or SQL Server.

Case 3: Tracking Town Contracts

A small town has hired you as a database specialist on contract. Your job is to develop a new database application to keep track of the jobs awarded to different contractors. All town hall employees will use the application. After several interviews with clerks and managers, you found out that a prior attempt at application development by a consulting company resulted in a draft of an entity-relationship (ER) diagram. The ER diagram depicts all the required information about the contractors and the awarded jobs. See Figure 10-2 for an illustration of the data model.

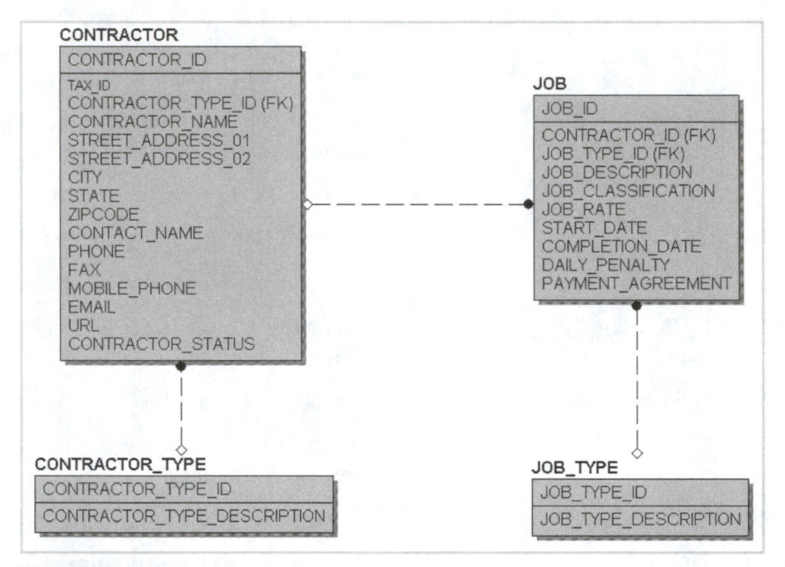

FIGURE 10-2 Contractor job data model for Case 3

During your meeting with the project manager for this application, you are asked to design an application with the following capabilities:

- Track all changes made to the application data.
- Obtain the approval of the project manager before accepting any contract job for more than $10,000.
- Alert the project manager whenever an awarded job is modified to a value greater than $10,000.
- Implement three levels of security:
  - The DEPARTMENT CLERK level allows clerks to add and update records.
  - The DEPARTMENT MANAGER level allows the manager to add, update, delete, and approve records.
  - The EXTERNAL CLERK level allows employees outside the department only to view data. Table 10-2 describes the columns for each table in the data model.

Table 10-2 Contractor job system column description

Table/Column	Description
CONTRACTOR	Table that contains data about contractors who are awarded jobs
CONTRACTOR_ID	(Primary key) A unique identifier generated by the application of a contractor
TAX_ID	The tax number of the contractor
CONTRACTOR_TYPE_ID	(Foreign key) Indication of the type of contractor
CONTRACTOR_NAME	The name of the individual contractor or the contractor's company name
STREET_ADDRESS_01	First address line; holds the contractor's permanent address
STREET_ADDRESS_02	Second address line; holds the contractor's permanent address
CITY	Contractor's permanent city
STATE	Contractor's permanent state
ZIPCODE	Contractor's permanent postal code
CONTACT_NAME	Name of the contact person for the contractor
PHONE	Contractor's office telephone number
FAX	Fax number for the contractor
MOBILE_PHONE	Cellular telephone number of the contractor's contact person
EMAIL	The e-mail address of the contractor
URL	The Web address of the contractor, if available
CONTRACTOR_STATUS	Indication of whether a contractor is active or inactive
CONTRACTOR_TYPE	Table that contains data about the contractor type, such as company, individual, or any other type
CONTRACTOR_TYPE_ID	(Primary key) Column that serves as a unique identifier of the contractor type, generated automatically by the application
CONTRACTOR_TYPE_DESCRIPTION	Description of the contractor type
JOB	Table that contains data about the awarded jobs
JOB_ID	(Primary key) Unique identifier of the awarded job, generated automatically by the system
CONTRACTOR_ID	(Foreign key) Contractor identification number

Table 10-2 Contractor job system column description (continued)

Table/Column	Description
JOB_TYPE_ID	(Foreign key) Job type identification number
JOB_DESCRIPTION	Description of the awarded job
JOB_CLASSIFICATION	Job classification description
JOB_RATE	Total amount the town is responsible to pay the awarded contractor upon completion of the job
START_DATE	Start date for the job
COMPLETION_DATE	Agreed upon completion date of the job
DAILY_PENALTY	Penalty amount to be charged against the contractor for each day (after the agreed upon completion date) that the contractor has not completed the job
PAYMENT_AGREEMENT	Column that indicates the number of days from the completion of the job that the town has before it must pay the contractor in full
JOB_TYPE	Table that contains contract job types, such as PLOWING, PAINTING, and so on
JOB_TYPE_ID	(Primary key) Unique identifier of the job type, generated automatically by the application
JOB_TYPE_DESCRIPTION	Description of the contract job type

Your assignment is to implement the security and auditing requirements. You may modify or add to the data model, as necessary. (*Note*: A SQL script, ch10case03.sql, is provided to create the data model for this case.)

Case 4: Tracking Database Changes

A friend recommended you to the company she works for. The company needs your help to solve a series of database and application violations. When you meet with the hiring manager, she explains that there has been a series of inexplicable, suspicious activities on the application and production databases. The company wants to know who accessed these databases, who modified data, and who changed the data structure. She also explains that they want to have an audit trail for all these activities but that the company was not interested in a historical data changes trail. As the consultant, your job is to design an audit model to meet these requirements. The following is a summary of the project requirements:

- Audit of database connections
- Audit trail of users that are performing DML operations
- Audit trail of users that are modifying structures of the application schema tables

You may use the two tables illustrated in Figure 10-3 as samples of application schema tables. (*Note*: A SQL script, ch10case04.sql, is provided to create this data model.)

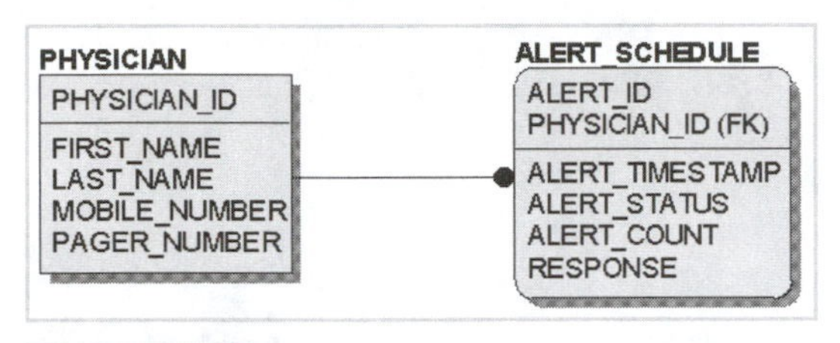

FIGURE 10-3 Sample data model for Case 4

Case 5: Developing a Secured Authorization Repository

A small retail company has asked you to provide them with database security services. The main requirement of this project is to create a security data model that will be used by the central authorization module. The model should include an auditing repository. This model will store application users, roles, applications, and application modules. Your mission is to create an authorization data model with a relevant auditing repository. The following is a summary of the project security requirements:

- There must be one database user account for the application schema owner.
- Database-assigned roles are not allowed.
- There must be application roles only.
- Each application user is assigned to application modules.
- Each application user is assigned a security level that indicates the type of operations the user can perform within the application. Operations are READ, WRITE, DELETE, and ADMINISTER.
- Passwords must be stored within the designed security module.
- Each user has a logon identification number that will be used to logon to the application.
- The security model should have the flexibility to logically lock, disable, and remove accounts.
- Application accounts must have an activation date and expiry date.

The security module must be coupled with an auditing module that meets these auditing requirements:

- It must have an audit trail of the date and time a user connects and disconnects from the application.
- It must have an audit trail of application operations that includes the date and time operations were performed by the application user.
- It must have an audit trail of all activities and operations performed on the security module.
- The auditing module must be coupled with the security module.

You are to provide only a design solution, not an implementation.

Database Security Checklist

Review the following security checklist to assess the level of security measures at your organization. When answering use the following scale:

- 0—Not implemented
- 1—Not fully implemented
- 2—Fully implemented

Item #	Database Security Check	Answer
1	Establish and publish security policies and standards, and raise awareness of these documents to all employees at all levels. Hold sessions to inform employees about security policies and the consequences of not abiding by these policies.	
2	Perform audits regularly and randomly to ensure that all security measures are working and enabled.	
3	Identify sensitive data in your database, and establish encryption mechanisms and security data models to hide data from being exposed to unauthorized persons.	
4	Provide only necessary privileges to users, and never grant any privilege unless the request is documented and approved.	
5	Follow up on all security violations no matter how small or large. Keep all investigation on security violations discreet.	
6	Make sure that all default database passwords are changed.	
7	Establish rules and guidelines for password complexity to be adopted by all departments in your organization.	
8	Never allow database development or testing to gain access to the production database through database links or linked servers.	
9	Prevent access to database files from all individuals except for database administrators and system administrators.	
10	Make sure that all database design changes are reviewed by the security architect and security officer to verify that the new design does not open any security gap that could make the database vulnerable to violations.	
11	Install only the database components, modules, or functionality that your application is intending to use. Any database functionality that is not used should be removed.	
12	Establish a disaster recovery strategy for every database application in production to ensure business continuity. Document all procedures that must be followed from the moment a disaster is declared by management to the moment the database applications are recovered.	
13	Change the default configuration for your application, database, and operating system in order to prevent easy access to data.	
14	Install antivirus and antispyware programs, an e-mail spam blocker, and a network firewall to prevent intrusions from malicious code.	
15	Establish a secure virtual private network for employees who need to access the database remotely.	

Item #	Database Security Check	Answer
16	Use encryption to protect passwords, credit card numbers, and other confidential and sensitive data.	
17	Implement good security design in your application to secure and protect data integrity and confidentiality.	
18	Employ secure measures to authenticate and authorize persons to access the operating system server and the database server.	
19	Enable password management that enforces password complexity. Employ password management software that assists administrators in saving all safely encrypted passwords in a repository. The password management program should also have the ability to generate random passwords that meet your company password policy.	
20	To protect the privacy and confidentiality of the data, make sure that production data is not used in a development or testing environment.	
21	Establish a position in your organizational structure that will be responsible for implementing and monitoring all security policies and standards.	
22	Make sure all security patches and service packs for the operating system and the database system are applied and up to date.	
23	Establish a database management change process and approval mechanism for any data or database modification. Explanation and justification of the change must be provided and approved by all parties who have a stake in the database.	
24	Ensure that all database production guides, operations procedures, and application manuals are always updated and current.	
25	Ensure that the network that connects to the database server is secured at all levels to prevent hackers and intruders from violating database integrity. This includes employment of firewalls, proxy servers, data encryption transmission, and other functionality that increase the security of your network.	

Add up your score and compare it to the score schedule listed below:

1. Score > 40 = ☆☆☆☆: Your organization is diligently working to ensure that database security is implemented at the highest level.
2. Score between 30 and 40 inclusively = ☆☆☆: Your organization should put more effort in implementing security measures to protect its data.
3. Score between 18 and 29 inclusively = ☆☆: Your organization lacks the vision to implement database security measures and your database security is at high risk.
4. Score < 18: Your organization is careless about database security and is highly vulnerable to security violations.

B

Database Auditing Checklist

Review the following auditing checklist to assess the level of implementation at your organization. When answering use the following scale:

- 0—Not implemented
- 1—Not fully implemented
- 2—Fully implemented

Item #	Database Auditing Check	Answer
1	Make sure that application procedures are being performed in compliance with the law, industry standards, and company policies.	
2	Use the auditing process as a means to improve your application quality.	
3	Initiate an audit whenever a security violation occurs and follow up with the audit results.	
4	Perform regular and random audits to verify that all security measures are enforced.	
5	Keep all auditing procedures up to date with legal requirements and industry standards. Review and revise application procedures and policies as needed.	
6	Inform all employees working with a database application that all confidential and sensitive data must be protected and that they could be audited to verify that data is secured.	
7	Track changes of sensitive data to produce an audit trail when needed.	
8	Implement an auditing design for your application that can be turned on or off when needed.	
9	Assign an internal employee to be responsible for compliance and auditing procedures.	
10	Review database and application configuration periodically to verify that changes to the database or application are not violating security measures.	
11	Ensure that there are no backdoors to application code and data.	
12	Employ tools that automatically perform audits and intelligently scan through reports for unusual activities.	
13	On a daily basis, check and review audit trail reports and logs for irregular activities.	
14	Use database auditing functionality whenever necessary to audit any suspicious activities.	
15	Randomly check application data input and output to verify data integrity.	
16	Log all database changes and establish procedures to roll back changes when needed.	
17	Randomly audit management activities, user activities, and application activities.	
18	Have your site certified to a standard to help you implement an auditing process.	

Item #	Database Auditing Check	Answer
19	Classify data into separate levels of accessibility and confidentiality, and assign employees to these levels based on their duties, trustworthiness, and position.	
20	Ensure all software licenses used for the application are legal and traceable.	
21	Investigate the cause of all database and application errors.	
22	Remove or lock all accounts owned by an employee when that employee is terminated.	
23	Randomly audit data transfer over the network to verify that data is encrypted when transmitted.	
24	Randomly audit user connections to the database and the type of programs used to connect to the database.	
25	Make sure that your application data model is normalized and all data constraints are implemented either in the database or the application.	

Add up your score and compare it to the score schedule listed below:

1. Score > 40 = ☆☆☆☆: Your organization is diligently enforcing auditing mechanisms to ensure that database security is implemented at the highest level.
2. Score between 30 and 40 inclusively = ☆☆☆: Your organization should put more effort in auditing processes and procedures.
3. Score between 18 and 29 inclusively = ☆☆: Your organization lacks the vision to implement security measures for database auditing procedures, and your database security is at high risk.
4. Score < 18: Your organization is careless about data.

Glossary

A

administration activity (one of three types of database activities) Encompasses commands issued by the database administrators or operators for maintenance and administration purposes; some of the commands and statements in this category are actually SQL statements

administration policy A type of documentation that includes all policies for handling new and terminated employees, managers, system and database administrators, database managers, operation managers, and human resources

adware A type of malicious code used to pop-up advertisements as well as capture keystrokes, Web sites visited, clicks on pages, and more; also called spyware

application activity (one of three types of database activities) Encompasses SQL statements issued against application tables

application administrator An application user who has application privileges to administer application users and their roles

application context A functionality in Oracle that allows a security context to be set, based on an application or user-defined environmental attribute

application owner A database user (schema owner) who owns application tables and objects

application role A role created in the database by a database administrator that is activated at the time of authorization

application user The user recorded in the application schema for authentication purposes and for enabling the user to perform tasks within the application

audit A process of examining and validating documents, data, processes, procedures, systems, or other activities to ensure that the audited entity is in compliance with its objectives

audit log A document generated by an automated system that contains all activities that are being audited

audit objectives A set of business rules, system controls, government regulations, or security policies against which the audited entity is measured to determine compliance

audit procedure A step-by-step instruction for performing an auditing process

audit report A document that contains audit findings and is generated by the individual conducting the audit

audit trail A chronological record of document changes, data changes, system activities, or operational events

audit, compliance Verifies that the system complies with industry standards, government regulations, or partner and client policies

audit, external An audit that is conducted by a party outside of the company that is being audited

audit, financial Ensures that all financial transactions are accounted for and comply with the law

audit, internal An examination, verification, and validation of documents, processes, procedures, systems, or activities conducted by staff members of the organization being audited

audit, internal An examination, verification, and validation of documents, processes, procedures, systems, or activities conducted by staff members of the organization being audited

audit, investigative Performed in response to an event, request, threat, or incident

audit, operational Verifies if an operation is working according to the policies of the company

audit, preventive Performed to identify problems before they occur

audit, product Performed to ensure that a product complies with industry standards

audit, security Evaluates the level of security of a system

auditing activities Tasks performed as a part of an audit, audit process, or audit plan

auditing environment Consists of objectives, procedures, people, and audited entities

auditing process Assures that the system is working and is compliant with the policies, standards, regulations, or laws set by the organization, industry, or government

auditing repository A set of tables used to store audited data and auditing meta-data

auditing, fine-grained (FGA) Column-level auditing

auditor A person with the proper qualifications and ethics who is authorized to examine, verify, and validate documents, data, processes, procedures, systems, or activities and to produce an audit report

authentication A fundamental service of the operating system that proves the identity of the user to permit access to the operating system

authorization A process that decides whether users are permitted to perform the functions they request

automatic audit An audit that is prompted and performed without human intervention

availability A major component of the C.I.A. triangle that defines a framework for information security; in an ideal state of availability, a system is always available and not vulnerable to unauthorized shutdowns

B

bug Software code that is faulty due to bad design, logic, or both

C

C.I.A triangle The C.I.A. triangle is a framework for protecting information: "C" stands for Confidentiality, "I" for Integrity, and "A" for Availability

confidentiality A major component of the C.I.A. triangle that defines a framework for information security; system information is classified into different levels of confidentiality to assure only those with "rights" can access the information

D

data audit A chronological record of data changes stored in a log file or a database table object

Data Definition Language (DDL) statements DDL statements are used to define the structure of a database object; examples of DDL statements include CREATE, ALTER and DROP commands

data manipulation history Audit trail

database administrator A user account who has database administration privileges enabling him or her to perform administration tasks

database auditing A chronological record of database activities, such as shutdown, startup, logins, and data structure changes of database objects

database auditing environment Consists of objectives, procedures, people, audited entities, and the database

database constraint The validation process by the database server that assures that data conforms to the condition and criteria defined in the constraints

database events (one of three types of database activities) Events that occur when a specific activity occurs; for example, when a user logs on or logs off, when the database is started or is shut down, or when an error is generated by a command or statement

database link (Oracle) Connection from one database to another database

database management system A collection of programs that manage the database and allow users to store, manipulate, and retrieve data efficiently

database security The degree to which all data is fully protected from tampering or unauthorized acts

database user A type of database user account that has roles and privileges assigned to it

DBMS See database management system

DDL statement See Data Definition Language statement

decision support system Used for tactical management tasks; deals with nonstructured problems and provides recommendations or answers to solve these problems

definer rights Indicates that the procedure is executed using the security credentials of the user that owns the procedure

denial-of-service-flood The act of flooding a Web site or network system with many requests to overload the system and force it to deny service to legitimate requests

digital authentication A process of verifying the identity of the user by means of a digital mechanism or software

digital card A security card or smart card; this card is similar to a credit card in dimensions, but instead of a magnetic strip, it has an electronic circuit that keeps user identification information such as name, ID, password, and other related data

digital certificate A type of authentication that is widely used in e-commerce; a digital certificate is a digital passport that identifies and verifies the holder of the certificate

digital token A small electronic device that users keep with them for authentication to a computer or network system

DML statements See Dynamic Markup Language

DSS See decision support systems

Dynamic Markup Language (DML) statement Statement that can be used to track data changes in the audit environment

E

e-mail spamming E-mail that is sent to many recipients without their permission

ES See expert systems

expert system A type of system used by top-level management for strategic management goals; a branch of artificial intelligence within the field of computer science studies

F

file permission An operating system method of granting read, write, or execute privileges to different users

file transfer A tool for sending files from one computer to another

File Transfer Protocol Allows the transfer of a file between the user hard drive and a remote server

fine-grained access (FGA) See virtual private database

fine-grained auditing An Oracle function that keeps an audit trail of all DML activities; it is an internal mechanism that allows the database administrator to set up auditing policies on tables; also known as column-level auditing

FTP See File Transfer Protocol

H

hybrid audit A combination of automatic and manual audits

I

information security The procedures and measures taken to protect each component of the information system that is involved in producing information; this includes protecting data, software, networks and people

information security architecture The overall design of a company's implementation of the C.I.A. triangle; the architecture's components range from physical equipment to logical security tools and utilities

information system Typically consists of data, procedures, hardware, software, networks, and people

integrity A major component of the C.I.A. triangle that defines a framework for information security; data is considered to have integrity if it is continually validated and is protected from the tampering of unauthorized persons

invoker rights Indicates that the procedure is executed using the security credentials of the caller, not the owner of the procedure

K

Kerberos Developed by the Project Athena team Massachusetts Institute of Technology (MIT) to enable two parties to exchange information over an open network by assigning a unique key, called a ticket, to each user

L

Lightweight Directory Access Protocol (LDAP) An authentication method developed by the University of Michigan that uses a centralized directory database that stores information about people, offices, and machines in a hierarchical manner

linked server (SQL Server 2000) A server that allows users to connect from one database to another database to issue SQL statements

logical table (within the context of SQL Server 2000 triggers) Similar to a pseudo-column, a logical table is generated by the database that contains columns and values that can be retrieved but not manipulated and are not stored permanently in the database; Microsoft SQL Server generates logical tables when a DML trigger is invoked

logon retries The practice of allowing a user to attempt to logon a set number of times, (usually three unsuccessful tries) before the account is locked and an administrator is contacted

M

malicious code A program, macro, or script that deliberately damages files or disrupts computer operations; there are several types of malicious code, such as viruses, worms, Trojan horses, and time bombs

manual audit An audit that is performed completely by humans

N

NT LAN Manager Developed and used by Microsoft, employs a challenge/response authentication protocol that uses an encryption and decryption mechanism to send and receive passwords over the network; this method is no longer used or supported by new versions of the Windows operating system

NTLM See NT LAN Manager

O

Object Linking and Embedding (OLE) database A set of COM-based interfaces that expose data from a variety of sources; OLE DB interfaces provide applications with uniform access to data stored in diverse information sources, or data stores

object privilege A privilege granted by the owner of the database object or a user who has been granted the grant privilege

OLE DB See Object Linking and Embedding (OLE) database (DB)

OLTP See transaction-processing system

operating system A collection of programs that allows the user to operate the computer hardware

P

password The key for opening a user account

password aging A process that indicates how long a password can be used before it expires

password complexity A set of guidelines used when selecting a password

password encryption A method that encrypts (scrambles) the password and stores it in a way that makes it impossible to be read directly

password history A practice related to password reuse that tells the system how many passwords it should maintain for an account; the password history can be used to determine if a password can be reused or not

password policy A set of guidelines that enhances the robustness of a password and reduces the likelihood of it being broken

password protection A practice of training employees about the dangers of not properly securing passwords

password reuse A password security practice that can be applied in three different ways: tells the system how many times you can reuse a password; indicates the number of days that must pass before you can reuse a password; or determines whether the system allows passwords to be reused

password storage A method of storing a password in an encrypted manner

peer-to-peer program A program that allows users to share files with other users over the Internet

performance monitoring process Monitors response time, after a product is commissioned into production

physical authentication Verifies identity to allows physical entrance into the company property

PKI See public key infrastructure

predefined roles description A full description of all predefined roles, outlining all tasks the role is responsible for and the role's relationship to other roles

privilege A method of permitting or denying access to data or the right to perform a database operation

procedure implementation scripts or programs The documentation of any script or program used to perform an administrative task

profile A security concept that describes the limitation of database resources for database users; a way of defining database user behavior to prevent users from wasting resources, such as memory or CPU consumption

proxy user A database user that has specific roles and privileges assigned to it; the proxy user works on behalf of an application user to isolate application users from the database; also called virtual user

pseudocolumn A database-generated logical column that contains a value that can be retrieved but cannot be manipulated; the value for the logical column is not stored permanently anywhere in the database; for example, Oracle generates a ROWNUM pseudocolumn that represents the row number for the returned result-set of a SELECT statement

public key encryption See public key infrastructure

public key infrastructure (PKI) An authentication method in which a user keeps a private key and the authentication company holds a public key; these two keys are used to encrypt and decrypt messages between the two parties; the private key is usually kept as a digital certificate on the user's system

Q

QA See quality assurance

quality assurance (QA) The process in software engineering that ensures that the system is bug-free and is functioning according to its specifications

R

RADIUS See Remote Authentication Dial-In User Service

Remote Authentications Dial-In User Service (RADIUS) An authentication method commonly used by network devices to provide a centralized authentication mechanism; RADIUS architecture is based on a client/server, which is dial-up server, virtual private network (VPN), or a wireless access point

remote user A user who logs onto a database using a different host machine than the machine where the database is located

role A concept used to organize and administer privileges in an easy manner; a role is like a user except it cannot own objects; a role can be assigned privileges and then assigned to users

rootkits and bots Malicious or legitimate software code that performs such functions as automatically retrieving and collecting information from computer systems

Row Level Security (RLS) See virtual private database

S

schema owner A database user that owns database objects

SDLC See system development life cycle

Secure Remote Password (SRP) SRP is an authentication protocol designed to integrate secure password authentication into existing networked applications

Secure Sockets Layers (SSL) A method in which authentication information is transmitted over the network in an encrypted form; this method is commonly used by Web sites to secure client communications

security access point A point at which security measures are needed to prevent unauthorized access

security procedure A step-by-step process for performing an administrative task according to company policies

security risk A known security gap that a company intentionally leaves open

security threat A security violation or attack that can occur because of a security vulnerability

security vulnerability A weakness in any of the information system components that can be exploited to violate the integrity, confidentiality, or accessibility of the system

services The services component of an operating system consists of functionality that the operating system offers as part of its core utilities; users employ these utilities to gain access to the operating system and to all the features that the users are authorized to work with

single sign-on A practice (that is not recommended) of allowing sign-on once to a system and then not requiring sign-on again for a user who moves to another system where he or she has an account

spoofing code Malicious code that looks like legitimate code

spyware See adware

SRP See Secure Remote Password

SSL See Secure Sockets Layers

system development life cycle (SDLC) A structured process for software development

system privilege Privilege granted only by a database administrator or users who have been granted the administration option

T

threat In the world of information security, a threat is defined as a security risk that has a high possibility of becoming a security breach

TPS See transaction-processing systems

transaction-processing systems A system for business transactions derived from business procedures, business rules, and policies. Also known as online transaction processing (OLTP)

trigger A stored PL/SQL procedure that fires (is called) automatically when a specific event occurs

trigger timing Oracle has six DML events, also known as trigger timings for INSERT, UPDATE, and DELETE

Trojan horse Malicious code that penetrates a computer system or network disguised as legitimate code

U

user administration A functionality used by administrators to create user accounts, set password policies, and grant privileges to users

V

virtual private database (VPD) A shared database schema containing data that belongs to different users with each user able to view and update only the data he or she owns

virtual user See proxy-user

VPD See virtual private database

Bibliography

Afyouni, H. A. *Oracle9i Performance Tuning: Optimizing Database Connectivity*. Boston, MA: Course Technology, 2004.

Castano, S., Fugini, M., Martella, G., and Samarati, P. *Database Security*. Addison-Wesley, 1994.

Clark, R., Holloway, S., and List, W. *The Security, Audit and Control of Databases*. UK: Ashgate, 1991.

Connolly, T., Begg, C., and Starchan, A. *Database Systems: A Practical Approach to Design, Implementation, and Management*. UK: Addison-Wesley Longman Limited, 1999.

Dewire, R. *Second-Generation Client/Server Computing*. McGraw-Hill, 1997.

Douglas, I. *Security and Audit of Database Systems*. UK: Blackwell, 1980.

Fernandez, E., Summers, R., and Wood, C. *Database Security and Integrity*. Reading, MA: Addison-Wesley, 1981.

Haag, S., Cummings, M., and Dawkins, J. *Management Information Systems for the Information Age*. Irwin/McGraw-Hill, 1998.

Helton, R., and Helton, J. *Java Security Solutions*. Wiley Publishing, 2002.

Lindstrom, P. "A Patch in Time." *Information Security. 7,* 27–37, 2004.

Maiwald, E. *Fundamentals of Network Security*. McGraw-Hill, 2004.

"Oracle Technology Network." Oracle 10*g* documentation. Oracle Corporation (*otn.oracle.com*).

Palmer, M. *Guide to Operating Systems Security* (1st ed.). Boston, MA: Course Technology, 2004.

Raftree, M. F. *MCSE Guide to Microsoft SQL Server 2000 Administration*. Boston, MA: Course Technology, 2002.

Rob, P., and Coronel, C. (2000). *Database Systems Design, Implementation, and Management* (4th ed.). Boston, MA: Course Technology.

Rosen, K. H., Rosinski, R. R., and Farber, J. M. (1990). *Unix System V Release 4: An Introduction: For New and Experienced Users*. Osborne McGraw-Hill.

Siddiqui, S., and NIIT. (2002). *Linux Security*. Boston: Premier Press.

Snyder, G., and Seebass, S. (1989). *Unix System Administration Handbook*. PTR Prentice Hall.

SQL Server documentation. Microsoft Corporation (*www.microsoft.com*).

Stair, R., and Reynolds, G. *Principles of Information Systems: A Managerial Approach.* Boston, MA: Course Technology, 1998.

Wells, N. *Guide to Linux Installation and Administration* (2nd ed.). Boston: Course Technology, 2003.

Whitman, M. E., and Mattord, H. J. *Principles of Information Security*. Boston: Course Technology, 2003.

Index

Bold page numbers indicate where a key term is defined in the text.

L

M

P

Q

R

S

T

V

W